Manual of Structural Kinesiology

R. T. Floyd EdD, ATC, CSCS

Director of Athletic Training and Sports Medicine
Professor of Physical Education and Athletic Training
Chair, Department of Physical Education and
    Athletic Training
    The University of West Alabama
    (formerly Livingston University)
    Livingston, Alabama

NINETEENTH EDITION

MANUAL OF STRUCTURAL KINESIOLOGY, NINTEENTH EDITION

Published by McGraw-Hill Education, 2 Penn Plaza, New York, NY 10121. Copyright © 2015 by McGraw-Hill Education. All rights reserved. Printed in the United States of America. Previous editions © 2012, 2009, and 2007. No part of this publication may be reproduced or distributed in any form or by any means, or stored in a database or retrieval system, without the prior written consent of McGraw-Hill Education, including, but not limited to, in any network or other electronic storage or transmission, or broadcast for distance learning.

Some ancillaries, including electronic and print components, may not be available to customers outside the United States.

This book is printed on acid-free paper.

1 2 3 4 5 6 7 8 9 0 DOW/DOW 1 0 9 8 7 6 5 4

ISBN 978-1-259-25389-8
MHID 1-259-25389-9

Contents

Preface

In this revision, I have attempted to update the information and improve the clarity of concepts and illustrations while maintaining the successful presentation approach the late Dr. Clem Thompson established from 1961 through 1989. I first used this book as an undergraduate and later in my teachings over the years. Having developed great respect for this text and Dr. Thompson's style, it is my intention to continue to preserve the effectiveness of this time-honored text, while adding material pertinent to the professions working with today's ever-growing physically active population. Hopefully, I have maintained a clear, concise, and simple presentation method supplemented with applicable information gained through my research and career experiences.

This text, now in its 67th year, has undergone many revisions over the years. My goal continues to be making the material as applicable as possible to physical activity and to make it more understandable and easier to use for the student and professional. While reading this text, I challenge kinesiology students and professionals to immediately apply the content to physical activities with which they are individually familiar. I hope that the reader will simultaneously palpate his or her own moving joints and contracting muscles to gain application. Concurrently, I encourage students to palpate the joints and muscles of fellow students to gain a better appreciation of the wide range of normal anatomy and, when possible, appreciate the variation from normal found in injured and pathological musculoskeletal anatomy. Additionally, with the tremendous growth of information and media available via the Internet and other technological means, I encourage careful and continuous exploration of these resources. These resources should be helpful, but must be reviewed with a critical eye, as all information should be.

Audience

This text is designed for students in an undergraduate structural kinesiology course after completing courses in human anatomy and physiology. While primarily utilized in physical education, exercise science, athletic training, physical therapy, and massage therapy curriculums, it is often used as a continuing reference by other clinicians and educators in addressing musculoskeletal concerns of the physically active. Applied kinesiologists, athletic trainers, athletic coaches, physical educators, physical therapists, occupational therapists, health club instructors, strength and conditioning specialists, personal trainers, massage therapists, physicians, and others who are responsible for evaluating, improving, and maintaining the muscular strength, endurance, flexibility, and overall health of individuals will benefit from this text.

With the ever-continuing growth in the number of participants of all ages in a spectrum of physical activity, it is imperative that medical, health, fitness, and education professionals involved in providing instruction and information to the physically active be correct and accountable for the teachings that they provide. The variety of exercise machines, techniques, strengthening and flexibility programs, and training programs is continuously expanding and changing, but the musculoskeletal system is constant in its design and architecture. Regardless of the goals sought or the approaches used in exercise activity, the human body is the basic ingredient and must be thoroughly understood and considered to maximize performance capabilities and minimize undesirable results. Most advances in kinesiology and exercise science continue to result from a better understanding of the body and how it functions. I believe that an individual in this field can never learn enough about the structure and function of the human body and that this is typically best learned through practical application.

Those who are charged with the responsibility of providing examination, instruction and consultation to the physically active will find this text a helpful and valuable resource in their never-ending quest for knowledge and understanding of human movement.

New to this edition

Some additional content has been added along with slight revisions in many areas. Tables and illustrations have been refined and updated, and a number of photographs and figures have been added or replaced to improve the visual quality and clarity. Some of the chapter worksheets have also been revised. Website addresses have been moved to the Online Learning Center at www.mhhe.com/floyd19e where they may be more easily accessed and updated as needed. Additional questions and exercises will continue to be added to the Online Learning Center. Finally, several new terms have been added to the Glossary.

Online Learning Center

www.mhhe.com/floyd19e
The Online Learning Center to accompany this text offers a number of additional resources for both students and instructors. Visit this website to find useful materials such as these:

For the instructor:
• Downloadable PowerPoint presentations
• Image bank
• Test bank questions
• End-of-chapter exercise and worksheet answers

For the student:
• Self-scoring multiple choice, matching, and video quizzes
• Anatomy flashcards and crossword puzzles for learning key terms and their definitions
• Student Success Strategies
• Glossary

Acknowledgments

I am very appreciative of the numerous comments, ideas, and suggestions provided by the eight reviewers. These reviews have been a most helpful guide in this revision and the suggestions have been incorporated to the extent possible when appropriate. These reviewers are:

Andrew J. Accacian, *University of Dubuque*

Jessica Adams, *Kean University*

Pam Brown, *The University of North Carolina at Greensboro*

Adam Bruenger, *University of Central Arkansas*

Phillip Morgan, *Washington State University*

Dean Smith, *Miami University*

Scott Strohmeyer, *University of Central Missouri*

Traci Worby, *Eastern Illinois University*

I would like to especially thank the kinesiology/athletic training students and faculty of the University of West Alabama for their suggestions, advice, and input throughout this revision. Their assistance and suggestions have been very helpful. I am particularly grateful to Britt Jones of Livingston, Alabama, for his outstanding photography. I also acknowledge John Hood and Lisa Floyd of Birmingham and Livingston, Alabama, respectively, for the fine photographs. Special thanks to Linda Kimbrough of Birmingham, Alabama, for her superb illustrations and insight. I appreciate the models for the photographs, Audrey Crawford, Fred Knighten, Darrell Locket, Amy Menzies, Matthew Phillips, Jay Sears, Marcus Shapiro, and David Whitaker. My thanks also go to Emily Nesheim and Erin Guendelsberger, Sara Jaeger, Adina Lonn and the McGraw-Hill staff who have been most helpful in their assistance and suggestions in preparing the manuscript for publication.

R. T. Floyd

About the Author

R. T. Floyd is in his fortieth year of providing athletic training services for the University of West Alabama. Currently, he serves as the Director of Athletic Training and Sports Medicine for the UWA Athletic Training and Sports Medicine Center, Program Director for UWA's CAATE accredited curriculum, and as a professor in the Department of Physical Education and Athletic Training, which he chairs. He has taught numerous courses in physical education and athletic training, including kinesiology, at both the undergraduate and graduate levels since 1980.

Floyd has maintained an active professional life throughout his career. He is currently serving as President of the National Athletic Trainers' Association (NATA) Research & Education Foundation after serving in multiple roles on the Board of Directors since 2002. He recently finished eight years of service on the NATA Board of Directors representing District IX, the Southeast Athletic Trainers' Association (SEATA). Previously, he served as the District IX representative to the NATA Educational Multimedia Committee from 1988 to 2002. He has served as the Convention Site Selection Chair for District IX from 1986 to 2004 and has directed the annual SEATA Competencies in Athletic Training Student Workshop since 1997. He has also served as a NATA BOC examiner for well over a decade and has served as a Joint Review Committee on Educational Programs in Athletic Training site visitor several times. He has provided over a hundred professional presentations at the local, state, regional, and national levels and has also had several articles and videos published related to the practical aspects of athletic training. He began authoring the *Manual of Structural Kinesiology* in 1992 with the twelfth edition after the passing of Dr. Clem W. Thompson, who authored the fourth through the eleventh editions. In 2010, much of the content of this text was incorporated into *Kinesiology for Manual Therapies,* which he co-authored with Nancy Dail and Tim Agnew.

Floyd is a certified member of the National Athletic Trainers' Association, a Certified Strength & Conditioning Specialist, and a Certified Personal Trainer in the National Strength and Conditioning Association. He is also a Certified Athletic Equipment Manager in the Athletic Equipment Managers' Association, a member of the American College of Sports Medicine, the American Orthopaedic Society for Sports Medicine, the American Osteopathic Academy of Sports Medicine, the American Sports Medicine Fellowship Society, and the American Alliance for Health, Physical Education, Recreation and Dance. Additionally, he is licensed in Alabama as an Athletic Trainer and an Emergency Medical Technician.

Floyd was presented the NATA Athletic Trainer Service Award in 1996, the Most Distinguished Athletic Trainer Award by the NATA in 2003, and received the NATA Sayers "Bud" Miller Distinguished Educator Award in 2007. In 2013 he was inducted into the NATA Hall of Fame. He received the District IX Award for Outstanding Contribution to the field of Athletic Training by SEATA in 1990 and the Award of Merit in 2001 before being inducted into the organization's Hall of Fame in 2008. He was named to Who's Who Among America's Teachers in 1996, 2000, 2004, and 2005. In 2001, he was inducted into the Honor Society of Phi Kappa Phi and the University of West Alabama Athletic Hall of Fame. He was inducted into the Alabama Athletic Trainers' Association Hall of Fame in May 2004.

To

my family,
Lisa, Robert Thomas, Jeanna, Rebecca, and Kate
who understand, support, and allow me to
pursue my profession

and to my parents,
Ruby and George Franklin,
who taught me the importance of a strong work ethic
with quality results

R.T.F.

CHAPTER 1

FOUNDATIONS OF STRUCTURAL KINESIOLOGY

Objectives

- To review the anatomy of the skeletal system

- To review and understand the terminology used to describe body part locations, reference positions, and anatomical directions

- To review the planes of motion and their respective axes of rotation in relation to human movement

- To describe and understand the various types of bones and joints in the human body and their functions, features, and characteristics

- To describe and demonstrate the joint movements

Online Learning Center Resources

Visit *Manual of Structural Kinesiology*'s **Online Learning Center** at **www.mhhe.com/floyd19e** for additional information and study material for this chapter, including:

- *Self-grading quizzes*
- *Anatomy flashcards*
- *Animations*
- *Related websites*

Kinesiology may be defined as the study of the principles of anatomy (active and passive structures), physiology, and mechanics in relation to human movement. The emphasis of this text is **structural kinesiology**—the study of muscles, bones, and joints as they are involved in the science of movement. To a much lesser degree, certain physiological and mechanical principles are addressed to enhance the understanding of the structures discussed.

Bones vary in size and shape, which factors into the amount and type of movement that occurs between them at the joints. The types of joint vary in both structure and function. Muscles also vary greatly in size, shape, and structure from one part of the body to another.

Anatomists, athletic trainers, physical therapists, occupational therapists, physicians, nurses, massage therapists, coaches, strength and conditioning specialists, performance enhancement specialists, personal trainers, physical educators, and others in health-related fields should have an adequate knowledge and understanding of all the large muscle groups so they can teach others how to strengthen, improve, and maintain these parts of the human body. This knowledge forms the basis of the exercise programs that should be followed to strengthen and maintain all the muscles. In most cases, exercises that involve the larger primary movers also involve the smaller muscles.

More than 600 muscles are found in the human body. In this book, an emphasis is placed on the larger muscles that are primarily involved in movement of the joints. Details related to many of the small muscles located in the hands, feet, and spinal column are provided to a lesser degree.

Fewer than 100 of the largest and most important muscles, primary movers, are considered in this text. Some small muscles in the human body, such as the multifidus, plantaris, scalenus, and serratus posterior, are omitted because they are exercised with other, larger primary movers. In addition, most small muscles of the hands and feet are not given the full attention provided to the larger muscles. Many small muscles of the spinal column are not considered in full detail.

Kinesiology students frequently become so engrossed in learning individual muscles that they lose sight of the total muscular system. They miss the "big picture"—that muscle groups move joints in given movements necessary for bodily movement and skilled performance. Although it is vital to learn the small details of muscle attachments, it is even more critical to be able to apply the information to real-life situations. Once the information can be applied in a useful manner, the specific details are usually much easier to understand and appreciate.

Reference positions

It is crucial for kinesiology students to begin with a reference point in order to better understand the musculoskeletal system, its planes of motion, joint classification, and joint movement terminology. Two reference positions can be used as a basis from which to describe joint movements. The **anatomical position** is the most widely used and is accurate for all aspects of the body. Fig. 1.1 demonstrates this reference position, with the subject standing in an upright posture, facing straight ahead, with feet parallel and close and palms facing forward. The **fundamental position** is essentially the same as the anatomical position, except that the arms are at the sides with the palms facing the body.

Reference lines

To further assist in understanding the location of one body part in relation to another, certain imaginary reference lines may be used. Some examples follow in Fig 1.2.

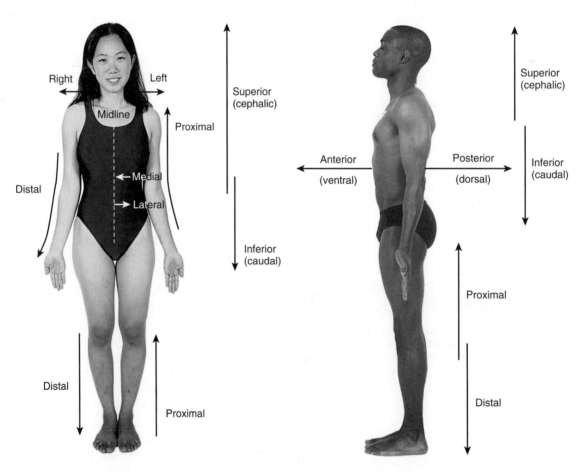

FIG. 1.1 • Anatomical position and anatomical directions. Anatomical directions refer to the position of one body part in relation to another.

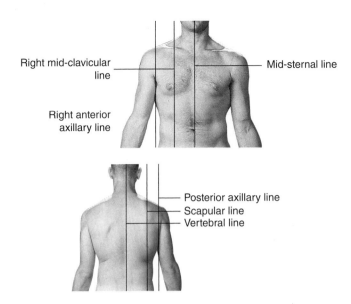

Right mid-clavicular line

Mid-sternal line

Right anterior axillary line

Posterior axillary line
Scapular line
Vertebral line

FIG. 1.2 • Reference lines.

Mid-axillary line: A line running vertically down the surface of the body passing through the apex of the axilla (armpit)

Mid-sternal line: A line running vertically down the surface of the body passing through the middle of the sternum

Anterior axillary line: A line that is parallel to the mid-axillary line and passes through the anterior axillary skinfold

Posterior axillary line: A line that is parallel to the mid-axillary line and passes through the posterior axillary skinfold

Mid-clavicular line: A line running vertically down the surface of the body passing through the midpoint of the clavicle

Mid-inguinal point: A point midway between the anterior superior iliac spine and the pubic symphysis

Scapula line: A line running vertically down the posterior surface of the body passing through the inferior angle of the scapula

Vertebral line: A line running vertically down through the spinous processes of the spine

Anatomical directional terminology FIGS. 1.1, 1.3, 1.4

It is important that we all be able to find our way around the human body. To an extent, we can think of this as similar to giving or receiving directions about how to get from one geographic location to another. Just as we use the terms *left, right, south, west, northeast,* etc. to describe geographic directions, we have terms such as *lateral, medial, inferior, anterior, inferomedial,* etc. to use for anatomical directions. With geographic directions we may use *west* to indicate the west end of a street

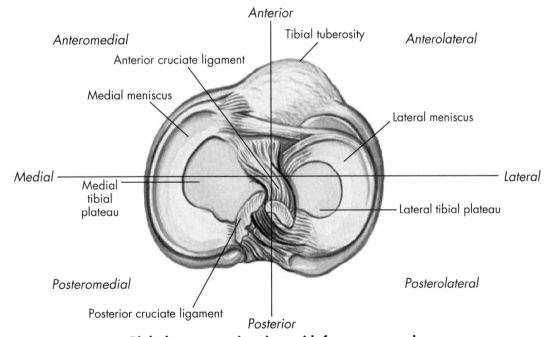

Anterior

Tibial tuberosity

Anteromedial

Anterolateral

Anterior cruciate ligament

Medial meniscus

Lateral meniscus

Medial

Medial tibial plateau

Lateral

Lateral tibial plateau

Posteromedial

Posterolateral

Posterior cruciate ligament

Posterior

Right knee, superior view with femur removed

FIG. 1.3 • Anatomical directional terminology.

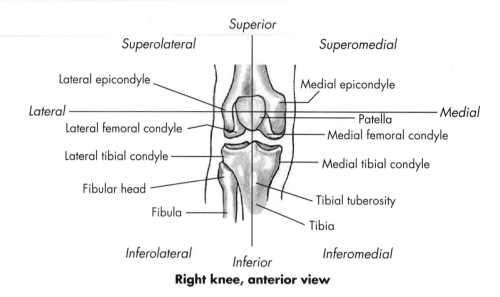

Right knee, anterior view

FIG. 1.4 • Anatomical directional terminology.

or the western United States. The same is true when we use anatomical directions. We may use *superior* to indicate the end of a bone in our lower leg closest to the knee, or we may be speaking about the top of the skull. It all depends on the context at the time. Just as we combine *south* and *east* to get *southeast* for the purpose of indicating somewhere in between these directions, we may combine *anterior* and *lateral* to get *anterolateral* for the purpose of describing the general direction or location "in the front and to the outside." Figs. 1.3 and 1.4 provide further examples.

Anterior: In front or in the front part

Anteroinferior: In front and below

Anterolateral: In front and to the outside

Anteromedial: In front and toward the inner side or midline

Anteroposterior: Relating to both front and rear

Anterosuperior: In front and above

Bilateral: Relating to the right and left sides of the body or of a body structure such as the right and left extremities

Caudal: Below in relation to another structure; inferior

Cephalic: Above in relation to another structure; higher, superior

Contralateral: Pertaining or relating to the opposite side

Deep: Beneath or below the surface; used to describe relative depth or location of muscles or tissue

Dexter: Relating to, or situated to the right or on the right side of, something

Distal: Situated away from the center or midline of the body, or away from the point of origin

Dorsal (dorsum): Relating to the back, being or located near, on, or toward the back, posterior part, or upper surface of; also relating to the top of the foot

Fibular: Relating to the fibular (lateral) side of the lower extremity

Inferior (infra): Below in relation to another structure; caudal

Inferolateral: Below and to the outside

Inferomedial: Below and toward the midline or inside

Ipsilateral: On the same side

Lateral: On or to the side; outside, farther from the median or midsagittal plane

Medial: Relating to the middle or center; nearer to the median or midsagittal plane

Median: Relating to, located in, or extending toward the middle; situated in the middle, medial

Palmar: Relating to the palm or volar aspect of the hand

Plantar: Relating to the sole or undersurface of the foot

Posterior: Behind, in back, or in the rear

Posteroinferior: Behind or in back and below

Posterolateral: Behind and to one side, specifically to the outside

Posteromedial: Behind and to the inner side

Posterosuperior: Behind or in back and above

Prone: Face-downward position of the body; lying on the stomach

Proximal: Nearest the trunk or the point of origin

Radial: Relating to the radial (lateral) side of the forearm or hand

Scapular plane: In line with the normal resting position of the scapula as it lies on the posterior rib cage; movements in the scapular plane are in line with the scapular, which is at an angle of 30 to 45 degrees from the frontal plane

Sinister: Relating to, or situated to the left or on the left side of, something

Superficial: Near the surface; used to describe relative depth or location of muscles or tissue

Superior (supra): Above in relation to another structure; higher, cephalic

Superolateral: Above and to the outside

Superomedial: Above and toward the midline or inside

Supine: Face-upward position of the body; lying on the back

Tibial: Relating to the tibial (medial) side of the lower extremity

Ulnar: Relating to the ulnar (medial) side of the forearm or hand

Ventral: Relating to the belly or abdomen, on or toward the front, anterior part of

Volar: Relating to palm of the hand or sole of the foot

Alignment variation terminology

Anteversion: Abnormal or excessive rotation forward of a structure, such as femoral anteversion

Kyphosis: Increased curving of the spine outward or backward in the sagittal plane

Lordosis: Increased curving of the spine inward or forward in the sagittal plane

Recurvatum: Bending backward, as in knee hyperextension

Retroversion: Abnormal or excessive rotation backward of a structure, such as femoral retroversion

Scoliosis: Lateral curving of the spine

Valgus: Outward angulation of the distal segment of a bone or joint, as in knock-knees

Varus: Inward angulation of the distal segment of a bone or joint, as in bowlegs

Planes of motion

When we study the various joints of the body and analyze their movements, it is helpful to characterize them according to specific planes of motion (Fig. 1.5). A plane of motion may be defined as an imaginary two-dimensional surface through which a limb or body segment is moved.

There are three specific, or **cardinal**, planes of motion in which the various joint movements can be classified. The specific planes that divide the body exactly into two halves are often referred to as cardinal planes. The cardinal planes are the sagittal, frontal, and transverse planes. There are an infinite number of planes within each half that are parallel to the cardinal planes. This is best understood in the following examples of movements in the sagittal plane. Sit-ups involve the spine and, as a result, are performed in the cardinal sagittal plane, which is also known as the **midsagittal** or **median** plane. Biceps curls and knee extensions are performed in **parasagittal** planes, which are parallel to the midsagittal plane. Even though these latter examples are not in the cardinal plane, they are thought of as movements in the sagittal plane.

Although each specific joint movement can be classified as being in one of the three planes of motion, our movements are usually not totally in one specific plane but occur as a combination of motions in more than one plane. These movements in the combined planes may be described as occurring in diagonal, or oblique, planes of motion.

Sagittal, anteroposterior, or AP plane

The sagittal, anteroposterior, or AP plane bisects the body from front to back, dividing it into right and left symmetrical halves. Generally, flexion and extension movements such as biceps curls, knee extensions, and sit-ups occur in this plane.

Frontal, coronal, or lateral plane

The frontal plane, also known as the coronal or lateral plane, bisects the body laterally from side to side, dividing it into front (ventral) and back (dorsal) halves. Abduction and adduction movements such as jumping jacks (shoulder and hip) and spinal lateral flexion occur in this plane.

Transverse, axial, or horizontal plane

The transverse plane, also known as the axial or horizontal plane, divides the body into superior (cephalic) and inferior (caudal) halves. Generally, rotational movements such as forearm pronation and supination and spinal rotation occur in this plane.

Diagonal or oblique plane FIG. 1.6

The diagonal or oblique plane is a combination of more than one plane of motion. In reality,

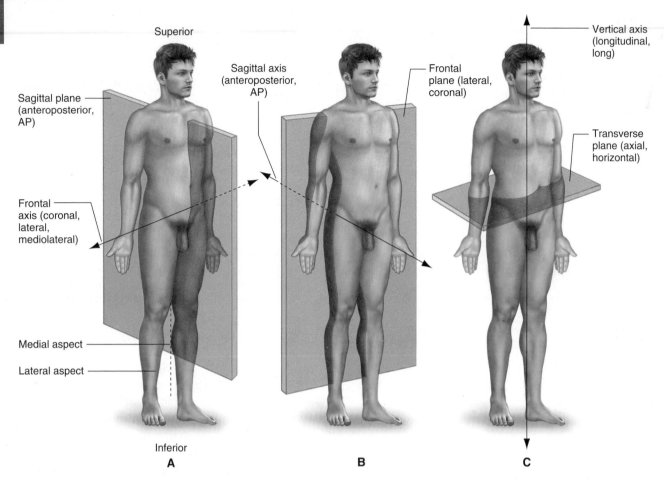

FIG. 1.5 • Planes of motion and axes of rotation. **A,** Sagittal plane with frontal axis; **B,** Frontal plane with sagittal axis; **C,** Transverse plane with vertical axis.

most of our movements in sporting activities fall somewhere between parallel and perpendicular to the previously described planes and occur in a diagonal plane. To further delineate, all movements in diagonal planes occur in a high diagonal plane or one of two low diagonal planes. The high diagonal plane is utilized for overhand movements in the upper extremity, whereas the two low diagonal planes are used to differentiate upper-extremity underhand movements from lower-extremity diagonal movements.

Axes of rotation

As movement occurs in a given plane, the joint moves or turns about an axis that has a 90-degree relationship to that plane. The axes are named in relation to their orientation (Fig. 1.5). Table 1.1 lists the planes of motion with their axes of rotation.

Frontal, coronal, lateral, or mediolateral axis

If the sagittal plane runs from anterior to posterior, then its axis must run from side to side. Since this axis has the same directional orientation as the frontal plane of motion, it is named similarly. As the elbow flexes and extends in the sagittal plane during a biceps curl, the forearm is actually rotating about a frontal axis that runs laterally through the elbow joint. The frontal axis may also be referred to as the bilateral axis.

Sagittal or anteroposterior axis

Movement occurring in the frontal plane rotates about a sagittal axis. This sagittal axis has the same directional orientation as the sagittal plane of motion and runs from front to back at a right angle to the frontal plane of motion. As the hip abducts and adducts during jumping jacks, the femur rotates about an axis that runs front to back through the hip joint.

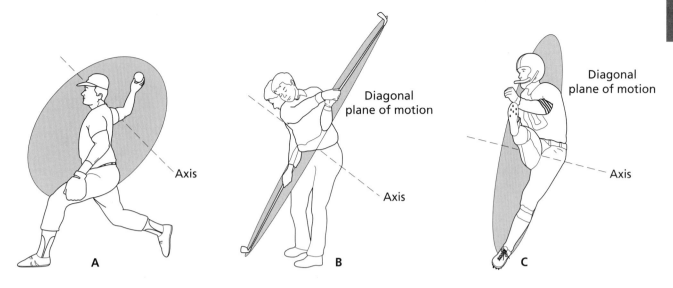

FIG. 1.6 • Diagonal planes and axes of rotation. **A,** Upper-extremity high diagonal plane movement and axis; **B,** Upper-extremity low diagonal plane movement and axis; **C,** Lower-extremity low diagonal plane movement and axis.

TABLE 1.1 • **Planes of motion and their axes of rotation**

Plane	Description of plane	Axis of rotation	Description of axis	Common movements
Sagittal (anteroposterior or AP)	Divides the body into right and left halves	Frontal (coronal, lateral, or mediolateral)	Runs medial/lateral	Flexion, extension
Frontal (coronal or lateral)	Divides the body into anterior and posterior halves	Sagittal (anteroposterior or AP)	Runs anterior/posterior	Abduction, adduction
Transverse (axial, horizontal)	Divides the body into superior and inferior halves	Vertical (longitudinal or long)	Runs superior/inferior	Internal rotation, external rotation

Vertical or longitudinal axis

The vertical axis, also known as the longitudinal or long axis, runs straight down through the top of the head and is at a right angle to the transverse plane of motion. As the head rotates or turns from left to right when indicating disapproval, the skull and cervical vertebrae are rotating around an axis that runs down through the spinal column.

Diagonal or oblique axis   FIG. 1.6

The diagonal axis, also known as the oblique axis, runs at a right angle to the diagonal plane. As the glenohumeral joint moves from diagonal abduction to diagonal adduction in overhand throwing,

its axis runs perpendicular to the plane through the humeral head.

Body regions

As mentioned later under the skeletal system, the body can be divided into axial and appendicular regions. Each of these regions may be further divided into different subregions, such as the cephalic, cervical, trunk, upper limbs, and lower limbs. Within each of these regions are many more subregions and specific regions. Table 1.2 details a breakdown of these regions and their common names, illustrated in Fig. 1.7.

TABLE 1.2 • **Body parts and regions**

	Region name	Common name	Subregion	Specific region name	Common name for specific region
Axial	Cephalic	Head	Cranial (skull)	Frontal	Forehead
				Occipital	Base of skull
			Facial (face)	Orbital	Eye
				Otic	Ear
				Nasal	Nose
				Buccal	Cheek
				Oral	Mouth
				Mental	Chin
	Cervical	Neck		Nuchal	Posterior neck
				Throat	Anterior neck
	Trunk	Thoracic	Thorax	Clavicular	Collar bone
				Pectoral	Chest
				Sternal	Breastbone
				Costal	Ribs
				Mammary	Breast
		Dorsal	Back	Scapula	Shoulder blade
				Vertebral	Spinal column
				Lumbar	Lower back or loin
		Abdominal	Abdomen	Celiac	Abdomen
				Umbilical	Navel
		Pelvic	Pelvis	Inguinal	Groin
				Pubic	Genital
				Coxal	Hip
				Sacral	Between hips
				Gluteal	Buttock
				Perineal	Perineum
Appendicular	Upper limbs	Shoulder		Acromial	Point of shoulder
				Omus	Deltoid
				Axillary	Armpit
				Brachial	Arm
				Olecranon	Point of elbow
				Cubital	Elbow
				Antecubital	Front of elbow
				Antebrachial	Forearm
				Carpal	Wrist
		Manual		Palmar	Palm
				Dorsal	Back of hand
				Digital	Finger
	Lower limbs			Femoral	Thigh
				Patella	Kneecap
				Popliteal	Back of knee
				Sural	Calf
				Crural	Leg
		Pedal	Foot	Talus	Ankle
				Calcaneal	Heel
				Dorsum	Top of foot
				Tarsal	Instep
				Plantar	Sole
				Digital	Toe

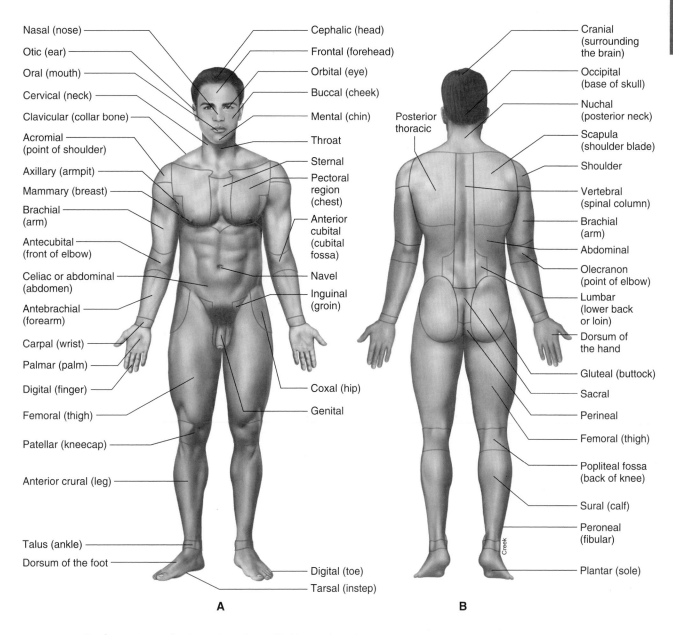

FIG. 1.7 ● Body regions. **A,** Anterior view; **B,** Posterior view.

Skeletal systems

Fig. 1.8 shows anterior and posterior views of the skeletal system. Some 206 bones make up the skeletal system, which provides support and protection for other systems of the body and provides for attachments of the muscles to the bones, by which movement is produced. Additional skeletal functions are mineral storage and hemopoiesis, which involves blood cell formation in the red bone marrow. The skeleton may be divided into the appendicular and the axial skeletons. The appendicular skeleton is composed of the appendages, or the upper and lower extremities, and the shoulder and pelvic girdles. The axial

skeleton consists of the skull, vertebral column, ribs, and sternum. Most students who take this course will have had a course in human anatomy, but a brief review is desirable before beginning the study of kinesiology. Later chapters provide additional information and more detailed illustrations of specific bones.

Osteology

The adult skeleton, consisting of approximately 206 bones, may be divided into the axial skeleton and the appendicular skeleton. The axial skeleton contains 80 bones, which include the skull, spinal column, sternum, and ribs. The appendicular skeleton contains

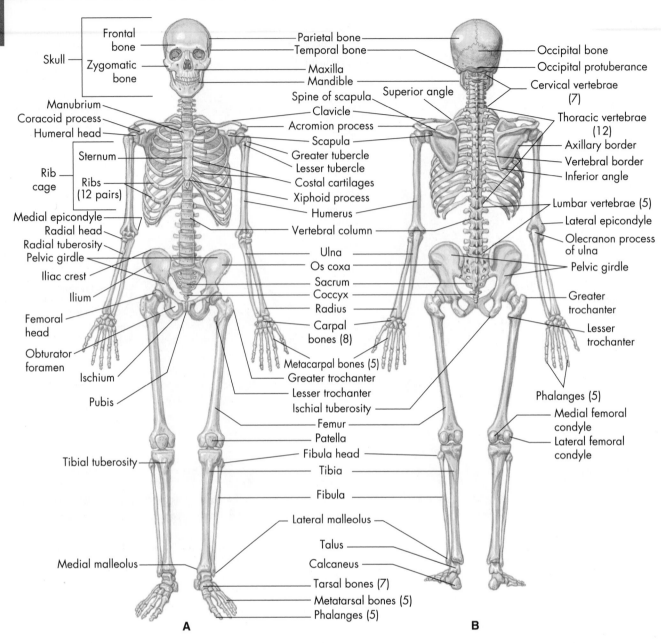

FIG. 1.8 • Skeleton. **A,** Anterior view; **B,** Posterior view.

126 bones, which include all the bones of the upper and lower extremities. The pelvis is sometimes classified as being part of the axial skeleton due to its importance in linking the axial skeleton with the lower extremities of the appendicular skeleton. The exact number of bones as well as their specific features occasionally varies from person to person.

Skeletal functions

The skeleton has five major functions:

1. Protection of vital soft tissues such as the heart, lungs, and brain

2. Support to maintain posture

3. Movement by serving as points of attachment for muscles and acting as levers

4. Storage for minerals such as calcium and phosphorus

5. Hemopoiesis, which is the process of blood formation that occurs in the red bone marrow located in the vertebral bodies, femur, humerus, ribs, and sternum

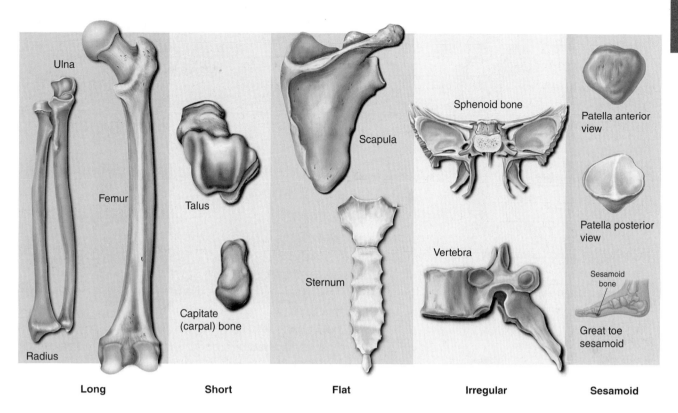

Long   Short   Flat   Irregular   Sesamoid

FIG. 1.9 • Classification of bones by shape.

Types of bones

Bones vary greatly in shape and size but can be classified in five major categories (Fig. 1.9).

Long bones: Composed of a long cylindrical shaft with relatively wide, protruding ends; serve as levers. The shaft contains the medullary cavity. Examples include phalanges, metatarsals, metacarpals, tibia, fibula, femur, radius, ulna, and humerus.

Short bones: Small cube-shaped, solid bones that usually have a proportionally large articular surface in order to articulate with more than one bone. Short bones provide some shock absorption and include the carpals and tarsals.

Flat bones: Usually having a curved surface and varying from thick (where tendons attach) to very thin. Flat bones generally provide protection and include the ilium, ribs, sternum, clavicle, and scapula.

Irregular bones: Irregular-shaped bones serve a variety of purposes and include the bones throughout the entire spine and the ischium, pubis, and maxilla.

Sesamoid bones: Small bones embedded within the tendon of a musculotendinous unit that provide protection as well as improve the mechanical advantage of musculotendinous units. In addition to the patella, there are small sesamoid bones within the flexor tendons of the great toe and the thumb. Sesamoid bones are sometimes referred to as accessory bones and, beyond those already mentioned, may occur in varying numbers from one individual to the next. They are most commonly found in smaller joints in the distal extremities of the foot, ankle, and hand.

Typical bony features

Long bones possess features that are typical of bones in general, as illustrated in Fig. 1.10. Long bones have a shaft or **diaphysis**, which is the long cylindrical portion of the bone. The diaphysis wall, formed from hard, dense, compact bone, is the **cortex**. The outer surface of the diaphysis is covered by a dense, fibrous membrane known as the **periosteum**. A similar fibrous membrane known as the **endosteum** covers the inside of the cortex. Between the walls of the diaphysis lies the **medullary** or marrow cavity, which contains yellow or fatty marrow. At each end of a long bone is the **epiphysis**, which is usually enlarged and shaped specifically to join with the epiphysis of an adjacent bone at a joint. The epiphysis is formed from spongy or **cancellous** or **trabecular** bone.

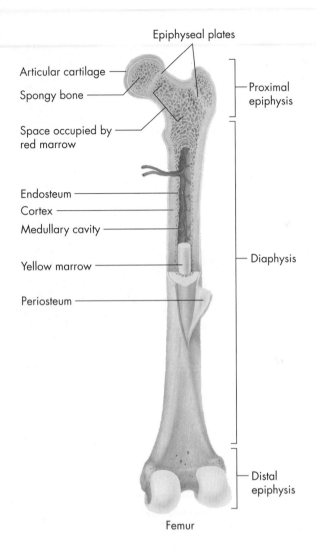

FIG. 1.10 • Major parts of a long bone.

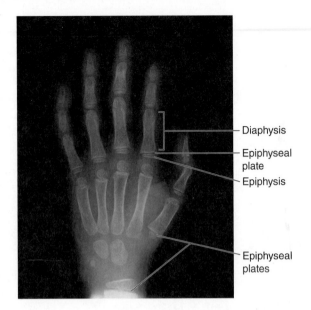

FIG. 1.11 • The presence of epiphyseal plates, as seen in a radiograph of a child's hand, indicates that the bones are still growing in length.

TABLE 1.3 • **Epiphyseal closure timetables**

Approximate age	Bones
7–8	Inferior rami of pubis and ischium (almost complete)
15–17	Scapula, lateral epicondyle of humerus, olecranon process of ulna
18–19	Medial epicondyle of humerus, head and shaft of radius
About 20	Humeral head, distal ends of radius and ulna, distal ends of femur and fibula, proximal end of tibia
20–25	Acetabulum in pelvis
25	Vertebrae and sacrum, clavicle, proximal end of fibula, sternum and ribs

Adapted from Goss CM: *Gray's anatomy of the human body,* ed 29, Philadelphia, 1973, Lea & Febiger.

During bony growth the diaphysis and the epiphysis are separated by a thin plate of cartilage known as the **epiphyseal plate**, commonly referred to as a growth plate (Fig. 1.11). As skeletal maturity is reached, on a timetable that varies from bone to bone as detailed in Table 1.3, the plates are replaced by bone and are closed. To facilitate smooth, easy movement at joints, the epiphysis is covered by **articular** or **hyaline** cartilage, which provides a cushioning effect and reduces friction.

Bone development and growth

Most of the skeletal bones of concern to us in structural kinesiology are **endochondral bones**, which develop from hyaline cartilage. As we develop from an embryo, these hyaline cartilage masses grow rapidly into structures shaped similarly to the bones they will eventually become.

This growth continues, and the cartilage gradually undergoes significant change to develop into long bone, as detailed in Fig. 1.12.

Bones continue to grow longitudinally as long as the epiphyseal plates are open. These plates begin closing around adolescence and disappear. Most close by age 18, but some may be open until age 25. Growth in diameter continues throughout life. This is done by an internal layer of periosteum

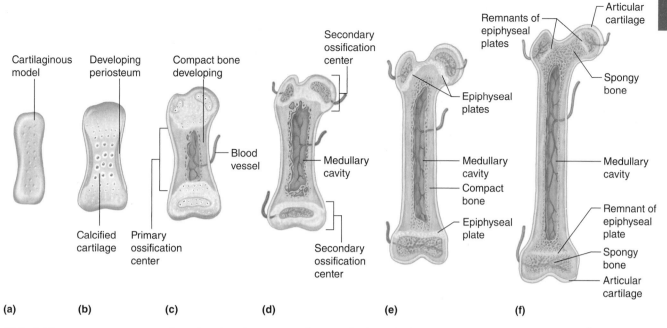

Cartilaginous model

Developing periosteum

Compact bone developing

Blood vessel

Secondary ossification center

Medullary cavity

Secondary ossification center

Calcified cartilage

Primary ossification center

Epiphyseal plates

Medullary cavity

Compact bone

Epiphyseal plate

Remnants of epiphyseal plates

Articular cartilage

Spongy bone

Medullary cavity

Remnant of epiphyseal plate

Spongy bone

Articular cartilage

(a)   (b)   (c)   (d)   (e)   (f)

FIG. 1.12 • Major stages **a–f** in the development of an endochondral bone (relative bone sizes not to scale).

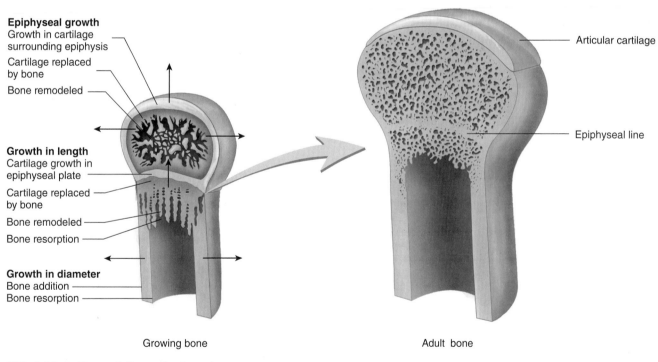

Epiphyseal growth
Growth in cartilage surrounding epiphysis

Cartilage replaced by bone

Bone remodeled

Growth in length
Cartilage growth in epiphyseal plate

Cartilage replaced by bone

Bone remodeled

Bone resorption

Growth in diameter
Bone addition
Bone resorption

Articular cartilage

Epiphyseal line

Growing bone

Adult bone

FIG. 1.13 • Remodeling of a long bone.

building new concentric layers on old layers. Simultaneously, bone around the sides of the medullary cavity is resorbed so that the diameter is continually increased. New bone is formed by specialized cells known as **osteoblasts**, whereas the cells that resorb old bone are **osteoclasts**. This bone remodeling, as depicted in Fig. 1.13, is necessary for continued bone growth, changes in bone shape, adjustment of bone to stress, and bone repair.

TABLE 1.4 • **Bone markings**

	Marking	Description	Examples	Page
Processes that form joints	Condyle	Large, rounded projection that usually articulates with another bone	Medial or lateral condyle of femur	276
	Facet	Small, flat or nearly flat surface	Articular facet of vertebra	331
	Head	Prominent, rounded projection of the proximal end of a bone, usually articulating	Head of femur, head of humerus	230, 232, 233, 113
Processes to which muscles, tendons, or ligaments attach	Angle	Bend or protruding angular projection	Superior and inferior angle of scapula	90, 91
	Border or margin	Edge or boundary line of a bone	Lateral and medial border of scapula	90, 91
	Crest	Prominent, narrow, ridgelike projection	Iliac crest of pelvis	230, 231, 232
	Epicondyle	Projection located above a condyle	Medial or lateral epicondyle of humerus	144
	Line	Ridge of bone less prominent than a crest	Linea aspera of femur	232
	Process	Any prominent projection	Acromion process of scapula, olecranon process of humerus	90, 91, 113, 114, 144
	Ramus	Part of an irregularly shaped bone that is thicker than a process and forms an angle with the main body	Superior and inferior ramus of pubis	230
	Spine (spinous process)	Sharp, slender projection	Spinous process of vertebra, spine of scapula	330, 331, 91
	Suture	Line of union between bones	Sagittal suture between parietal bones of skull	16
	Trochanter	Very large projection	Greater or lesser trochanter of femur	230, 232
	Tubercle	Small, rounded projection	Greater and lesser tubercles of humerus	113
	Tuberosity	Large, rounded or roughened projection	Radial tuberosity, tibial tuberosity	144, 276
Cavities (depressions)	Facet	Flattened or shallow articulating surface	Intervertebral facets in cervical, thoracic, and lumbar spine	331
	Foramen	Rounded hole or opening in bone	Obturator foramen in pelvis	230, 231
	Fossa	Hollow, depressed, or flattened surface	Supraspinatus fossa, iliac fossa	90, 230
	Fovea	Very small pit or depression	Fovea capitis of femur	233
	Meatus	Tubelike passage within a bone	External auditory meatus of temporal bone	343
	Notch	Depression in the margin of a bone	Trochlear and radial notch of the ulna	144
	Sinus	Cavity or hollow space within a bone	Frontal sinus	
	Sulcus (groove)	Furrow or groovelike depression on a bone	Intertubercular (bicipital) groove of humerus	113

Bone properties

Calcium carbonate, calcium phosphate, collagen, and water are the basis of bone composition. About 60% to 70% of bone weight is made up of calcium carbonate and calcium phosphate, with water making up approximately 25% to 30% of bone weight. Collagen provides some flexibility and strength in resisting tension. Aging causes progressive loss of collagen and increases bone brittleness, resulting in increased likelihood of fractures.

Most outer bone is cortical; cancellous bone is underneath. Cortical bone is harder and more compact, with only about 5% to 30% of its volume being porous, with nonmineralized tissue. In contrast, cancellous bone is spongy, with around 30% to 90% of its volume being porous. Cortical bone is stiffer; it can withstand greater stress, but less strain, than cancellous bone. Due to its sponginess, cancellous bone can undergo greater strain before fracturing.

Bone size and shape are influenced by the direction and magnitude of forces that are habitually applied to them. Bones reshape themselves based on the stresses placed upon them, and their mass increases over time with increased stress.

This concept of bone adaptation to stress is known as **Wolff's law**, which essentially states that bone in a healthy individual will adapt to the loads it is placed under. When a particular bone is subjected to increased loading, the bone will remodel itself over time to become stronger to resist that particular type of loading. As a result, the external cortical portion of the bone becomes thicker. The opposite is also true: when the loading on a bone decreases, the bone will become weaker.

Bone markings

Bones have specific markings that exist to enhance their functional relationship with joints, muscles, tendons, nerves, and blood vessels. Many of these markings serve as important bony landmarks in determining muscle location and attachment and joint function. Essentially, all bone markings may be divided into

1. Processes (including elevations and projections), which either form joints or serve as a point of attachment for muscles, tendons, or ligaments, and
2. Cavities (depressions), which include openings and grooves that contain tendons, vessels, nerves, and spaces for other structures.

Detailed descriptions and examples of many bony markings are provided in Table 1.4.

Types of joints

The articulation of two or more bones allows various types of movement. The extent and type of movement determine the name applied to the joint. Bone structure limits the kind and amount of movement in each joint. Some joints or **arthroses** have no movement, others are only very slightly movable, and others are freely movable with a variety of movement ranges. The type and range of movements are similar in all humans; but the freedom, range, and vigor of movements are limited by the configuration of the bones where they fit together, and by ligaments and muscles.

Articulations may be classified according to the structure or function. Classification by structure places joints into one of three categories: fibrous, cartilaginous, or synovial. Functional classification also results in three categories: synarthrosis (synarthrodial), amphiarthrosis (amphiarthrodial), and diarthrosis (diarthrodial). There are subcategories in each classification. Due to the strong relationship between structure and function, there is significant overlap between the classification systems. That is, there is more similarity than difference between the two members in each of the following pairs: fibrous and synarthrodial joints, cartilaginous and amphiarthrodial joints, and synovial and diarthrodial joints. However, not all joints fit neatly into both systems. Table 1.5 provides a detailed listing of all joint types according to both classification systems. Since this text is concerned primarily with movement, the more functional system (synarthrodial, amphiarthrodial, and diarthrodial joints) will be used throughout, following a brief explanation of structural classification.

Fibrous joints are joined together by connective tissue fibers and are generally immovable. Subcategories are suture and gomphosis, which are immovable, and syndesmosis, which allows a slight amount of movement. Cartilaginous joints are joined together by hyaline cartilage or fibrocartilage, which allows very slight movement. Subcategories include synchondrosis and symphysis. Synovial joints are freely movable and generally are diarthrodial. Their structure and subcategories are discussed in detail under diarthrodial joints.

The articulations are grouped into three classes based primarily on the amount of movement possible, with consideration given to their structure.

TABLE 1.5 • Joint classification by structure and function

		Structural classification		
		Fibrous	Cartilaginous	Synovial
Functional classification	Synarthrodial	Gomphosis Suture	———	———
	Amphiarthrodial	Syndesmosis	Symphysis Synchondrosis	———
	Diarthrodial	———	———	Arthrodial Condyloidal Enarthrodial Ginglymus Sellar Trochoidal

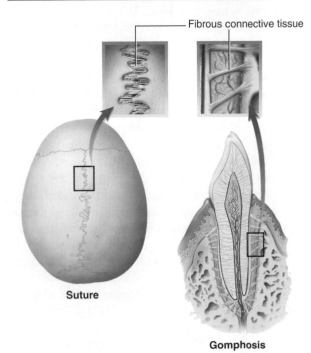

FIG. 1.14 • Synarthrodial joints.

Synarthrodial (immovable) joints   FIG. 1.14

Structurally, these articulations are divided into two types:

Suture

Found in the sutures of the cranial bones. The sutures of the skull are truly immovable beyond infancy.

Gomphosis

Found in the sockets of the teeth. The socket of a tooth is often referred to as a gomphosis (type of joint in which a conical peg fits into a socket). Normally, there should be essentially no movement of the teeth in the mandible or maxilla.

Amphiarthrodial (slightly movable) joints   FIG. 1.15

Structurally, these articulations are divided into three types:

Syndesmosis

Type of joint held together by strong ligamentous structures that allow minimal movement between the bones. Examples are the coracoclavicular joint and the inferior tibiofibular joint.

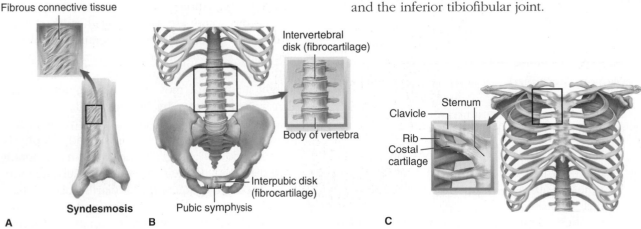

FIG. 1.15 • Amphiarthrodial joints. **A,** Syndesmosis joint; **B,** Symphysis joint; **C,** Synchondrosis joint.

Symphysis

Type of joint separated by a fibrocartilage pad that allows very slight movement between the bones. Examples are the symphysis pubis and the intervertebral disks.

Synchondrosis

Type of joint separated by hyaline cartilage that allows very slight movement between the bones. Examples are the costochondral joints of the ribs with the sternum.

Diarthrodial (freely movable) joints   FIG. 1.16

Diarthrodial joints, also known as synovial joints, are freely movable. A sleevelike covering of ligamentous tissue known as the **joint capsule** surrounds the bony ends forming the joints. This ligamentous capsule is lined with a thin vascular synovial capsule that secretes synovial fluid to lubricate the area inside the joint capsule, known as the **joint cavity**. In certain areas the capsule is thickened to form tough,

nonelastic **ligaments** that provide additional support against abnormal movement or joint opening. These ligaments vary in location, size, and strength depending upon the particular joint. Ligaments, in connecting bones to bones, provide static stability to joints.

In many cases, additional ligaments, not continuous with the joint capsule, provide further support. In some cases, these additional ligaments may be contained entirely within the joint capsule; or intraarticularly, such as the anterior cruciate ligament in the knee; or extraarticularly, such as the fibular collateral ligament of the knee, which is outside the joint capsule.

The articular surfaces on the ends of the bones inside the joint cavity are covered with layers of articular or **hyaline cartilage** that helps protect the ends of the bones from wear and damage. This cartilage is quite resilient because it is slightly compressible and elastic, which enables it to absorb compressive and shear forces. The articular surface, thanks in part to lubrication

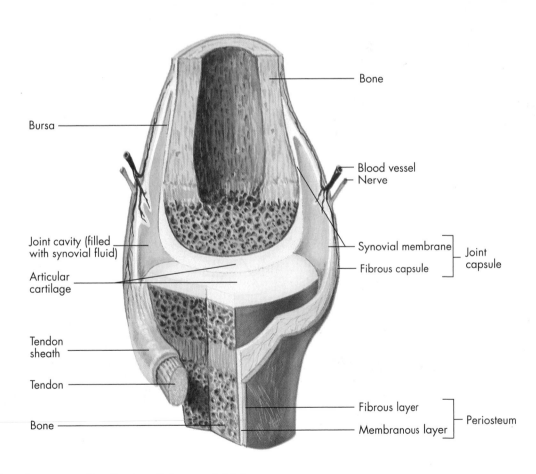

FIG. 1.16 • Structure of a diarthrodial synovial joint.

from synovial fluid, has a very low amount of friction and is very durable. When the joint surfaces are unloaded or distracted, this articular cartilage slowly absorbs a slight amount of the joint synovial fluid, only to slowly secrete it during subsequent weight bearing and compression. Articular cartilage has a very limited blood supply and as a result depends on joint movement to provide its nutrition through this synovial flow. Therefore, maintaining and utilizing a joint through its normal range of motion are important to sustaining joint health and function.

Additionally, some diarthrodial joints have a fibrocartilage disk between their articular surfaces to provide additional shock absorption and further enhance joint stability. Examples are the knee's medial and lateral menisci and the acetabular and glenoid labrum of the hip and shoulder joints, respectively. Structurally, this type of articulation can be divided into six groups, as shown in Fig. 1.17.

Diarthrodial joints have motion possible in one or more planes. Those joints having motion possible in one plane are said to have one degree of freedom of motion, whereas joints having motion in two and three planes of motion are described as having two and three degrees of freedom of motion, respectively. Refer to Table 1.6 for a comparison of diarthrodial joint features by subcategory.

Arthrodial (gliding, plane) joint

This joint type is characterized by two flat, or plane, bony surfaces that butt against each other. This type of joint permits limited gliding movement. Examples are the carpal bones of the wrist and the tarsometatarsal joints of the foot.

Condyloidal (ellipsoid, ovoid, biaxial ball-and-socket) joint

This is a type of joint in which the bones permit movement in two planes without rotation. Examples are the wrist (radiocarpal joint) between the radius and the proximal row of the carpal bones or the second, third, fourth, and fifth metacarpophalangeal joints.

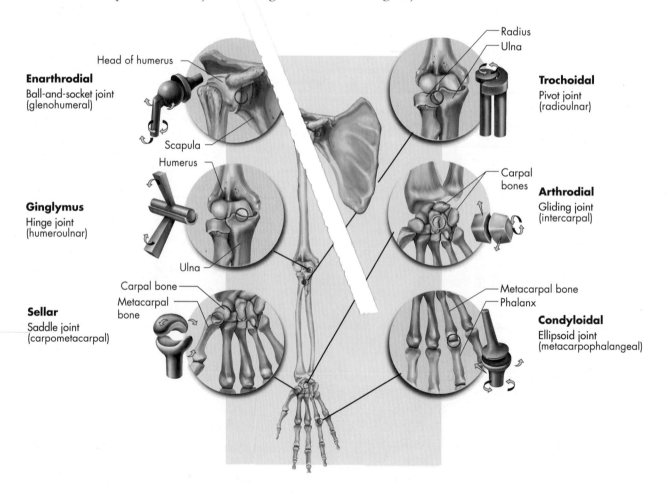

FIG. 1.17 • Types of diarthrodial or synovial joints.

TABLE 1.6 • Diarthrodial joint classification

Classification name	Number of axes	Degrees of freedom	Typical movements	Joint examples	Plane for examples	Axis for examples
Ginglymus (hinge)	Uniaxial	One	Flexion, extension	Elbow joint (humeroulnar) Ankle joint (talocrural)	Sagittal	Frontal
Trochoidal (pivot, screw)			Internal rotation, external rotation	Proximal and distal radioulnar joint Atlantoaxial joint	Transverse	Vertical
Condyloidal (ellipsoid, ball-and-socket, ovoid)	Biaxial	Two	Flexion, extension, abduction, adduction	Wrist (radiocarpal) 2nd–5th metacarpophalangeal joints	Sagittal Frontal	Frontal Sagittal
Arthrodial (gliding, plane)	Multiaxial	Three	Flexion, extension, abduction, adduction, internal rotation, external rotation	Transverse tarsal joint Vertebral facets in spine Intercarpal joints in wrist	Variable Frontal Variable	Variable Sagittal Variable
Enarthrodial (ball-and-socket, spheroidal)				Glenohumeral joint Hip joint (acetabularfemoral)	Sagittal Frontal Transverse	Frontal Sagittal Vertical
Sellar (saddle)				1st carpometacarpal joint	Sagittal Frontal Transverse	Frontal Sagittal Vertical

Enarthrodial (spheroidal, multiaxial ball-and-socket) joint

This type of joint is most like a true ball-and-socket in that it permits movement in all planes. Examples are the shoulder (glenohumeral) and hip (acetabularfemoral) joints.

Ginglymus (hinge) joint

This is a type of joint that permits a wide range of movement in only one plane. Examples are the elbow (humeroulnar), ankle (talocrural), and knee (tibiofemoral) joints.

Sellar (saddle) joint

This type of reciprocal reception is found only in the thumb at the carpometacarpal joint and permits ball-and-socket movement, with the exception of slight rotation.

Trochoidal (pivot, screw) joint

This is a type of joint with a rotational movement around a long axis. An example is the rotation of the radius on the ulna at the proximal and distal radioulnar joints.

Stability and mobility of diarthrodial joints

Generally, the more mobile a joint is, the less stable it is, and vice versa. This is true when comparing the same joints between individuals, but also when comparing one joint versus another in the same individual. Both heredity and developmental factors (Wolff's law for bone and Davis's law for soft tissue) contribute to these variances. In a manner similar to the adaption of bone to loading, as previously discussed in Wolff's law, soft tissue also adapts to stress or the lack thereof. This corollary to Wolff's law is known as **Davis's law**, which essentially states that ligaments, muscle, and other soft tissue when placed under appropriate tension will adapt over time by lengthening, and conversely, when maintained in a loose or shortened state over a period of time will gradually shorten.

Five major factors affect the total stability, and consequently the mobility, of a joint (see Fig. 1.18).

- Bones—Although bones are usually very similar in bilateral comparisons within an individual, the actual anatomical configuration at the joint surfaces in terms of depth and shallowness may vary significantly between individuals.
- Cartilage—The structures of both hyaline cartilage and specialized cartilaginous structures, such as the knee menisci, glenoid labrum, and acetabular labrum, further assist in joint congruency and stability. As with bones, these

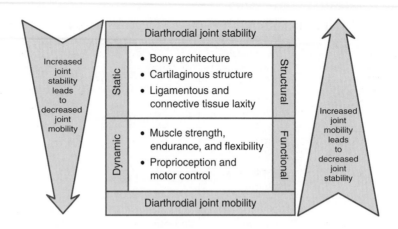

FIG. 1.18 • Factors affecting diarthrodial joint stability.

structures normally are the same in bilateral comparisons within an individual, but may vary between individuals in size and configuration.

- Ligaments and connective tissue—Ligaments and connective tissue provide static stability to joints. As with bones and cartilage, variances exist between individuals in the degree of restrictiveness of ligamentous tissue. An individual's amount of hypo- or hyperlaxity is primarily due to the proportional amount of elastin versus collagen within the joint structures. Simply put, individuals with proportionally higher elastin-to-collagen ratios are hyperlax, or "loose-jointed," whereas individuals with proportionally lower ratios are tighter.
- Muscles—Muscles provide dynamic stability to joints when actively contracting. Without active tension via contraction, muscles provide minimal static stability. Consequently, strength and endurance are significant factors in stabilizing joints, whereas muscle flexibility may affect the total range of joint motion possible.
- Proprioception and motor control— Proprioception is the subconscious mechanism by which the body is able to regulate posture and movements by responding to stimuli originating in the proprioceptors embedded in joints, tendons, muscles, and the inner ear. Motor control is the process by which bodily actions and movements are organized and executed. To determine the appropriate amount of muscular forces and joint activations needed, sensory information from the environment and the body must be integrated and then coordinated in a cooperative manner between the central nervous system and the musculoskeletal system. Muscle strength and endurance are not very useful

in providing joint stability unless they can be activated precisely when needed.

The integrity of any of these structures may be affected by acute or chronic injury. These structures adapt over time both positively and negatively to the specific biomechanical demands placed upon them. When any of the above factors are compromised, additional demands are placed on the remaining structures to provide stability, which in turn may compromise their integrity, resulting in abnormal mobility. This abnormal mobility, whether hypermobility or hypomobility, may lead to further pathological conditions such as tendinitis, bursitis, arthritis, internal derangement, and joint subluxations.

Movements in joints

In many joints, several different movements are possible. Some joints permit only flexion and extension; others permit a wide range of movements, depending largely on the joint structure. We refer to the area through which a joint may normally be freely and painlessly moved as the **range of motion (ROM)**. The specific amount of movement possible in a joint or range of motion may be measured by using an instrument known as a **goniometer** to compare the change in joint angles. The goniometer has a moving arm, a stationary arm, and an axis or fulcrum. Measuring the available range of motion in a joint or the angles created by the bones of a joint is known as **goniometry**.

The goniometer axis, or hinge point, is placed even with the axis of rotation at the joint line. The stationary arm is held in place either along or parallel to the long axis of the more stationary bone (usually the more proximal bone), and

FIG. 1.19 ● Goniometric measurement of knee joint flexion.

the moving arm is placed either along or parallel to the long axis of the bone that moves the most (usually the more distal bone). The joint angle can then be read from the goniometer, as shown in Fig. 1.19. As an example, we could measure the angle between the femur and the trunk in the anatomical position (which would usually be zero), and then ask the person to flex the hip as far as possible. If we measured the angle again at full hip flexion, we would find a goniometer reading of around 130 degrees.

Depending on the size of the joint and its movement potential, different goniometers may be more or less appropriate. Fig. 1.20 depicts a variety of goniometers that may be utilized to determine the range of motion for a particular joint. Inclinometers may also be used to measure range of motion, particularly in the spine.

Please note that the normal range of motion for a particular joint varies to some degree from

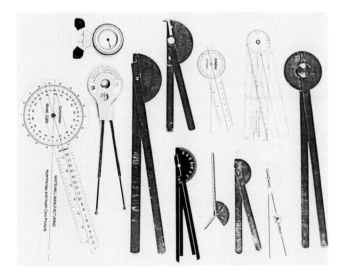

FIG. 1.20 ● Various goniometers used for measuring joint range of motion.

person to person. Appendixes 1 and 2 provide the average normal ranges of motion for all joints.

When using movement terminology, it is important to understand that the terms are used to describe the actual change in position of the bones relative to each other. That is, the angles between the bones change, whereas the movement occurs between the articular surfaces of the joint. We may say, in describing knee movement, "flex the leg at the knee"; this movement results in the leg moving closer to the thigh. Some describe this as leg flexion occurring at the knee joint and may say "flex the leg," meaning flex the knee. Additionally, movement terms are utilized to describe movement occurring throughout the full range of motion or through a very small range. Using the knee flexion example again, we may flex the knee through the full range by beginning in full knee extension (zero degrees of knee flexion) and flexing it fully, so that the heel comes in contact with the buttocks; this would be approximately 140 degrees of flexion. We may also begin with the knee in 90 degrees of flexion and then flex it 30 degrees more; this movement results in a knee flexion angle of 120 degrees, even though the knee flexed only 30 degrees. In both examples, the knee is in different degrees of flexion. We may also begin with the knee in 90 degrees of flexion and extend it 40 degrees, which would result in a flexion angle of 50 degrees. Even though we extended the knee, it is still flexed, only less so than before.

In this example, we more commonly move the distal extremity in relation to the proximal extremity, which is usually more stationary. However, there are examples in every joint where the distal segment may be more stationary and we move the proximal segment in relation to it. An example is the knee in doing a squat from the standing position. As the squat occurs, the thigh moves toward the stabler leg, still resulting in knee flexion that could be stated as flexing the thigh at the knee.

Some movement terms may be used to describe motion at several joints throughout the body, whereas other terms are relatively specific to a joint or group of joints (Fig. 1.21). Rather than list the terms alphabetically, we have chosen to group them according to the body area and pair them with opposite terms where applicable. Additionally, the prefixes **hyper-** and **hypo-** may be combined with these terms to emphasize motion beyond and below normal, respectively. Of these combined terms, **hyperextension** is the most commonly used.

A **B** **C**

FIG. 1.21 • Joint movements. **A,** Examples of sagittal plane movements: extension of left toes, ankle (plantar flexion), knee, hip, shoulder, elbow, wrist, fingers, lumbar and cervical spine; flexion of right toes, ankle (dorsiflexion), knee, hip, shoulder, elbow, wrist, and fingers. **B,** Examples of frontal plane movements: abduction of left transverse tarsal/subtalar joints (eversion), shoulder, wrist, fingers, and shoulder girdle (upward rotation), lumbar (lateral flexion to right) and cervical spine (lateral flexion to left), and right hip; adduction of right transverse tarsal/subtalar joints (inversion), shoulder, wrist, fingers, and shoulder girdle (downward rotation). **C,** Examples of transverse plane movements: internal rotation of right hip, left shoulder, radioulnar joints (pronation); external rotation of left knee, hip, right shoulder, radioulnar joints (supination), and lumbar (right rotation) and cervical spine (right rotation).

Terms describing general movements

Abduction: Lateral movement away from the midline of the trunk in the frontal plane. An example is raising the arms or legs to the side horizontally.

Adduction: Movement medially toward the midline of the trunk in the frontal plane. An example is lowering the arm to the side or the thigh back to the anatomical position.

Flexion: Bending movement that results in a decrease of the angle in a joint by bringing bones together, usually in the sagittal plane. An example is the elbow joint when the hand is drawn to the shoulder.

Extension: Straightening movement that results in an increase of the angle in a joint by moving bones apart, usually in the sagittal plane. Using the elbow, an example is when the hand moves away from the shoulder.

Circumduction: Circular movement of a limb that delineates an arc or describes a cone. It is a combination of flexion, extension, abduction, and adduction. Sometimes referred to as circumflexion. An example is when the shoulder joint or the hip joint moves in a circular

fashion around a fixed point, either clockwise or counterclockwise.

Diagonal abduction: Movement by a limb through a diagonal plane away from the midline of the body, such as in the hip or glenohumeral joint.

Diagonal adduction: Movement by a limb through a diagonal plane toward and across the midline of the body, such as in the hip or glenohumeral joint.

External rotation: Rotary movement around the longitudinal axis of a bone away from the midline of the body. Occurs in the transverse plane and is also known as rotation laterally, outward rotation, and lateral rotation.

Internal rotation: Rotary movement around the longitudinal axis of a bone toward the midline of the body. Occurs in the transverse plane and is also known as rotation medially, inward rotation, and medial rotation.

Terms describing ankle and foot movements

Eversion: Turning the sole of the foot outward or laterally in the frontal plane; abduction. An example is standing with the weight on the inner edge of the foot.

Inversion: Turning the sole of the foot inward or medially in the frontal plane; adduction. An example is standing with the weight on the outer edge of the foot.

Dorsal flexion (dorsiflexion): Flexion movement of the ankle that results in the top of the foot moving toward the anterior tibia in the sagittal plane.

Plantar flexion: Extension movement of the ankle that results in the foot and/or toes moving away from the body in the sagittal plane.

Pronation: A position of the foot and ankle resulting from a combination of ankle dorsiflexion, subtalar eversion, and forefoot abduction (toe-out).

Supination: A position of the foot and ankle resulting from a combination of ankle plantar flexion, subtalar inversion, and forefoot adduction (toe-in).

Terms describing radioulnar joint movements

Pronation: Internally rotating the radius in the transverse plane so that it lies diagonally across the ulna, resulting in the palm-down position of the forearm.

Supination: Externally rotating the radius in the transverse plane so that it lies parallel to the ulna, resulting in the palm-up position of the forearm.

Terms describing shoulder girdle (scapulothoracic) movements

Depression: Inferior movement of the shoulder girdle in the frontal plane. An example is returning to the normal position from a shoulder shrug.

Elevation: Superior movement of the shoulder girdle in the frontal plane. An example is shrugging the shoulders.

Protraction (abduction): Forward movement of the shoulder girdle in the horizontal plane away from the spine. Abduction of the scapula.

Retraction (adduction): Backward movement of the shoulder girdle in the horizontal plane toward the spine. Adduction of the scapula.

Rotation downward: Rotary movement of the scapula in the frontal plane with the inferior angle of the scapula moving medially and downward. Occurs primarily in the return from upward rotation. The inferior angle may actually move upward slightly as the scapula continues in extreme downward rotation.

Rotation upward: Rotary movement of the scapula in the frontal plane with the inferior angle of the scapula moving laterally and upward.

Terms describing shoulder joint (glenohumeral) movements

Horizontal abduction: Movement of the humerus or femur in the horizontal plane away from the midline of the body. Also known as horizontal extension or transverse abduction.

Horizontal adduction: Movement of the humerus or femur in the horizontal plane toward the midline of the body. Also known as horizontal flexion or transverse adduction.

Scaption: Movement of the humerus away from the body in the scapular plane. Glenohumeral abduction in a plane 30 to 45 degrees between the sagittal and frontal planes.

Terms describing spine movements

Lateral flexion (side bending): Movement of the head and/or trunk in the frontal plane laterally away from the midline. Abduction of the spine.

Reduction: Return of the spinal column in the frontal plane to the anatomic position from lateral flexion. Adduction of the spine.

Terms describing wrist and hand movements

Dorsal flexion (dorsiflexion): Extension movement of the wrist in the sagittal plane with the dorsal or posterior side of the hand moving toward the posterior side of the forearm.

Palmar flexion: Flexion movement of the wrist in the sagittal plane with the volar or anterior side of the hand moving toward the anterior side of the forearm.

Radial flexion (radial deviation): Abduction movement at the wrist in the frontal plane of the thumb side of the hand toward the lateral forearm.

Ulnar flexion (ulnar deviation): Adduction movement at the wrist in the frontal plane of the little finger side of the hand toward the medial forearm.

Opposition of the thumb: Diagonal movement of the thumb across the palmar surface of the hand to make contact with the fingers.

Reposition of the thumb: Diagonal movement of the thumb as it returns to the anatomical position from opposition with the hand and/or fingers.

These movements are considered in detail in the chapters that follow as they apply to the individual joints.

Combinations of movements can occur. Flexion or extension can occur with abduction, adduction, or rotation.

Movement icons (pedagogical feature)

Throughout this text a series of movement icons will be utilized to represent different joint movements. These icons will be displayed in the page margins to indicate the joint actions of the muscles displayed on that page. As further explained in Chapter 2, the actions displayed represent the movements that occur when the muscle contracts concentrically. Table 1.7 provides a complete list of the icons. Refer to them as needed when reading Chapters 4, 5, 6, 7, 9, 10, 11, and 12.

TABLE 1.7 • **Movement icons representing joint actions**

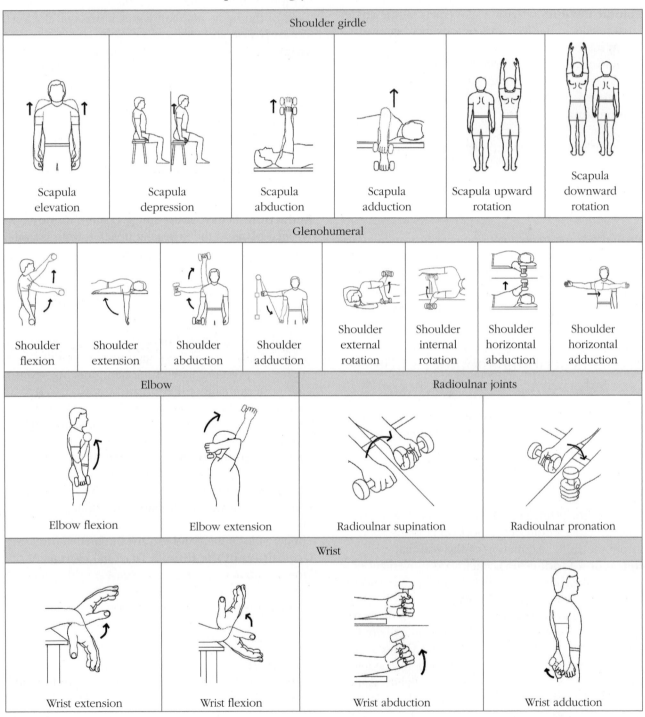

TABLE 1.7 (continued) • Movement icons representing joint actions

Thumb carpometacarpal joint			Thumb metacarpophalangeal joint		Thumb interphalangeal joint	
Thumb CMC flexion	Thumb CMC extension	Thumb CMC abduction	Thumb MCP flexion	Thumb MCP extension	Thumb IP flexion	Thumb IP extension

2nd, 3rd, 4th, and 5th MCP, PIP, and DIP joints		2nd, 3rd, 4th, and 5th MCP and PIP joints	2nd, 3rd, 4th, and 5th metacarpophalangeal joints		2nd, 3rd, 4th, and 5th PIP joints	2nd, 3rd, 4th, and 5th DIP joints
2nd–5th MCP, PIP, and DIP flexion	2nd–5th MCP, PIP, and DIP extension	2nd–5th MCP and PIP flexion	2nd–5th MCP flexion	2nd–5th MCP extension	2nd–5th PIP flexion	2nd–5th DIP flexion

Hip					
Hip flexion	Hip extension	Hip abduction	Hip adduction	Hip external rotation	Hip internal rotation

Knee			
Knee flexion	Knee extension	Knee external rotation	Knee internal rotation

Ankle		Transverse tarsal and subtalar joints	
Ankle plantar flexion	Ankle dorsal flexion	Transverse tarsal and subtalar inversion	Transverse tarsal and subtalar eversion

Great toe metatarsophalangeal and interphalangeal joints		2nd–5th metatarsophalangeal, proximal interphalangeal, and distal interphalangeal joints	
Great toe MTP and IP flexion	Great toe MTP and IP extension	2nd–5th MTP, PIP, and DIP flexion	2nd–5th MTP, PIP, and DIP extension
Cervical spine			
Cervical flexion	Cervical extension	Cervical lateral flexion	Cervical rotation unilaterally
Lumbar spine			
Lumbar flexion	Lumbar extension	Lumbar lateral flexion	Lumbar rotation unilaterally

Physiological movements versus accessory motions

Movements such as flexion, extension, abduction, adduction, and rotation occur by the bones moving through planes of motion about an axis of rotation at the joint. These movements may be referred to as physiological movements. The motion of the bones relative to the three cardinal planes resulting from these physiological movements is referred to as **osteokinematic motion**. In order for these osteokinematic motions to occur, there must be movement between the actual articular surfaces of the joint. This motion between the articular surfaces is known as **arthrokinematics**, and it includes three specific types of **accessory motions**. These accessory motions, named specifically to describe the actual change in relationship between the articular surface of one bone relative to another, are **spin**, **roll**, and **glide** (Fig. 1.22).

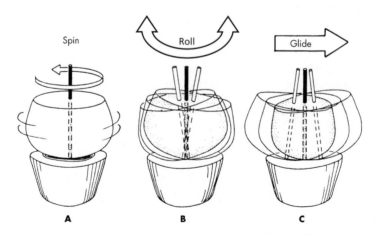

FIG. 1.22 • Joint arthrokinematics. **A,** Spin; **B,** Roll; **C,** Glide.

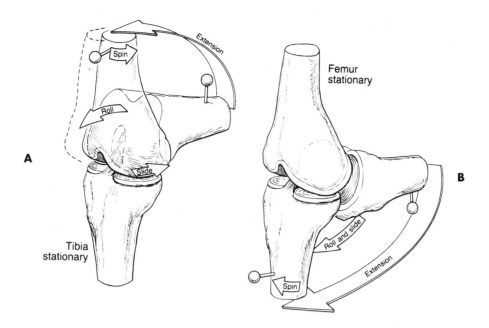

FIG. 1.23 • Knee joint arthrokinematics. **A,** Standing from squatting; **B,** Flexing from non-weight-bearing position.

Roll is sometimes referred to as rock or rocking, whereas glide is sometimes referred to as slide or translation. If accessory motion is prevented from occurring, then physiological motion cannot occur to any substantial degree other than by joint compression or distraction. Because most diarthrodial joints in the body are composed of a concave surface articulating with a convex surface, roll and glide must occur together to some degree. For example, as illustrated in Fig. 1.23, as a person stands from a squatting position, in order for the knee to extend, the femur must roll forward and simultaneously slide backward on the tibia. If not for the slide, the femur would roll off the front of the tibia, and if not for the roll, the femur would slide off the back of the tibia.

Spin may occur in isolation or in combination with roll and glide, depending upon the joint structure. To some degree, spin occurs at the knee as it flexes and extends. In the squatting to standing example, the femur spins medially or internally rotates as the knee reaches full extension. Table 1.8 provides examples of accessory motion.

Roll (rock): A series of points on one articular surface contacts a series of points on another articular surface.

Glide (slide, translation): A specific point on one articulating surface comes in contact with a series of points on another surface.

Spin: A single point on one articular surface rotates about a single point on another articular surface. Motion occurs around some stationary longitudinal mechanical axis in either a clockwise or a counterclockwise direction.

TABLE 1.8 • **Accessory motion**

Accessory motion	Anatomical joint example	Analogy	
Roll (rocking)	Knee extension occurring from femoral condyles rolling forward on tibia as a person stands from squatting position	Tire rolling across a road surface, as in normal driving with good traction	*Combination of roll and glide:* Tire spinning on slick ice (i.e., poor traction) but still resulting in movement across the road surface
Glide (slide or translation)	Knee extension occurring from femoral condyles sliding backward on tibia as a person stands from squatting position	Tire skidding across a slick surface with the brakes locked	
Spin	Radioulnar pronation/supination occurring from spinning of radial head against humeral capitulum	Point of a toy top spinning around in one spot on the floor	

REVIEW EXERCISES

1. Complete the blanks in the following paragraphs using each word from the list below only once except for the ones marked with two asterisks,**, which are used twice. The number of dashes indicates the number of letters of the word for each blank.

a. anterior**	s. medial
b. anteroinferior	t. palmar
c. anterolateral	u. plantar
d. anteromedial	v. posterior**
e. anteroposterior	w. posteroinferior
f. anterosuperior	x. posterolateral
g. bilateral	y. posteromedial
h. caudal	z. posterosuperior
i. cephalic	aa. prone
j. contralateral	bb. proximal
k. deep	cc. superficial
l. distal	dd. superior
m. dorsal	ee. superolateral**
n. inferior	ff. superomedial
o. inferolateral	gg. supine
p. inferomedial	hh. ventral
q. ipsilateral	ii. volar
r. lateral	

When Jacob greeted Stephanie at the beach, he reached out with the _____ surface of his hand to grasp the _____ surface of her hand for a handshake. As the _____ aspects of their bodies faced each other, Jacob noticed that the hair located on the most _____ part of Stephanie's head appeared to be a different color than he remembered. He then asked her to turn around so that he could see it from a(n) _____ view. As she did so, it became obvious to him that she had blonde streaks running from her _____ region in a(n) _____ direction all the way down to her _____ region.

Stephanie then asked Jacob if the sunburn on the _____ portions of his shoulders was due to the exposure that his tank-top shirt provided. He replied yes but that it was only a(n) _____ burn and did not go too ____. He then said, "I wish I had had my shirt off so that I would have gotten some more sun on the _____ portion of my shoulders up to my neck."

Stephanie said that she recently got sunburned on her back while lying _____ at the beach. She then flipped her hair around the _____ side of her neck toward the _____ portion of her trunk to expose her _____ region. Jacob remarked, "Wow, instead of the bikini you have on today with straps over your shoulders running from your _____ chest to your _____ shoulders, you must have been wearing one with crossing straps as I see you have tan lines running in a(n) _____ direction to your _____ low back from the _____ aspect of your _____ shoulders. You should have spent more time lying _____." She replied, "Well, I did lie partially on my back and my right side for a while. See where the _____ portion of my right thigh and the _____ portion of my left thigh are tanned just right, but unfortunately in that position the _____ right thigh and _____ thigh received relatively little exposure." Jacob commented, "Yep, when you lie on one side most of the time, you get all the sun on the _____ side and none on the _____ side. It looks like you must have had a towel covering your feet and ankles since your _____ lower extremities are not nearly as tan as your _____ lower extremities." Stephanie replied, "You are correct. I kept the bottom of my lower legs covered almost all of the time while lying on both sides so that the sensitive skin on my _____ and _____ shins would not burn. But I did get a good _____ tan on my _____ trunk, except for the _____ aspect of my right elbow I was resting on." As Jacob slipped his sandals on to protect the _____ aspect of his feet from the hot sand, he said, "Well, nice to see you. I have to go by the doctor's office and get a(n) _____ chest X-ray to make sure my pneumonia has cleared up."

2. **Joint movement terminology chart**

The specific body area joint movement terms arise from the basic motions in the three specific planes: flexion/extension in the sagittal plane, abduction/adduction in the frontal plane, and rotation in the transverse plane. With this in mind, complete the chart by writing the basic motion in the right column for each specific motion listed in the left column by using either *flexion, extension, abduction, adduction,* or *rotation (external* or *internal)*.

Specific motion	Basic motion
Eversion	
Inversion	
Dorsal flexion	
Plantar flexion	
Pronation (radioulnar)	
Supination (radioulnar)	
Lateral flexion	

Specific motion	Basic motion
Reduction	
Radial flexion	
Ulnar flexion	

3. Bone typing chart

Utilizing Fig. 1.8 and other resources, place an "X" in the appropriate column to indicate its classification.

Bone	Long	Short	Flat	Irregular	Sesamoid
Frontal					
Zygomatic					
Parietal					
Temporal					
Occipital					
Maxilla					
Mandible					
Cervical vertebrae					
Clavicle					
Scapula					
Humerus					
Ulna					
Radius					
Carpal bones					
Metacarpals					
Phalanges					
Ribs					
Sternum					
Lumbar vertebrae					
Ilium					
Ischium					
Pubis					
Femur					
Patella					
Fabella					
Tibia					
Fibula					
Talus					
Calcaneus					
Navicular					
Cuneiforms					
Metatarsals					

4. What are the five functions of the skeleton?
5. List the bones of the upper extremity.
6. List the bones of the lower extremity.
7. List the bones of the shoulder girdle.
8. List the bones of the pelvic girdle.
9. Describe and explain the differences and similarities between the radius and ulna.
10. Describe and explain the differences and similarities between the humerus and femur.
11. Using body landmarks, how would you suggest determining the length of each lower extremity for comparison to determine whether someone had a true total leg length discrepancy?
12. Explain why the fibula is more susceptible to fractures than the tibia.
13. Why is the anatomical position so important in understanding anatomy and joint movements?
14. Label the parts of a long bone.

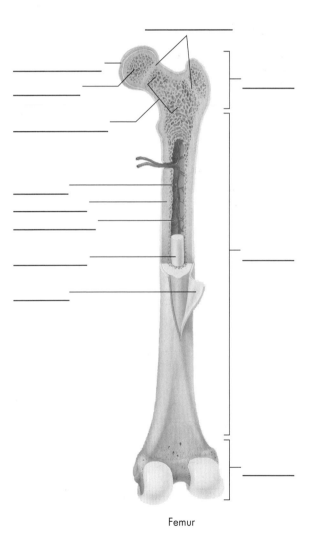

Femur

15. Joint type, movement, and plane of motion chart

Complete the chart by filling in the type of diarthrodial joint and then listing the movements of the joint under the plane of motion in which they occur.

Joint	Type	Planes of motion		
		Sagittal	Lateral	Transverse
Scapulothoracic joint				
Sternoclavicular				
Acromioclavicular				
Glenohumeral joint				
Elbow				
Radioulnar joint				
Wrist				
1st carpometacarpal joint				
1st metacarpophalangeal joint				
Thumb interphalangeal joint				
2nd, 3rd, 4th, and 5th metacarpophalangeal joints				
2nd, 3rd, 4th, and 5th proximal interphalangeal joints				
2nd, 3rd, 4th, and 5th distal interphalangeal joints				
Cervical spine C1–C2				
Cervical spine C2–C7				
Lumbar spine				
Hip				
Knee (tibiofemoral joint)				
Knee (patellofemoral joint)				
Ankle				
Transverse tarsal and subtalar joints				
Metatarsophalangeal joints				
Great toe interphalangeal				
2nd, 3rd, 4th, and 5th proximal interphalangeal joints				
2nd, 3rd, 4th, and 5th distal interphalangeal joints				

16. Joint position chart

Using proper terminology, complete the chart by listing the name of each joint involved and its position upon completion of the multiple joint movement.

Multiple joint movement	Joints and respective position of each
Reach straight over the superior aspect of your head to touch the contralateral ear	
Place the toe of one foot against the posterior aspect of the contralateral calf	
Reach behind the back and use your thumb to touch a spinous process	
Pull the knee as far as possible to the ipsilateral shoulder	
Place the plantar aspect of both feet against each other	

17. Plane of motion and axis of rotation chart

For each joint motion listed in the chart, list the plane of motion in which the motion occurs and its axis of rotation.

Motion	Plane of motion	Axis of rotation
Cervical rotation		
Shoulder girdle elevation		
Glenohumeral horizontal adduction		
Elbow flexion		
Radioulnar pronation		
Wrist radial deviation		
Metacarpophalangeal abduction		
Lumbar lateral flexion		
Hip internal rotation		
Knee extension		
Ankle inversion		
Great toe extension		

18. List two sport skills that involve movements more clearly seen from the side. List the primary movements that occur in the ankle, knee, hip, spine, glenohumeral joint, elbow, and wrist. In which plane are these movements occurring primarily? What axis of rotation is involved primarily?

19. List two sport skills that involve movements more clearly seen from the front or rear. List the primary movements that occur in the transverse tarsal/subtalar joint, hip, spine, glenohumeral joint, and wrist. Which plane are these movements occurring in primarily? What axis of rotation is involved primarily?

20. List the similarities between the ankle/foot/toes and the wrist/hand/fingers regarding the bones, joint structures, and movements. What are the differences?

21. Compare and contrast the glenohumeral and acetabulofemoral joints. Which one is more susceptible to dislocations and why?

22. Compare and contrast the elbow and knee joints. Considering the bone and joint structures and their functions, what are the similarities and differences?

LABORATORY EXERCISES

1. Choose several different locations on your body at random and specifically describe the locations, using the correct anatomical directional terminology.

2. Determine which joints have movements possible in each of the following planes:
   a. Sagittal
   b. Frontal
   c. Transverse

3. List all the diarthrodial joints of the body that are capable of the following paired movements:
   a. Flexion/extension
   b. Abduction/adduction
   c. Rotation (left and right)
   d. Rotation (internal and external)

4. Determine the planes in which the following activities occur. Also, use a pencil to visualize the axis for each of the following activities.
   a. Walking up stairs
   b. Turning a knob to open a door
   c. Nodding the head to agree

d. Shaking the head to disagree

e. Shuffling the body from side to side

f. Looking over your shoulder to see behind you

5. Individually practice the various joint movements, on yourself or with another subject.

6. Locate the various types of joints on a human skeleton and palpate their movements on a living subject.

7. Stand in the anatomical position facing a closed door. Reach out and grasp the knob with your right hand. Turn it and open the door widely toward you. Determine all of the joints involved in this activity and list the movements for each joint.

8. Utilize a goniometer to measure the joint ranges of motion for several students in your class for each of the following movements. Compare your results with the average ranges provided in Appendixes 1 and 2.

a. External and internal rotation of the shoulder with the shoulder in 90 degrees of abduction while supine

b. Elbow flexion in the supine position

c. Wrist extension with the forearm in neutral and the elbow in 90 degrees of flexion

d. Hip external and internal rotation in the sitting position with the hip and knee each in 90 degrees of flexion

e. Knee flexion in the prone position

f. Ankle dorsiflexion with the knee in 90 degrees of flexion versus knee in full extension

9. Discuss the following joints among your classmates and place them in order from the least total range of motion to the most. Be prepared to defend your answer.

a. Ankle   d. Hip

b. Elbow   e. Knee

c. Glenohumeral   f. Wrist

10. Is there more inversion or more eversion possible in the transverse tarsal and subtalar joints? Explain this occurrence based on anatomy.

11. Is there more abduction or more adduction possible in the wrist joint? Explain this occurrence based on anatomy.

References

Anthony C, Thibodeau G: *Textbook of anatomy and physiology,* ed 10, St. Louis, 1979, Mosby.

Booher JM, Thibodeau GA: *Athletic injury assessment,* ed 4, New York, 2000, McGraw-Hill.

Goss CM: *Gray's anatomy of the human body,* ed 29, Philadelphia, 1973, Lea & Febiger.

Hamilton N, Weimar W, Luttgens K: *Kinesiology: scientific basis of human motion,* ed 12, New York, 2012, McGraw-Hill.

Lindsay DT: *Functional human anatomy,* St. Louis, 1996, Mosby.

Logan GA, McKinney WC: *Anatomic kinesiology,* ed 3, Dubuque, IA, 1982, Brown.

National Strength and Conditioning Association; Baechle TR, Earle RW: *Essentials of strength training and conditioning,* ed 2, Champaign, IL, 2000, Human Kinetics.

Neumann, DA: *Kinesiology of the musculoskeletal system: foundations for physical rehabilitation,* ed 2, St. Louis, 2010, Mosby.

Northrip JW, Logan GA, McKinney WC: *Analysis of sport motion: anatomic and biomechanic perspectives,* ed 3, Dubuque, IA, 1983, Brown.

Prentice WE: *Principles of athletic training: a competency based approach,* ed 15, New York, 2014, McGraw-Hill.

Prentice WE: *Rehabilitation techniques in sports medicine,* ed 5, New York, 2011, McGraw-Hill.

Seeley RR, Stephens TD, Tate P: *Anatomy & physiology,* ed 8, New York, 2008, McGraw-Hill.

Shier D, Butler J, Lewis R: *Hole's essentials of human anatomy and physiology,* ed 10, New York, 2009, McGraw-Hill.

Stedman TL: *Stedman's medical dictionary,* ed 28, Baltimore, 2005, Lippincott Williams & Wilkins.

Steindler A: *Kinesiology of the human body,* Springfield, IL, 1970, Thomas.

Van De Graaff KM: *Human anatomy,* ed 6, New York, 2002, McGraw-Hill.

Van De Graaff KM, Fox SI, LaFleur KM: *Synopsis of human anatomy & physiology,* Dubuque, IA, 1997, Brown.

For additional resources and a list of related websites, visit **www.mhhe.com/floyd19e**.

Worksheet Exercises

For in- or out-of-class assignments, or for testing, utilize this tear-out worksheet.

Anterior skeletal worksheet

On the anterior skeletal worksheet, label the bones and their prominent features by filling in the blanks.

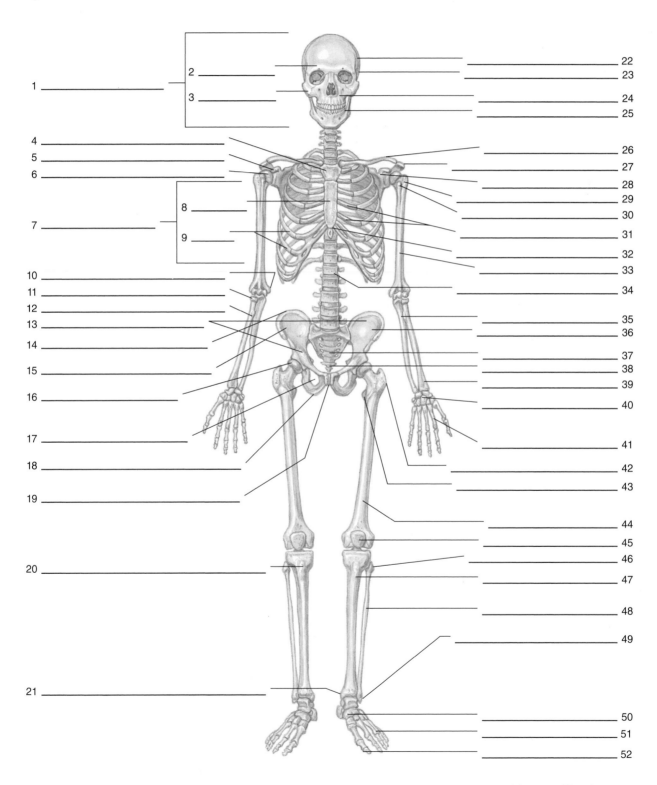

1 _____

2 _____

3 _____

4 _____

5 _____

6 _____

7 _____

8 _____

9 _____

10 _____

11 _____

12 _____

13 _____

14 _____

15 _____

16 _____

17 _____

18 _____

19 _____

20 _____

21 _____

22 _____

23 _____

24 _____

25 _____

26 _____

27 _____

28 _____

29 _____

30 _____

31 _____

32 _____

33 _____

34 _____

35 _____

36 _____

37 _____

38 _____

39 _____

40 _____

41 _____

42 _____

43 _____

44 _____

45 _____

46 _____

47 _____

48 _____

49 _____

50 _____

51 _____

52 _____

Worksheet Exercises

For in- or out-of-class assignments, or for testing, utilize this tear-out worksheet.

Posterior skeletal worksheet

On the posterior skeletal worksheet, label the bones and their prominent features by filling in the blanks.

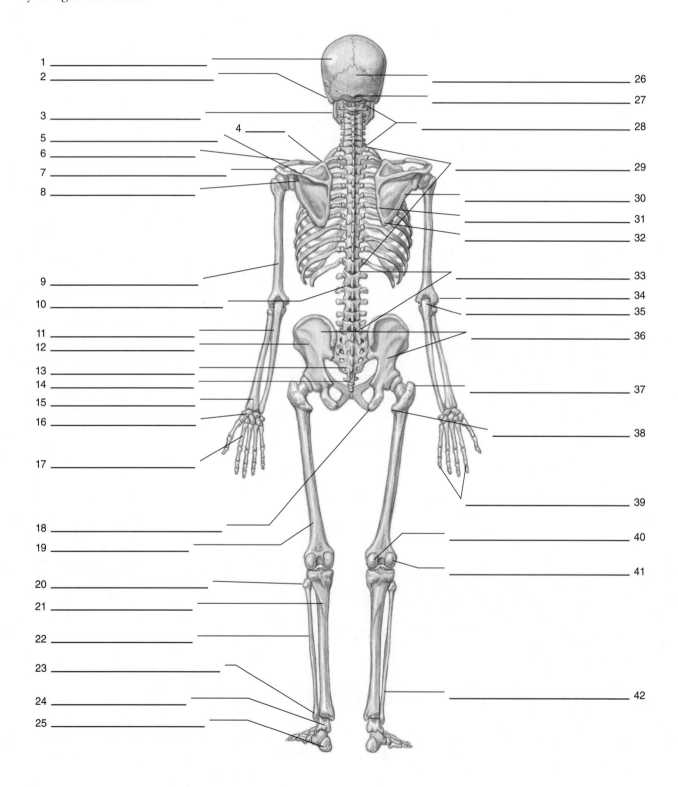

1 _____
2 _____

3 _____
5 _____
6 _____
7 _____
8 _____

4 _____

9 _____
10 _____

11 _____
12 _____

13 _____
14 _____
15 _____
16 _____

17 _____

18 _____
19 _____

20 _____
21 _____

22 _____

23 _____

24 _____
25 _____

26 _____
27 _____
28 _____

29 _____

30 _____
31 _____
32 _____

33 _____
34 _____
35 _____

36 _____

37 _____

38 _____

39 _____

40 _____

41 _____

42 _____

CHAPTER 2

NEUROMUSCULAR FUNDAMENTALS

Skeletal muscles are responsible for movement of the body and all its joints. Muscle contraction produces the force that causes joint movement in the human body. In addition to the function of movement, muscles also provide both dynamic stability of joints and protection, contribute to posture and support, and produce a major portion of total body heat. There are over 600 skeletal muscles, which constitute approximately 40% to 50% of body weight. Of these, there are 215 pairs of skeletal muscles. These pairs of muscles usually work in cooperation with each other to perform opposite actions at the joints they cross. In most cases, muscles work in groups rather than independently to achieve a given joint motion. This is known as **aggregate muscle action**.

Muscle nomenclature

In attempting to learn the skeletal muscles, it is helpful to have an understanding of how they are named. Muscles are usually named because of one or more distinctive characteristics, such as their visual appearance, anatomical location, or function. Examples of skeletal muscle naming are as follows:

Shape—deltoid, rhomboid
Size—gluteus maximus, teres minor
Number of divisions—triceps brachii
Direction of its fibers—external abdominal oblique
Location—rectus femoris, palmaris longus
Points of attachment—coracobrachialis, extensor hallucis longus, flexor digitorum longus
Action—erector spinae, supinator, extensor digiti minimi

Action and shape—pronator quadratus
Action and size—adductor magnus
Shape and location—serratus anterior
Location and attachment—brachioradialis
Location and number of divisions—biceps femoris

In discussions regarding the muscles, they are often grouped together for brevity of conversation and clearer understanding. The naming of muscle groups follows a similar pattern. Here are some

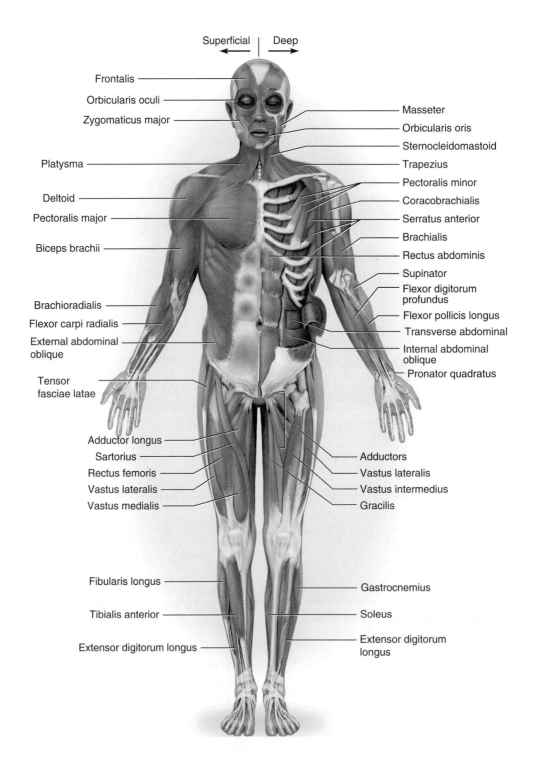

Superficial | Deep

Frontalis

Orbicularis oculi

Zygomaticus major

Platysma

Deltoid

Pectoralis major

Biceps brachii

Brachioradialis

Flexor carpi radialis

External abdominal oblique

Tensor fasciae latae

Adductor longus

Sartorius

Rectus femoris

Vastus lateralis

Vastus medialis

Fibularis longus

Tibialis anterior

Extensor digitorum longus

Masseter

Orbicularis oris

Sternocleidomastoid

Trapezius

Pectoralis minor

Coracobrachialis

Serratus anterior

Brachialis

Rectus abdominis

Supinator

Flexor digitorum profundus

Flexor pollicis longus

Transverse abdominal

Internal abdominal oblique

Pronator quadratus

Adductors

Vastus lateralis

Vastus intermedius

Gracilis

Gastrocnemius

Soleus

Extensor digitorum longus

FIG. 2.1 • Superficial and deep muscles of the human body, anterior view.

muscle groups assembled according to different naming rationales:

Shape—hamstrings
Number of divisions—quadriceps, triceps surae
Location—peroneals, abdominal, shoulder girdle
Action—hip flexors, rotator cuff

Figs. 2.1 and 2.2 depict the muscular system from both a superficial and a deep point of view.

Muscles shown in these figures, and many other muscles, will be studied in more detail as each joint of the body is considered in later chapters.

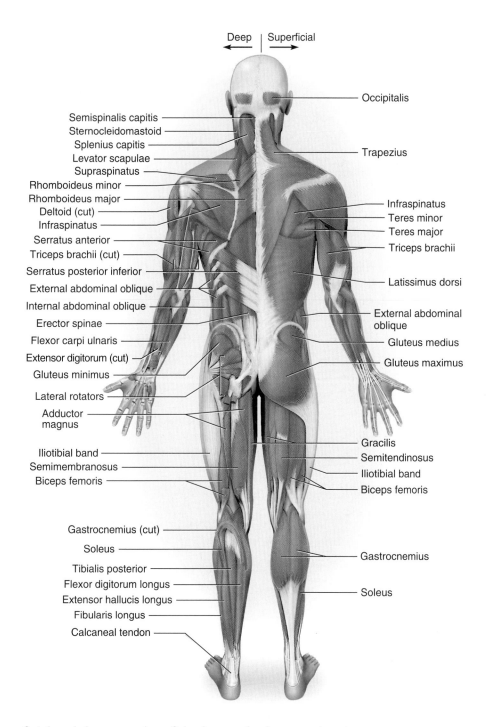

FIG. 2.2 • Superficial and deep muscles of the human body, posterior view.

Shape of muscles and fiber arrangement

Various muscles have different shapes, and their fibers may be arranged differently in relation to each other and to the tendons that connect them to bone. The shape and fiber arrangement play a role in the muscle's ability to exert force and in the range through which it can effectively exert force on the bones to which it is attached. A factor in the ability of a muscle to exert force is its cross-section diameter. Keeping all other factors constant, a muscle with a greater cross-section diameter will be able to exert a greater force. A factor in the ability of a muscle to move a joint through a large range of motion is its ability to shorten. Generally, longer muscles can shorten through a greater range and therefore are more effective in moving joints through large ranges of motion.

Essentially, all skeletal muscles may be grouped into two major types of fiber arrangements: parallel and pennate. Each may be subdivided further according to shape.

Parallel muscles have their fibers arranged parallel to the length of the muscle. Generally, parallel muscles will produce a greater range of movement than similar-size muscles with a pennate arrangement. Parallel muscles are categorized into the following shapes:

Flat muscles are usually thin and broad, originating from broad, fibrous, sheetlike aponeuroses that allow them to spread their forces over a broad area. Examples include the rectus abdominis and external oblique.

Fusiform muscles are spindle-shaped with a central belly that tapers to tendons on each end; this allows them to focus their power on small, bony targets. Examples are the brachialis and the brachioradialis.

Strap muscles are more uniform in diameter with essentially all their fibers arranged in a long parallel manner. This also enables a focusing of power on small, bony targets. The sartorius is an example.

Radiate muscles are also sometimes described as being triangular, fan-shaped, or convergent. They have the combined arrangement of flat and fusiform muscles, in that they originate on a broad surface or an aponeurosis and converge onto a tendon. Examples include the pectoralis major and trapezius.

Sphincter or circular muscles are technically endless strap muscles that surround openings and function to close them upon contraction. An example is the orbicularis oris, surrounding the mouth.

Pennate muscles have shorter fibers that are arranged obliquely to their tendons in a structure similar to that of a feather. This arrangement increases the cross-sectional area of the muscle, thereby increasing its force production capability. Pennate muscles are categorized on the basis of the exact arrangement between the fibers and the tendon, as follows:

Unipennate muscle fibers run obliquely from a tendon on one side only. Examples are seen in the biceps femoris, extensor digitorum longus, and tibialis posterior.

Bipennate muscle fibers run obliquely from a central tendon on both sides, as in the rectus femoris and flexor hallucis longus.

Multipennate muscles have several tendons with fibers running diagonally between them, as in the deltoid.

Bipennate and unipennate muscles produce the strongest contractions. Review Table 2.1 regarding muscle shapes and fiber arrangements.

Muscle tissue properties

Skeletal muscle tissue has four properties related to its ability to produce force effecting movement about joints. **Irritability** or **excitability** is the muscle property of being sensitive or responsive to chemical, electrical, or mechanical stimuli. When an appropriate stimulus is provided, muscle responds by developing tension. **Contractility** is the ability of muscle to contract and develop tension or internal force against resistance when stimulated. The ability of muscle tissue to develop tension or contract is unique in that other body tissues do not have this property. **Extensibility** is the ability of muscle to be passively stretched beyond its normal resting length. As an example, the triceps brachii displays extensibility when it is stretched beyond its normal resting length by the biceps brachii and other elbow flexors contracting to achieve full elbow flexion. **Elasticity** is the ability of muscle to return to its original resting length following stretching. To continue with the elbow example, the triceps brachii displays elasticity by returning to its original resting length when the elbow flexors cease contracting and relax.

Muscle terminology

Locating the muscles, their proximal and distal attachments, and their relationship to the joints they cross is critical to determining the effects that muscles have on the joints. It is also necessary to understand certain terms as body movement is considered.

TABLE 2.1 • Muscle shape and fiber arrangement

Fiber arrangement	Advantage	Shape	Appearance	Characteristics/description	Examples
Parallel (fibers arranged parallel to the length of the muscle)	Produces greater range of movement than similar-size pennate muscles; long excursion (contract over a great distance); good endurance	Flat		Usually thin and broad, originating from broad, fibrous, sheetlike aponeuroses that allow them to spread their forces over a broad area	Rectus abdominis, external oblique
		Fusiform	Tendon — Belly — Tendon	Spindle-shaped with central belly that tapers to tendons on each end; can focus their power on small, bony targets	Biceps brachii, brachialis
		Strap		More uniform in diameter with essentially all their fibers arranged in a long parallel manner; can focus their power on small, bony targets	Sartorius
		Radiate (triangular, fan-shaped, convergent)		Combined arrangement of flat and fusiform muscles; originate on broad aponeuroses and converge to a single point of attachment via a tendon	Pectoralis major, trapezius
		Sphincter (circular)		Fibers concentrically arranged around a body opening; technically endless strap muscles surround openings and function to close them upon contraction	Orbicularis oris, orbicularis oculi
Pennate (shorter fibers, arranged obliquely to their tendons)	Produces greater power than similar-size parallel muscles due to increased cross-sectional area; strong muscles; short excursion	Unipennate		Run obliquely from a tendon on one side only	Biceps femoris, extensor digitorum longus, tibialis posterior
		Bipennate		Run obliquely from a central tendon on both sides	Rectus femoris, flexor hallucis longus
		Multipennate		Several tendons with fibers running diagonally between them	Deltoid

Modified from Saladin, KS: *Anatomy & physiology: the unity of form and function,* ed 4, New York, 2007, McGraw-Hill; and Seeley RR, Stephens TD, Tate P: *Anatomy & physiology*, ed 7, New York, 2008, McGraw-Hill.

Intrinsic

Pertaining usually to muscles within or belonging solely to the body part on which they act. The small intrinsic muscles found entirely within the hand are examples. See page 197.

Extrinsic

Pertaining usually to muscles that arise or originate outside of (proximal to) the body part on which they act. The forearm muscles that attach proximally on the distal humerus and insert on the fingers are examples of extrinsic muscles of the hand. See Chapter 7.

Action

Action is the specific movement of the joint resulting from a concentric contraction of a muscle that crosses the joint. An example is the biceps brachii, which has the action of flexion at the elbow. In most cases a particular action is caused by a group of muscles working together. Any of the muscles in the group can be said to cause the action, even though it is usually an effort of the entire group. A particular muscle may cause more than one action either at the same joint or at a different joint, depending upon the characteristics of the joints crossed by the muscle and the exact location of the muscle and its attachments in relation to the joint(s).

Innervation

Innervation occurs in the segment of the nervous system responsible for providing a stimulus to muscle fibers within a specific muscle or portion of a muscle. A particular muscle may be innervated by more than one nerve, and a particular nerve may innervate more than one muscle or portion of a muscle.

Amplitude

The amplitude is the range of muscle fiber length between maximal and minimal lengthening.

Gaster (belly or body)

The **gaster** is the central, fleshy portion of the muscle. This contractile portion of the muscle generally increases in diameter as the muscle contracts.

When a particular muscle contracts, it tends to pull both ends toward the gaster, or middle, of the muscle. Consequently, if neither of the bones to which a muscle is attached were stabilized, both bones would move toward each other upon contraction. The more common case, however, is that one bone is more stabilized by a variety of factors, and as a result the less stabilized bone usually moves toward the more stabilized bone upon contraction.

Tendon

Tendons are tough yet flexible bands of fibrous connective tissue, often cordlike in appearance, that connect muscles to bones and other structures. By providing this connection, tendons transmit the force generated by the contracting muscle to the bone. In some cases, two muscles may share a common tendon, such as the Achilles tendon of the gastrocnemius and soleus muscles. In other cases a muscle may have multiple tendons connecting it to one or more bones, such as the three proximal attachments of the triceps brachii.

Aponeurosis

An aponeurosis is a tendinous expansion of dense fibrous connective tissue that is sheet- or ribbonlike in appearance and resembles a flattened tendon. Aponeuroses serve as a fascia to bind muscles together or as a means of connecting muscle to bone.

Fascia

Fascia is a sheet or band of fibrous connective tissue that envelopes, separates, or binds together parts of the body such as muscles, organs, and other soft-tissue structures of the body. In certain places throughout the body, such as around joints like the wrist and ankle, fascial tissue forms a **retinaculum** to retain tendons close to the body.

Origin

From a structural perspective, the proximal attachment of a muscle or the part that attaches closest to the midline or center of the body is usually considered to be the origin. From a functional and historical perspective, the least movable part or attachment of the muscle has generally been considered to be the origin.

Insertion

Structurally, the distal attachment, or the part that attaches farthest from the midline or center of the body, is considered the insertion. Functionally and historically, the most movable part is generally considered the insertion.

As an example, in the biceps curl exercise, the biceps brachii muscle in the arm has its origin on the scapula (least movable bone) and its insertion on the radius (most movable bone). In some movements this process can be reversed. An example of this reversal can be seen in the pull-up, where the radius is relatively stable and the scapula moves up. Even though in this example the most movable bone is reversed, the proximal attachment of the biceps brachii is always on the scapula and is still considered to be the origin, and the insertion is still on the radius. The biceps brachii would be an extrinsic muscle of the elbow, whereas the brachialis would be intrinsic to the elbow. For each muscle studied, the origin and insertion are indicated.

Types of muscle contraction (action)

When tension is developed in a muscle as a result of a stimulus, it is known as a contraction. The term *muscle contraction* may be confusing, because in some types of contractions the muscle does not shorten in length as the term *contraction* indicates. As a result, it has become increasingly common to refer to the various types of muscle contractions as *muscle actions* instead.

Muscle contractions can be used to *cause, control,* or *prevent* joint movement. To elaborate, muscle contractions can be used to initiate or accelerate the movement of a body segment, to slow down or decelerate the movement of a body segment, or to prevent movement of a body segment by external forces. All muscle contractions or actions can be classified as either isometric or isotonic. An isometric contraction occurs when tension is developed within the muscle but the joint angles remain constant. Isometric contractions may be thought of as **static** contractions, because a significant amount of active tension may be developed in the muscle to maintain the joint angle in a relatively static or stable position. Isometric contractions may be used to stabilize a body segment to prevent it from being moved by external forces.

Isotonic contractions involve the muscle developing tension to either cause or control joint movement. They may be thought of as **dynamic** contractions, because the varying degrees of active tension in the muscles are either causing the joint angles to change or controlling the joint angle change that is caused by external forces. The isotonic type of muscle contraction is classified further as either concentric or eccentric on the basis of whether shortening or lengthening occurs. Concentric contractions involve the muscle developing active tension as it shortens, whereas eccentric contractions involve the muscle lengthening under active tension. In Fig. 2.3, *A, B, E,* and *F* illustrate isotonic contractions, while *C* and *D* demonstrate isometric contractions.

It is also important to note that movement may occur at any given joint without any muscle contraction whatsoever. Such movement is referred to as passive and is due solely to external forces, such as those applied by another person, object, or resistance, or to the force of gravity in the presence of muscle relaxation.

Concentric contraction

Concentric contractions involve the muscle developing active tension as it shortens and occur when the muscle develops enough force to overcome the applied resistance. Concentric contractions may be thought of as causing movement against gravity or resistance and are described as positive contractions. The force developed by the muscle is greater than that of the resistance. This results in the joint angle being changed in the direction of the applied muscular force and causes the body part to move against gravity or external forces. Concentric contractions are used to accelerate the movement of a body segment from a lower speed to a higher speed.

Eccentric contraction (muscle action)

Eccentric contractions involve the muscle lengthening under active tension and occur when the muscle gradually lessens in tension to control the descent of the resistance. The weight or resistance may be thought of as overcoming the muscle contraction, but not to the point that the muscle cannot control the descending movement. Eccentric muscle actions control movement with gravity or resistance and are described as negative contractions. The force developed by the muscle is less than that of the resistance; this results in a change in the joint angle in the direction of the resistance or external force and allows the body part to move with gravity or external forces (resistance). Eccentric contractions are used to decelerate the movement of a body segment from a faster speed to a slower speed or stop the movement of a joint already in motion. Because the muscle is lengthening as opposed to shortening, the relatively recent change in terminology from muscle contraction to muscle action is becoming more commonly accepted.

Movement differentiation

Some confusion exists regarding body movement and the factors affecting it. Joint movement may occur with muscle groups on either or both sides of the joint actively contracting or even without any muscles contracting. Similarly, when no movement is occurring there may or may not be muscle contraction present, depending on the external forces acting on the joint. To further add to the confusion, a variety of terms and descriptive phrases are used by different authorities to

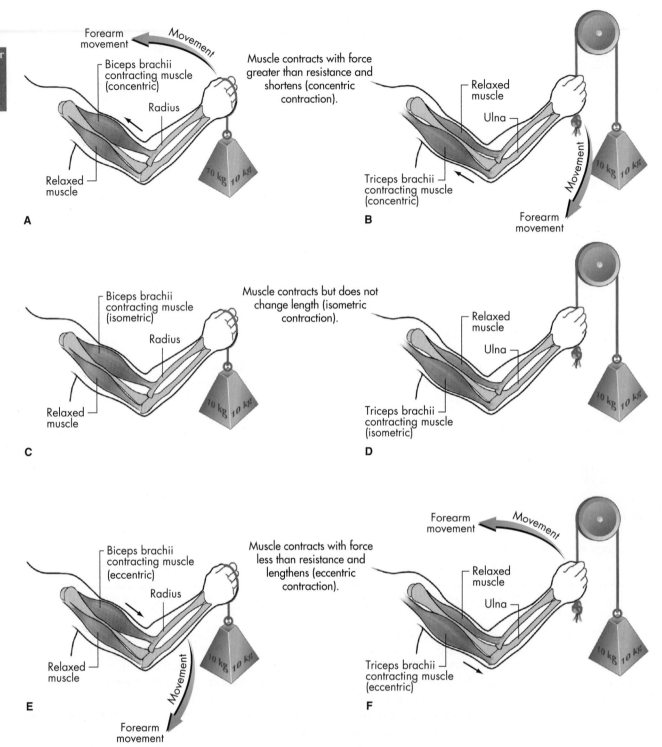

FIG. 2.3 • Agonist–antagonist relationship with isotonic and isometric contractions. **A,** Biceps is agonist in flexing the elbow with a concentric contraction, and triceps is antagonist; **B,** Triceps is agonist in extending the elbow with a concentric contraction, and biceps is antagonist; **C,** Biceps is maintaining the elbow in a flexed position with an isometric contraction, and triceps is antagonist; **D,** Triceps is maintaining the elbow in a flexed position with an isometric contraction, and biceps is antagonist; **E,** Biceps is controlling elbow extension with an eccentric contraction, and triceps is antagonist; **F,** Triceps is controlling elbow flexion with an eccentric contraction, and biceps is antagonist.

TABLE 2.2 • **Muscle contraction and movement matrix**

| Definitive and descriptive factors | Type of contraction (muscle action) | | | Movement without contraction |
| | Isometric | Isotonic | | |
		Concentric	Eccentric	
Agonist muscle length	No appreciable change	Shortening ➡️⬅️	Lengthening ⬅️➡️	Dictated solely by gravity and/or external forces
Antagonist muscle length	No appreciable change	Lengthening ⬅️➡️	Shortening ➡️⬅️	Dictated solely by gravity and/or external forces
Joint angle changes	No appreciable change	In direction of applied muscular force	In direction of external force (resistance)	Dictated solely by gravity and/or external forces
Direction of body part	Against immovable object or matched external force (resistance)	Against gravity and/or other external force (resistance)	With gravity and/or other external force (resistance)	Consistent with gravity and/or other external forces
Motion	Prevents motion; pressure (force) applied, but no resulting motion	Causes motion	Controls motion	Either no motion or passive motion as a result of gravity and/or other external forces
Description	Static; fixating	Dynamic shortening; positive work	Dynamic lengthening; negative work	Passive; relaxation
Applied muscle force versus resistance	Force = resistance	Force > resistance	Force < resistance	No force, all resistance
Speed relative to gravity or applied resistance including inertial forces	Equal to speed of applied resistance	Faster than the inertia of the resistance	Slower than the speed of gravity or applied inertial forces	Consistent with inertia of applied external forces or the speed of gravity
Acceleration/ deceleration	Zero acceleration	Acceleration ↗️	Deceleration ↘️	Either zero or acceleration consistent with applied external forces
Descriptive symbol	(=)	(+)	(−)	(0)
Practical application	Prevents external forces from causing movement	Initiates movement or speeds up the rate of movement	Slows down the rate of movement or stops movement, "braking action"	Passive motion by force from gravity and/or other external forces

describe these phenomena. Table 2.2 attempts to provide an exhaustive explanation of the various types of contraction and resulting joint movements. The varying terminology utilized in defining and describing these actions is included. Appendix 5 provides an algorithm for determining if a muscle or muscle group is contracting and, if so, the type of contraction.

Various exercises may use any one or all of these contraction types for muscle development. Development of exercise machines has resulted in another type of muscle exercise known as **isokinetics**. Isokinetics is not another type of contraction, as some authorities have mistakenly described; rather, it is a specific technique that may use any or all of the different types of contractions. Isokinetics is a type of dynamic exercise usually using concentric and/or eccentric muscle contractions in which the speed (or velocity) of movement is constant and muscular contraction (ideally, maximum contraction) occurs throughout the movement. Biodex, Cybex, and other types of apparatuses are engineered to allow this type of exercise.

Students well educated in kinesiology should be qualified to prescribe exercises and activities for the development of large muscles and of muscle groups in the human body. They should be able to read the description of an exercise or observe an exercise and immediately know the most important muscles being used. Terms describing how muscles function in joint movements follow.

Role of muscles

When a muscle contracts, it simply attempts to pull the bones to which both of its ends are attached toward each other. Usually this does not happen, however, because one of the bones is usually more stable than the other. As a result, the less stable bone moves toward the more stable bone. When a muscle that is capable of performing multiple actions contracts, it attempts to perform all of its actions unless other forces, such as those provided by other muscles, prevent the undesired actions.

Agonist   FIG. 2.3

Agonist muscles, when contracting concentrically, cause joint motion through a specified plane of motion. Any concentrically contracting muscle that causes the same joint motion is an agonist for the motion. However, some muscles, because of their relative location, size, length, or force generation capacity, are able to contribute significantly more to the joint movement than other agonists. These muscles are known as prime or **primary movers** or as muscles most involved. Agonist muscles that contribute significantly less to the joint motion are commonly referred to as assisters or assistant movers. Consensus among all authorities regarding which muscles are primary movers and which are weak assistants does not exist in every case. This text will emphasize the primary movers. The remaining agonists or assistants, when listed, will be referred to as weak contributors to the motion involved. As an example, the hamstrings (semitendinosus, semimembranosus, biceps femoris), sartorius, gracilis, popliteus, and gastrocnemius are all agonists in knee flexion, but most kinesiologists regard only the hamstrings as the prime movers.

Antagonist   FIG. 2.3

Antagonist muscles have the opposite concentric action from agonists. Referred to as contralateral muscles, antagonists are located on the opposite side of the joint from the agonist and work in cooperation with agonist muscles by relaxing and allowing movement; but when contracting concentrically, they perform the joint motion opposite to that of the agonist. Using the previous example, the quadriceps muscles are antagonists to the hamstrings in knee flexion.

Stabilizers

Stabilizers surround the joint or body part and contract to fixate or stabilize the area to enable another limb or body segment to exert force and move. Known as fixators, they are essential in establishing a relatively firm base for the more distal joints to work from when carrying out movements. In a biceps curling example, the muscles of the scapula and glenohumeral joint must contract in order to maintain the shoulder complex and humerus in a relatively static position so that the biceps brachii can more effectively perform the curls. The antagonists for each motion of the proximal joint co-contract or contract against each other to prevent motion. This is an example of proximal stabilization to enhance the effectiveness of distal joint motion, which occurs commonly with the upper extremity.

Synergist

Muscles that assist in the action of an agonist but are not necessarily prime movers for the action, known as guiding muscles, assist in refined movement and rule out undesired motion. Synergist muscles may be either helping synergists or true synergists. **Helping synergists** have an action in common but also have actions antagonistic to each other. They help another muscle move the joint in the desired manner and simultaneously prevent undesired actions. An example involves the anterior and posterior deltoid. The anterior deltoid acts as an agonist in glenohumeral flexion, while the posterior deltoid acts as an extensor. Helping each other, they work in synergy with the middle deltoid to accomplish abduction. **True synergists** contract to prevent an undesired joint action of the agonist and have no direct effect on the agonist action. The finger flexors are provided true synergy by the wrist extensors when one is grasping an object. The finger flexors originating on the forearm and humerus are agonists in both wrist flexion and finger flexion. The wrist extensors contract to prevent wrist flexion by the finger flexors. This allows the finger flexors to maintain more of their length and therefore utilize more of their force in flexing the fingers.

Neutralizers

Neutralizers counteract or neutralize the action of other muscles to prevent undesirable movements such as inappropriate muscle substitutions. They contract to resist specific actions of other muscles. As an example, when only the supination action of the biceps brachii is desired, the triceps brachii contracts to neutralize the flexion action of the biceps brachii. Another example may be seen in the biceps curl, when only the flexion force of the biceps brachii is desired. When the biceps brachii contracts, it normally attempts to both flex the elbow and supinate the forearm. In this case the pronator teres contracts to neutralize the supination component of the biceps.

Force couples

Force couples occur when two or more forces are pulling in different directions on an object, causing the object to rotate about its axis. Fig. 2.4, *A* depicts a force couple consisting of one hand on each side of a steering wheel. One hand pulls the wheel up and to the right, and the other hand pulls it down and to the left.

Coupling of muscular forces in the body can result in a more efficient movement. Figure 2.4, *B* illustrates a force couple in which the middle trapezius, lower trapezius, and serratus anterior all attach at different points on the scapula. Each muscle pulls on the scapula from a different direction to produce the combined result of upward rotation. Another example of muscular force couples is seen in standing, when the hip flexors (iliopsoas and rectus femoris) are used to pull the front of the pelvis downward and the erector spinae are used to pull the posterior pelvis upward, resulting in anterior pelvic rotation.

Tying the roles of muscles together

When a muscle with multiple agonist actions contracts, it attempts to perform all its actions. Muscles cannot determine which actions are appropriate for the task at hand. The resulting actions actually performed depend upon several factors, such as the motor units activated, joint position at the time of contraction, planes of motion allowed in the joint, axis of rotation

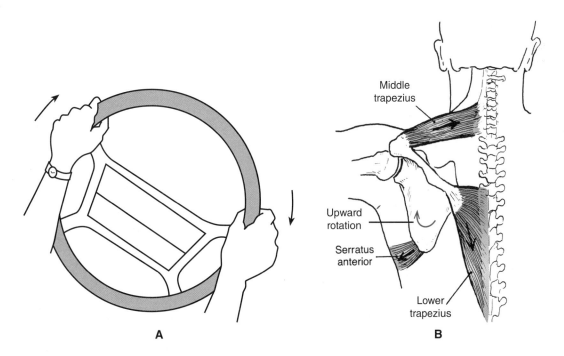

FIG. 2.4 • Force couples. **A,** When a person steers with two hands, the hands act as a force couple; **B,** Two force couples act on the scapula to rotate it upward. The middle trapezius and lower serratus anterior are excellent examples. The middle trapezius and lower trapezius also tend to act as a force couple, although their pulls are not in opposite directions.

possible in the joint, muscle length, and the relative contraction or relaxation of other muscles acting on the joint. In certain instances, two muscles may work in synergy by counteracting their opposing actions to accomplish a common action.

As discussed, agonist muscles are primarily responsible for a given movement, such as those of hip flexion and knee extension while kicking a ball. In this example, the hamstrings are antagonistic and relax to allow the kick to occur. This does not mean that all other muscles in the hip area are uninvolved. The preciseness of the kick depends on the involvement of many other muscles. As the lower extremity swings forward, its route and subsequent angle at the point of contact depend on a certain amount of relative contraction or relaxation in the hip abductors, adductors, internal rotators, and external rotators. These muscles act in a synergistic fashion to guide the lower extremity in a precise manner. That is, they are not primarily responsible for knee extension and hip flexion, but they do contribute to the accuracy of the total movement. These guiding muscles assist in refining the kick and preventing extraneous motions. Additionally, the muscles in the contralateral hip and pelvic area must be under relative tension to help fixate or stabilize the pelvis on that side in order to provide a relatively stable pelvis for the hip flexors on the involved side to contract against. In kicking the ball, the pectineus and tensor fascia latae are adductors and abductors, respectively, in addition to flexors. The actions of adduction and abduction are neutralized by each other, and the common action of the two muscles results in hip flexion.

From a practical point of view, it is not essential that individuals know the exact force exerted by each of the elbow flexors—biceps, brachialis, and brachioradialis—in chinning. It is important to understand that this muscle group is the agonist or primary mover responsible for elbow joint flexion. Similarly, it is important to understand that these muscles contract concentrically when the chin is pulled up to the bar and that they contract eccentrically when the body is lowered slowly. Antagonistic muscles produce actions opposite those of the agonist. For example, the muscles that produce extension of the elbow joint are antagonistic to the muscles that produce flexion of the elbow joint. It is important to understand that specific exercises need to be prescribed for the development of each antagonistic muscle group. The return movement to the hanging position at the elbow joint after chinning is elbow joint extension, but the triceps and anconeus are not being strengthened. A concentric contraction of the elbow joint flexors occurs, followed by an eccentric contraction of the same muscles.

Reversal of muscle function

A muscle group that is described to perform a given function can contract to control the exact opposite motion. Fig. 2.3, *A* illustrates how the biceps is an agonist by contracting concentrically to flex the elbow. The triceps is an antagonist to elbow flexion, and the pronator teres is considered to be a synergist to the biceps in this example. If the biceps were to slowly lengthen and control elbow extension, as in Fig. 2.3, *E*, it would still be the agonist, but it would be contracting eccentrically. Fig. 2.3, *B* illustrates how the triceps is an agonist by contracting concentrically to extend the elbow. The biceps is an antagonist to elbow extension in this example. If the triceps were to slowly lengthen and control elbow flexion, as in Fig. 2.3, *F*, it would still be the agonist, but it would be eccentrically contracting. In both of these examples, the deltoid, trapezius, and various other shoulder muscles are serving as stabilizers of the shoulder area.

Determination of muscle action

The specific action of a muscle may be determined through a variety of methods. These include considering anatomical lines of pull, anatomical dissection, palpation, models, electromyography, and electrical stimulation.

With an understanding of a muscle's line of pull relative to a joint, one may determine the muscle's action at the joint. (See lines of pull below.) Although not available to all students, cadaver dissection of muscles and joints is an excellent way to further understand muscle action.

For most of the skeletal muscles, **palpation** is a very useful way to determine muscle action. It is done through using the sense of touch to feel or examine a muscle as it contracts. Palpation is limited to superficial muscles but is helpful in furthering an understanding of joint mechanics. Models such as long rubber bands may be used to facilitate understanding of lines of pull and to simulate muscle lengthening or shortening as joints move through various ranges of motion.

Electromyography (EMG) utilizes either surface electrodes that are placed over the muscle or fine wire/needle electrodes placed into the muscle. As the subject then moves the joint and contracts the muscles, the EMG unit detects the action potentials of the muscles and provides an electronic readout of the contraction intensity and duration. EMG is the most accurate way of detecting the presence and extent of muscle activity.

Electrical muscle stimulation is somewhat a reverse approach of electromyography. Instead of electricity being used to detect muscle action, it is used to cause muscle activity. Surface electrodes are placed over a muscle, and then the stimulator causes the muscle to contract. The joint's actions may then be observed to see the effect of the muscle's contraction on it.

Lines of pull   FIG. 2.5

Combining the knowledge of a particular joint's functional design and diarthrodial classification with an understanding of the specific location of a musculotendinous unit as it crosses a joint is extremely helpful in understanding its action on the joint. For example, knowing that the rectus femoris has its origin on the anterior inferior iliac spine and its insertion on the tibial tuberosity via the patella, you can then determine that the muscle must have an anterior relationship to the knee and hip. Combining this knowledge with the knowledge that both joints are capable of sagittal plane movements such as flexion/extension, you can then determine that when the rectus femoris contracts concentrically, it should cause the knee to extend and the hip to flex.

Furthermore, knowing that the semitendinosus, semimembranosus, and biceps femoris all originate on the ischial tuberosity and that the semitendinosus and semimembranosus cross the knee posteromedially before inserting on the tibia, but that the biceps femoris crosses the knee posterolaterally before inserting on the fibula head, you may determine that all three muscles have posterior relationships to the hip and knee, which would enable them to be hip extensors and knee flexors upon concentric contraction. The specific knowledge related to their distal attachments and the knee's ability to

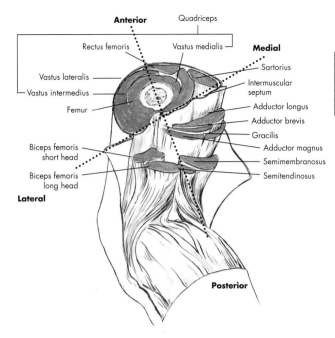

FIG. 2.5 ● Lines of pull in relation to the left knee. Biceps femoris with a posterolateral relationship enables it to externally rotate the knee; semitendinosus and semimembranosus have a posteromedial relationship enabling them to internally rotate the knee; hamstrings (biceps femoris, semitendinosus, and semimembranosus) all have a posterior relationship enabling them to flex the knee; quadriceps muscles have an anterior relationship enabling them to extend the knee.

rotate when flexed would allow you to determine that the semitendinosus and semimembranosus will cause internal rotation, whereas the biceps femoris will cause external rotation. Knowledge that the knee's axes of rotation are only frontal and vertical, but not sagittal, enables you to determine that even though the semitendinosus and semimembranosus have a posteromedial line of pull and the biceps femoris has a posterolateral line of pull, they are not capable of causing knee adduction and abduction, respectively.

You can also apply this concept in reverse. For example, if the only action of a muscle such as the brachialis is known to be elbow flexion, then you should be able to determine that its line of pull must be anterior to the joint. Additionally, you would know that the origin of the brachialis must be somewhere on the anterior humerus and the insertion must be somewhere on the anterior ulna.

Consider all the following factors and their relationships as you study movements of the body to gain a more thorough understanding.

1. Exact locations of bony landmarks to which muscles attach proximally and distally and their relationship to joints
2. The planes of motion through which a joint is capable of moving
3. The muscle's relationship or line of pull relative to the joint's axes of rotation
4. As a joint moves through a particular range of motion, the ability of the line of pull of a particular muscle to change and even result in the muscle having a different or opposite action than in the original position
5. The potential effect of other muscles' relative contraction or relaxation on a particular muscle's ability to cause motion
6. The effect of a muscle's relative length on its ability to generate force (See muscle length–tension relationship, p. 57, and active and passive insufficiency, p. 62.)
7. The effect of the position of other joints on the ability of a biarticular or multiarticular muscle to generate force or allow lengthening (See uniarticular, biarticular, and multiarticular muscles, p. 61.)

Neural control of voluntary movement

When we discuss muscular activity, we should really state it as neuromuscular activity, since muscle cannot be active without nervous innervation. All voluntary movement is a result of the muscular and the nervous systems working together. All muscle contraction occurs as a result of stimulation from the nervous system. Ultimately, every muscle fiber is innervated by a somatic motor neuron, which, when an appropriate stimulus is provided, results in a muscle contraction. Depending upon a variety of factors, this stimulus may be processed in varying degrees at different levels of the **central nervous system (CNS)**. The CNS, for the purposes of this discussion, may be divided into five levels of control. Listed in order from the most general level of control and the most superiorly located to the most specific level of control and the most inferiorly located, these levels are the cerebral cortex, the basal ganglia, the cerebellum, the brain stem, and the spinal cord.

The **cerebral cortex**, the highest level of control, provides for the creation of voluntary movement as aggregate muscle action but not as specific muscle activity. Sensory stimuli from the body also are interpreted here, to a degree, for the determination of needed responses.

At the next level, the **basal ganglia** control the maintenance of postures and equilibrium and learned movements such as driving a car. Sensory integration for balance and rhythmic activities is controlled here.

The **cerebellum** is a major integrator of sensory impulses and provides feedback relative to motion. It controls the timing and intensity of muscle activity to assist in the refinement of movements.

Next, the **brain stem** integrates all central nervous system activity through excitation and inhibition of desired neuromuscular actions and functions in arousal or maintaining a wakeful state.

Finally, the **spinal cord** is the common pathway between the CNS and the **peripheral nervous system (PNS)**, which contains all the remaining nerves throughout the body. It has the most specific control and integrates various simple and complex spinal reflexes, as well as cortical and basal ganglia activity.

Functionally, the PNS can be divided into sensory and motor divisions. The sensory or **afferent nerves** bring impulses from receptors in the skin, joints, muscles, and other peripheral aspects of the body to the CNS, while the motor or **efferent nerves** carry impulses to the outlying regions of the body.

The spinal nerves, illustrated in Fig. 2.6, also provide both motor and sensory function for their respective portions of the body and are named for the locations from which they exit the vertebral column. From each side of the spinal column, there are 8 cervical nerves, 12 thoracic nerves, 5 lumbar nerves, 5 sacral nerves, and 1 coccygeal nerve. Cervical nerves 1 through 4 form the cervical plexus, which is generally responsible for sensation from the upper part of the shoulders to the back of the head and front of the neck. The cervical plexus supplies motor innervation to several muscles of the neck. Cervical nerves 5 through 8, along with thoracic nerve 1, form the brachial plexus, which supplies motor and sensory function to the upper extremity and most of the scapula. Thoracic nerves 2 through 12 run directly to specific anatomical locations in the thorax. All of the lumbar, sacral, and coccygeal nerves form the lumbosacral plexus, which supplies sensation and motor

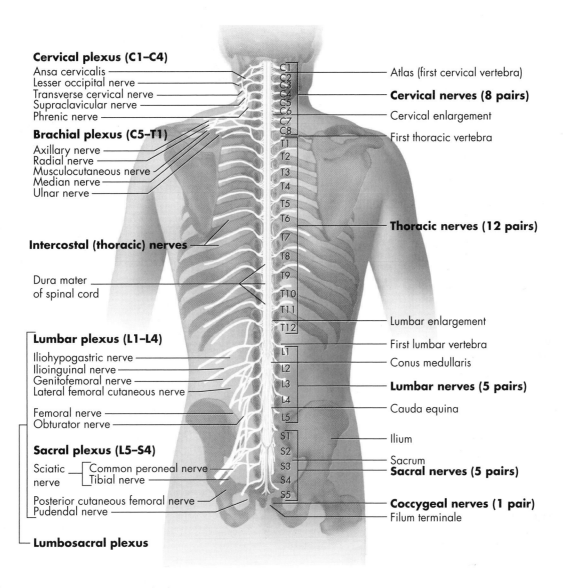

Cervical plexus (C1–C4)
Ansa cervicalis
Lesser occipital nerve
Transverse cervical nerve
Supraclavicular nerve
Phrenic nerve

Brachial plexus (C5–T1)
Axillary nerve
Radial nerve
Musculocutaneous nerve
Median nerve
Ulnar nerve

Intercostal (thoracic) nerves

Dura mater
of spinal cord

Lumbar plexus (L1–L4)
Iliohypogastric nerve
Ilioinguinal nerve
Genitofemoral nerve
Lateral femoral cutaneous nerve
Femoral nerve
Obturator nerve

Sacral plexus (L5–S4)
Sciatic — Common peroneal nerve
nerve — Tibial nerve
Posterior cutaneous femoral nerve
Pudendal nerve

Lumbosacral plexus

Atlas (first cervical vertebra)
Cervical nerves (8 pairs)
Cervical enlargement
First thoracic vertebra

Thoracic nerves (12 pairs)

Lumbar enlargement
First lumbar vertebra
Conus medullaris
Lumbar nerves (5 pairs)
Cauda equina
Ilium
Sacrum
Sacral nerves (5 pairs)
Coccygeal nerves (1 pair)
Filum terminale

FIG. 2.6 • Spinal nerve roots and plexuses.

function to the lower trunk and the entire lower extremity and perineum.

One aspect of the sensory function of spinal nerves is to provide feedback to the CNS regarding skin sensation. A defined area of skin supplied by a specific spinal nerve is known as a **dermatome** (Fig. 2.7). Regarding motor function of spinal nerves, a **myotome** is defined as a muscle or group of muscles supplied by a specific spinal nerve. Certain spinal nerves are also responsible for reflexes. Table 2.3 summarizes the specific spinal nerve functions.

The basic functional units of the nervous system responsible for generating and transmitting impulses are nerve cells known as **neurons**.

Neurons consist of a **neuron cell body**; one or more branching projections known as **dendrites**, which transmit impulses to the neuron and cell body; and an **axon**, an elongated projection that transmits impulses away from neuron cell bodies. As shown in Fig. 2.8, neurons are classified into three types, according to the direction in which they transmit impulses. **Sensory neurons** transmit impulses to the spinal cord and brain from all parts of the body, whereas **motor neurons** transmit impulses away from the brain and spinal cord to muscle and glandular tissue. **Interneurons** are central or connecting neurons that conduct impulses from sensory neurons to motor neurons.

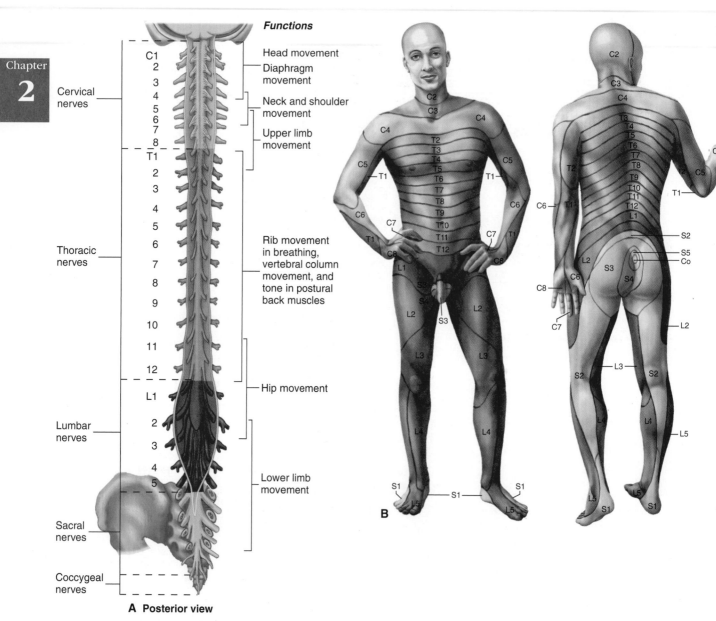

Functions

Cervical nerves
C1
2
3
4
5
6
7
8

Thoracic nerves
T1
2
3
4
5
6
7
8
9
10
11
12

Lumbar nerves
L1
2
3
4
5

Sacral nerves

Coccygeal nerves

A Posterior view

Head movement
Diaphragm movement
Neck and shoulder movement
Upper limb movement

Rib movement in breathing, vertebral column movement, and tone in postural back muscles

Hip movement

Lower limb movement

B

FIG. 2.7 • Spinal cord and dermatomal map. **A,** Nerves and functions of the spinal cord (regions color-coded); **B,** Letters and numbers indicate the spinal nerves innervating a given region of skin.

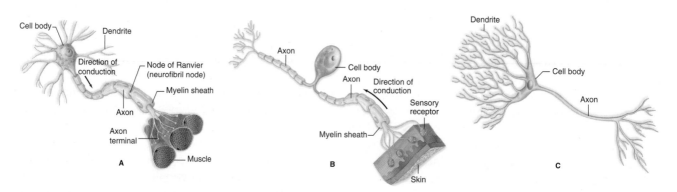

FIG. 2.8 • Neuron anatomy. **A,** Motor neuron. Note the branched dendrites and the single long axon, which branches only near its tip; **B,** Sensory neuron with dendritelike structures projecting from the peripheral end of the axon; **C,** Interneuron (from the cortex of the cerebellum) with very highly branched dendrites.

TABLE 2.3 • Spinal nerve root dermatomes, myotomes, reflexes, and functional applications

	Nerve root	Dermatome afferent (sensory)	Myotome efferent (motor)	Reflexes	Functional application
Cervical plexus	C1	Touch: Vertex of skull	Upper neck muscles	None	Capital flexion and extension
	C2	Touch: Temple, forehead, occiput	Upper neck muscles	None	Sensation behind the ear and posterior skull. Capital and upper cervical movements
	C3	Touch: Entire neck, posterior cheek, temporal area, under mandible	Trapezius, splenius, capitis	None	Scapula retraction, neck extension. Sensation to cheek and side of neck
	C4	Touch: Shoulder area, clavicular area, upper scapular area	Trapezius, levator scapulae	None	Scapula retraction and elevation. Sensation to clavicle and upper scapula
Brachial plexus	C5	Touch: Deltoid area, anterior aspect of entire arm to base of thumb	Supraspinatus, infraspinatus, deltoid, biceps brachii	Biceps brachii	Shoulder abduction. Sensation to lateral side of arm and elbow
	C6	Touch: Anterior arm, radial side of hand to thumb and index finger	Biceps, supinator, wrist extensors	Biceps brachii, brachioradialis	Elbow flexion, wrist extension. Sensation to lateral side of forearm including thumb and index fingers
	C7	Touch: Lateral arm and forearm to index, long, and ring fingers	Triceps brachii, wrist flexors	Triceps brachii	Elbow extension, wrist flexion. Sensation to middle of anterior forearm and long finger
	C8	Touch: Medial side of forearm to ring and little fingers	Ulnar deviators, thumb extensors, thumb adductors (rarely triceps)	None	Wrist ulnar deviation, thumb extension. Sensation to posterior elbow and medial forearm to little fingers
	T1	Touch: Medial arm and forearm to wrist	Intrinsic muscles of the hand except for opponens pollicis and abductor pollicis brevis	None	Abduction and adduction of fingers. Sensation to medial arm and elbow
Thoracic	T2	Touch: Medial side of upper arm to medial elbow, pectoral, and midscapular areas	Intercostal muscles	None	Sensation to medial upper arm, upper chest, and midscapular area
	T3–T12	Touch: T3–T6, upper thorax; T5–T7, coastal margin; T8–T12, abdomen and lumbar region	Intercostal muscles, abdominal muscles	None	Sensation to chest, abdomen, and low back

Proprioception and kinesthesis

The performance of various activities is significantly dependent upon neurological feedback from the body. Very simply, we use the various senses to determine a response to our environment, as when we use sight to know when to lift our hand to catch a fly ball. We are familiar with the senses of smell, touch, sight, hearing, and taste. We are also aware of other sensations, such as pain, pressure, heat, and cold, but we often take for granted the sensory feedback provided by proprioceptors during neuromuscular activity. Proprioceptors are internal receptors located in the skin, joints, muscles, and tendons that provide feedback relative to the tension, length, and contraction state of muscle, the position of the body and limbs, and movements of the joints. These proprioceptors in combination with the other sense organs of the body are vital in **kinesthesis**, the conscious awareness of the position and movement of the body in space. For example, if standing on one leg with the other knee flexed, you do not have to look at your

	Nerve root	Dermatome afferent (sensory)	Myotome efferent (motor)	Reflexes	Functional application
Lumbosacral plexus	L1	Touch: Lower abdomen, groin, lumbar region from 2nd to 4th vertebrae, upper and outer aspect of buttocks	Quadratus lumborum	None	Sensation to low back, over trochanter and groin
	L2	Touch: Lower lumbar region, upper buttock, anterior aspect of thigh	Iliopsoas, quadriceps	None	Hip flexion Sensation to back, front of thigh to knee
	L3	Touch: Medial aspect of thigh to knee, anterior aspect of lower 1/3 of the thigh to just below patella	Psoas, quadriceps	Patella or knee extensors	Hip flexion and knee extension Sensation to back, upper buttock, anterior thigh and knee, medial lower leg
	L4	Touch: Medial aspect of lower leg and foot, inner border of foot, great toe	Tibialis anterior, extensor hallucis and digitorum longus, peroneals	Patella or knee extensors	Ankle dorsiflexion, transverse tarsal/subtalar inversion Sensation to medial buttock, lateral thigh, medial leg, dorsum of foot, great toe
	L5	Touch: Lateral border of leg, anterior surface of lower leg, top of foot to middle three toes	Extensor hallucis and digitorum longus, peroneals, gluteus maximus and medius, dorsiflexors	None	Great toe extension, transverse tarsal/subtalar eversion Sensation to upper lateral leg, anterior surface of the lower leg, middle three toes
	S1	Touch: Posterior aspect of the lower 1/4 of the leg, posterior aspect of the foot, including the heel, lateral border of the foot and sole	Gastrocnemius, soleus, gluteus maximus and medius, hamstrings, peroneals	Achilles reflex	Ankle plantar flexion, knee flexion, transverse tarsal/subtalar eversion Sensation to lateral leg, lateral foot, lateral two toes, plantar aspect of foot
	S2	Touch: Posterior central strip of the leg from below the gluteal fold to 3/4 of the way down the leg	Gastrocnemius, soleus, gluteus maximus, hamstrings	None	Ankle plantar flexion and toe flexion Sensation to posterior thigh and upper posterior leg
	S3	Touch: Groin, medial thigh to knee	Intrinsic foot muscles	None	Sensation to groin and adductor region
	S4	Touch: Perineum, genitals, lower sacrum	Bladder, rectum	None	Urinary and bowel control Sensation to saddle area, genitals, anus

non-weight-bearing leg to know the approximate number of degrees that you may have it flexed. The proprioceptors in and around the knee provide information so that you are kinesthetically aware of your knee position. Muscle spindles and Golgi tendon organs (GTO) are proprioceptors specific to the muscles, whereas Meissner's corpuscles, Ruffini's corpuscles, Pacinian corpuscles, and Krause's end-bulbs are proprioceptors specific to the joints and skin.

While kinesthesis is concerned with the conscious awareness of the body's position, **proprioception** is the subconscious mechanism by which the body is able to regulate posture and movement by responding to stimuli originating in the proprioceptors embedded in the joints, tendons, muscles, and inner ear. When we unexpectedly step on an unlevel or unstable surface, if we have good proprioception the muscles in and about our lower extremity may respond very quickly by contracting appropriately to prevent a fall or injury. This protective response of the body occurs without our having time to make a conscious decision about how to respond.

Muscle spindles (Fig. 2.9), concentrated primarily in the muscle belly between the fibers, are sensitive to stretch and rate of stretch. Specifically, they insert into the connective tissue within the muscle and run parallel with the muscle fibers. The number of spindles in a particular muscle varies depending upon the level of control needed for the area. Consequently, the concentration of muscle spindles in the hands is much greater than in the thigh.

When rapid stretch occurs, an impulse is sent to the CNS. The CNS then activates the motor neurons of the muscle and causes it to contract. All muscles possess this **myotatic** or **stretch reflex**, but it is most remarkable in the extensor muscles of the extremities. The knee jerk or patellar tendon reflex is an example, as shown in Fig. 2.10. When the reflex hammer strikes the patellar tendon, it causes a quick stretch of the musculotendinous unit of the quadriceps. In response, the quadriceps fires and the knee extends. To an extent, the more sudden the tap of the hammer, the more

significant the reflexive contraction. A more practical example is seen in maintaining posture, as when a student begins to doze off in class. As the head starts to nod forward, a sudden stretch is placed on the neck extensors, which activates the muscle spindles and ultimately results in a sudden jerk back to an extended position.

The stretch reflex provided by the muscle spindle may be utilized to facilitate a greater response, as in the case of a quick, short squat before attempting a jump. The quick stretch placed upon the muscles in the squat enables the same muscles to generate more force in the subsequent jump off the floor.

The Golgi tendon organs (Fig. 2.11), serially located in the tendon close to the muscle–tendon junction, are continuously sensitive to both muscle tension and active contraction. The GTO is much less sensitive to stretch than muscle spindles are and requires a greater stretch to be activated. Tension in tendons and consequently in the GTO increases as the muscle contracts, which in turn activates the GTO. When the GTO stretch threshold is reached, an impulse is sent to the CNS, which in turn causes the muscle to relax and facilitates activation of the antagonists as a protective mechanism. That is, the GTO, through this inverse stretch reflex, protects us from excessive contraction by causing the muscle it supplies to relax. As an example, when a weight lifter attempts a very heavy resistance in the biceps curl and reaches the point of extreme overload, the GTO is activated, the biceps suddenly relaxes, and the triceps contracts. This is why it appears as if the lifter is throwing the weight down.

Pacinian corpuscles, concentrated around joint capsules, ligaments, and tendon sheaths and beneath the skin, are activated by rapid changes in the joint angle and by pressure changes affecting the capsule. This activation lasts only briefly and is not effective in detecting constant pressure. Pacinian corpuscles are helpful in providing feedback regarding the location of a body part in space following quick movements such as running or jumping.

Ruffini's corpuscles, located in deep layers of the skin and the joint capsule, are activated by strong and sudden joint movements as well as by pressure changes. Compared to Pacinian corpuscles, their reaction to pressure changes is slower to develop, but their activation is continued as long as pressure is maintained. They are essential in detecting even minute joint position changes and providing information as to the exact joint angle.

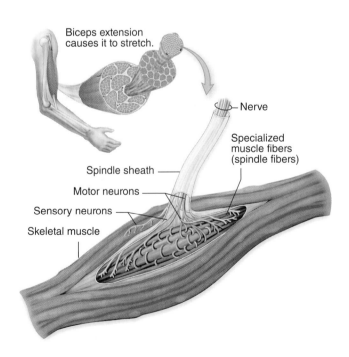

Biceps extension causes it to stretch.

Nerve

Specialized muscle fibers (spindle fibers)

Spindle sheath

Motor neurons

Sensory neurons

Skeletal muscle

FIG. 2.9 • Muscle spindles.

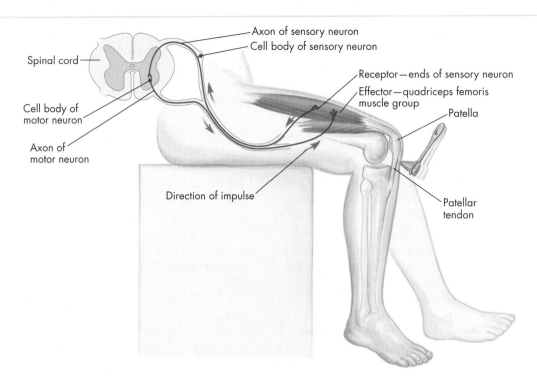

FIG. 2.10 • Knee jerk, or patellar tendon reflex. A sudden tap on the patellar tendon causes a quick stretch of the quadriceps, which activates the muscle spindle. The information regarding the stretch is sent via the axon of the sensory neuron to the spinal cord, where it synapses with a motor neuron that, in turn, carries via its axon a motor response for the quadriceps to contract.

1. Golgi tendon organs detect tension applied to a tendon.

2. Sensory neurons conduct action potentials to the spinal cord.

3. Sensory neurons synapse with inhibitory interneurons that synapse with alpha motor neurons.

4. Inhibition of the alpha motor neurons causes muscle relaxation, relieving the tension applied to the tendon.

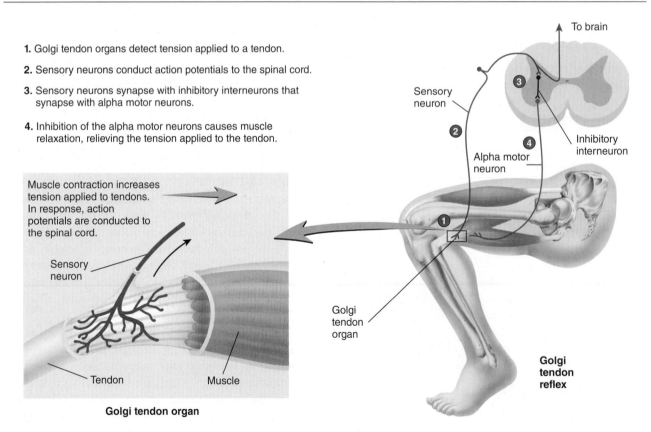

FIG. 2.11 • The Golgi tendon organ. Golgi tendon organs are located in series with muscle and serve as "tension monitors" that act as a protective device for muscle.

TABLE 2.4 • Sensory receptors

Receptors	Sensitivity	Location	Response
Muscle spindles	Subconscious muscle sense, muscle length changes	In skeletal muscles among muscle fibers in parallel with fibers	Initiate rapid contraction of stretched muscle Inhibit development of tension in antagonistic muscles
Golgi tendon organs	Subconscious muscle sense, muscle tension changes	In tendons, near muscle–tendon junction in series with muscle fibers	Inhibit development of tension in stretched muscles Initiate development of tension in antagonistic muscles
Pacinian corpuscles	Rapid changes in joint angles, pressure, vibration	Subcutaneous, submucosa, and subserous tissues around joints and external genitals, mammary glands	Provide feedback regarding location of body part in space following quick movements
Ruffini's corpuscles	Strong, sudden joint movements, touch, pressure	Skin and subcutaneous tissue of fingers, collagenous fibers of the joint capsule	Provide feedback as to exact joint angle
Meissner's corpuscles	Fine touch, vibration	In skin	Provide feedback regarding touch, two-point discrimination
Krause's end-bulbs	Touch, thermal change	Skin, subcutaneous tissue, lip and eyelid mucosa, external genitals	Provide feedback regarding touch

Meissner's corpuscles and Krause's end-bulbs are located in the skin and in subcutaneous tissues. They are important in receiving stimuli from touch, but they are not so relevant to our discussion of kinesthesis. See Table 2.4 for further comparisons of sensory receptors.

The quality of movement and how we react to position change are significantly dependent upon proprioceptive feedback from the muscles and joints. Like the other factors involving body movement, proprioception may be enhanced through specific training that utilizes the proprioceptors to a high degree, such as balancing and functional activities. Attempting to maintain your balance on one leg, first with the eyes open on a level surface, may serve as an initial low-level proprioceptive activity, which may eventually progress to a much higher level such as balancing on an unlevel, unstable surface with your eyes closed. There are numerous additional proprioceptive training activities that are limited only by the imagination and the level of proprioception.

Neuromuscular concepts

Motor units and the all or none principle

Each muscle cell is connected to a motor neuron at the **neuromuscular junction**. A **motor unit**, shown in Fig. 2.12, consists of a single

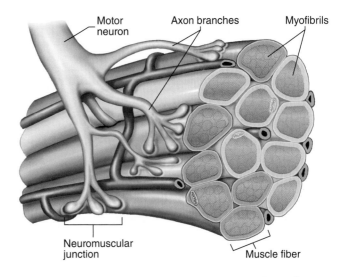

FIG. 2.12 • Motor unit. A motor unit consists of a single motor neuron and all the muscle fibers its branches innervate.

motor neuron and all the muscle fibers it innervates. Motor units function as a single unit. When a particular muscle contracts, the contraction actually occurs at the muscle fiber level within a particular motor unit. In a typical muscle contraction, the number of motor units responding and consequently the number of muscle fibers contracting within the muscle may vary significantly, from relatively few to virtually all of the muscle fibers, depending on the number of muscle fibers within each activated motor unit and the number of motor units activated (Fig. 2.13). Regardless of the number involved, the individual muscle fibers within a given motor unit will fire and contract either maximally or not at all. This is referred to as the **all or none principle**.

Muscle fiber type

Most agree that humans have three types of muscle fiber—two subtypes of fast fibers known as type IIa and type IIb (more recently referred to as type IIx), and a slow fiber known as type I. Fast fibers can produce greater forces due to a greater shortening velocity but fatigue more quickly than slow fibers. Slow muscle fibers have a higher resistance to fatigue, but generally produce less tension than fast fibers.

Factors affecting muscle tension development

The difference between a particular muscle contracting to lift a minimal resistance and the same muscle contracting to lift a maximal resistance lies in the number of muscle fibers recruited. The number of muscle fibers recruited may be increased by activating those motor units containing a greater number of muscle fibers, by activating more motor units, or by increasing the frequency of motor unit activation. The number of muscle fibers per motor unit varies significantly from fewer than 10 in muscles requiring a very precise and detailed response, such as the muscles of the eye, to as many as a few thousand in large muscle groups, such as the quadriceps, that perform less complex activities.

Additionally, recruitment of motor units containing fast muscle fibers is helpful in developing greater tension. Finally, recruitment of muscle fibers that are at optimal length can help generate greater muscle tension. Tension development is minimized in muscle fibers that are shortened to around 60% of their resting length, and muscle fibers stretched beyond 130% of their resting length are significantly compromised in their tension development capability. (See Muscle length–tension relationship, p. 57.)

For the muscle fibers in a particular motor unit to contract, the motor unit must first receive a stimulus via an electrical signal known as an **action potential** from the brain and spinal cord through its axons. If the stimulus is not strong enough to cause an action potential, it is known as a **subthreshold stimulus**

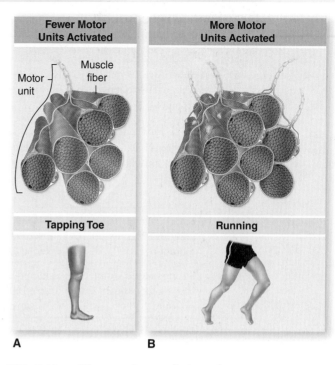

FIG. 2.13 • The number and size of motor units. **A,** Precise muscle contractions require smaller motor units; **B,** Large muscle movements require larger motor units.

and does not result in a contraction. When the stimulus becomes strong enough to produce an action potential in a single motor unit axon, it is known as a **threshold stimulus**, and all the muscle fibers in the motor unit contract. Stimuli that are stronger to the point of producing action potentials in additional motor units are known as **submaximal stimuli**. For action potentials to be produced in all the motor units of a particular muscle, a **maximal stimulus** is required. As the strength of the stimulus increases from threshold up to maximal, more motor units are recruited, and the overall force of the muscle contraction increases in a graded fashion. Increasing the stimulus beyond maximal has no effect. The effect of increasing the number of motor units activated is detailed in Fig. 2.14.

Greater contraction forces may also be achieved by increasing the frequency of motor unit activation. To simplify the phases of a single muscle fiber contraction or twitch, a stimulus is provided and followed by a brief **latent period** of a few milliseconds. Then the second phase, known as the **contraction phase**, begins and the muscle fiber starts shortening. The contraction phase lasts about 40 milliseconds and is

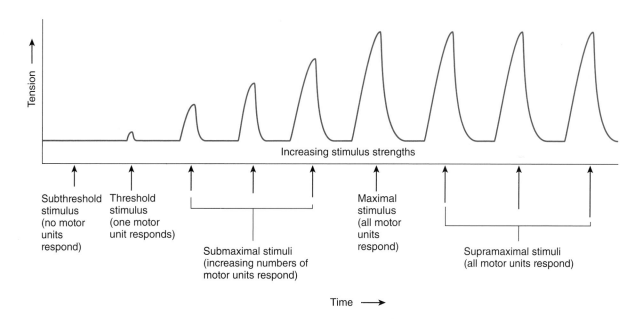

FIG. 2.14 ● Achieving threshold stimulus and the effect on increasing tension of recruiting more motor units. If the stimulus does not reach threshold, there is no motor unit response. As the stimulus strength increases, more motor units are recruited until eventually all motor units are recruited and maximal tension of the muscle is generated. Increasing stimulus strength beyond this point has no effect.

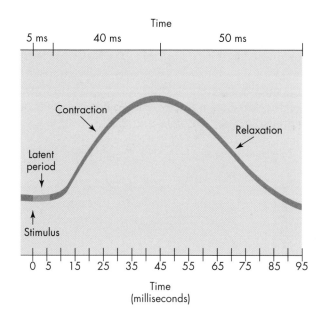

FIG. 2.15 ● A recording of a simple twitch. Note the three time periods (latent period, contraction, and relaxation) following the stimulus.

followed by the **relaxation phase**, which lasts approximately 50 milliseconds. This sequence is illustrated in Fig. 2.15. When successive stimuli are provided before the relaxation phase of the first twitch is complete, the subsequent twitches combine with the first to produce a sustained contraction. This **summation** of contractions generates greater tension than a single contraction would produce on its own. As the frequency of stimuli increases, the resultant summation increases accordingly, producing increasingly greater total muscle tension. If the stimuli are provided at a frequency high enough that no relaxation can occur between contractions, then **tetanus** results. Fig. 2.16 illustrates the effect of increasing the rate of stimulation to gain increased muscle tension.

Treppe, another phenomenon of muscle contraction, occurs when multiple maximal stimuli are provided to rested muscle at a low enough frequency to allow complete relaxation between contractions. Slightly greater tension is produced by the second stimulus than by the first. A third stimulus produces even greater tension than the second. This staircase effect, illustrated in Fig. 2.17, occurs only with the first few stimuli, with the resultant contractions after the initial ones resulting in equal tension being produced.

Muscle length–tension relationship

Tension in a muscle can be thought of as a pulling force. Tension may be either passive through external applied forces or active via muscle

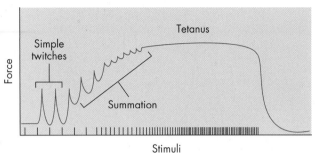

FIG. 2.16 • Recording showing the change from simple twitches to summation and finally tetanus. Peaks to the left represent simple twitches. Increasing the frequency of the stimulus results in a summation of the twitches and finally tetanus.

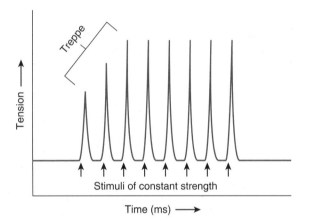

FIG. 2.17 • Treppe. When a rested muscle is stimulated repeatedly with a maximal stimulus at a frequency that allows complete relaxation between stimuli, the second contraction produces a slightly greater tension than the first, and the third contraction produces greater tension than the second. After a few contractions, the tension produced by all contractions is equal.

contraction. **Passive tension** is developed as a muscle is stretched beyond its normal resting length. As it becomes stretched further, its passive tension increases similarly to that of a rubber band being stretched. **Active tension** is dependent on the number of motor units and their respective muscle fibers recruited in a given contraction. However, the length of the muscle during the contraction is a factor in the amount of active tension the muscle may be able to generate.

Generally, depending on the particular muscle involved, the greatest amount of active tension can be developed when a muscle is stretched to between 100% and 130% of its resting length. As a

muscle is stretched beyond this point, the amount of active tension it can generate decreases significantly. Likewise, there is a proportional decrease in the ability to develop tension as a muscle is shortened. When a muscle is shortened to around 50% to 60% of its resting length, its ability to develop contractile tension is essentially reduced to zero.

In the preparatory phase of most sporting activities, we generally place an optimum stretch on the muscles we intend to contract forcefully in the subsequent movement or action phase of the skill. The various phases of performing a movement skill are discussed in much greater detail in Chapter 8. This principle may be seen at work when we squat slightly to stretch the calf, hamstrings, and quadriceps before contracting them concentrically to jump. If we do not first lengthen these muscles through squatting slightly, they are unable to generate enough contractile force to allow us to jump very high. If we squat fully and lengthen the muscles too much, we lose the ability to generate as much force and cannot jump as high.

We can take advantage of this principle by effectively reducing the contribution of some muscles in a group by placing them in a shortened state so that we can isolate the work to those muscle(s) remaining in the lengthened state. For example, in hip extension, we may isolate the work of the gluteus maximus as a hip extensor by maximally shortening the hamstrings with flexion of the knee to reduce their ability to act as hip extensors. See Figs. 2.18 and 2.19.

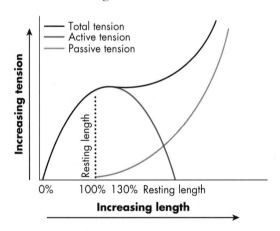

FIG. 2.18 • Muscle length–tension relationship. As the length increases, the amount of active tension that can be developed increases until approximately 130% of the muscle's resting length is reached. After this point, further increases in length result in decreased ability to generate active tension. Passive tension begins to increase on the muscle lengthened beyond its resting length.

A   **B**   **C**

FIG. 2.19 • Practical application of the muscle length–tension relationship involving the calf, hamstrings, and quadriceps muscle groups in jumping. **A,** The muscles are in a relatively shortened position and consequently are not able to generate much tension upon contraction; **B,** The muscles are in a more optimally lengthened position to generate significant tension to jump high; **C,** The muscles are lengthened too much and are not able to generate as much force as in *B*.

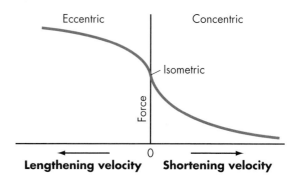

FIG. 2.20 • Muscle–force velocity relationship. From right to left: When little force is needed to move relatively light loads, the muscles may contract concentrically at a relatively high velocity. As the amount of force needed increases with greater loads, the velocity of the concentric contraction decreases proportionally. The amount of force needed continues to increase as the load increases until eventually the load cannot be moved, resulting in zero velocity and an isometric contraction. When the muscle can no longer generate the amount of force needed to maintain the load in a static position, the muscle begins eccentrically contracting to control the velocity, and it can do so at a relatively slow velocity. As the amount of force needed increases to control greater loads, the velocity increases proportionally.

Muscle force–velocity relationship

When the muscle is either concentrically or eccentrically contracting, the rate of length change is significantly related to the amount of force potential. When contracting concentrically against a light resistance, the muscle is able to contract at a high velocity. As the resistance increases, the maximal velocity at which the muscle is able to contract decreases. Eventually, as the load increases, the velocity decreases to zero, resulting in an isometric contraction.

As the load increases even further beyond that which the muscle can maintain with an isometric contraction, the muscle begins to lengthen, resulting in an eccentric contraction or action. A slight increase in the load will result in a relatively low velocity of lengthening. As the load increases even further, the velocity of lengthening will increase as well. Eventually, the load may increase to the point where the muscle can no longer resist. This will result in uncontrollable lengthening or, more likely, dropping of the load.

From this explanation you can see that there is an inverse relationship between concentric velocity and force production. As the force needed

to cause movement of an object increases, the velocity of concentric contraction decreases. Furthermore, there is a somewhat proportional relationship between eccentric velocity and force production. As the force needed to control the movement of an object increases, the velocity of eccentric lengthening increases, at least until the point at which control is lost. This is illustrated in Fig. 2.20.

Stretch-shortening cycle

In addition to the previously discussed factors affecting the force generation capabilities of muscle, the sequencing and timing of contractions can enhance the total amount of force produced. When a muscle is suddenly stretched, resulting in an eccentric contraction that is followed by a concentric contraction of the same muscle, the total force generated in the concentric contraction is greater than that of an isolated concentric contraction. This is often referred to as the **stretch-shortening cycle** and functions by integration of the Golgi tendon organ (GTO) and the muscle spindle. Elastic energy is stored during the eccentric stretch phase, transitioned, and utilized in the concentric contraction phase. A stretch reflex is elicited in the eccentric phase of the motion, which subsequently increases the activation of the muscle that was stretched, resulting in a more forceful concentric contraction. For this to be effective, the transition phase must be immediate or the potential energy gained in the eccentric phase will be lost as heat. The shorter the transition phase, the more effective the force production. This is the basis of plyometric training. An example may be seen when a jumper moves quickly downward immediately prior to jumping upward, resulting in a greater jumping height.

Reciprocal inhibition or innervation

As stated earlier, antagonist muscle groups must relax and lengthen when the agonist muscle group contracts. This effect, called reciprocal innervation, occurs through reciprocal inhibition of the antagonists. Activation of the motor units of the agonists causes a reciprocal neural inhibition of the motor units of the antagonists. This reduction in neural activity of the antagonists allows them to subsequently lengthen under less tension. This may be demonstrated by comparing the ease with which one can stretch the hamstrings while simultaneously contracting the quadriceps with the difficulty of attempting to stretch the hamstrings without the quadriceps contracted. See Fig. 2.21.

Angle of pull

Another factor of considerable importance in using the leverage system is the angle of pull of the muscles on the bone. The angle of pull may be defined as the angle between the line of pull of the muscle and the bone on which it inserts. For the sake of clarity and consistency, we need to specify that the actual angle referred to is the angle toward the joint. With every degree of joint motion, the angle of pull changes. Joint movements and insertion angles involve mostly small angles of pull. The angle of pull decreases as the bone moves away from its anatomical position through the contraction of the local muscle group. This range of movement depends on the type of joint and bony structure.

Most muscles work at a small angle of pull, generally less than 50 degrees. The amount of muscular force needed to cause joint movement is affected by the angle of pull. Three components of muscular force are involved. The **rotary**

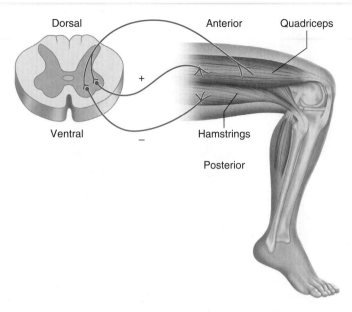

FIG 2.21 ● Reciprocal inhibition. A contraction of the agonist (quadriceps) will produce relaxation in the antagonist (hamstrings).

component, also referred to as the vertical component, acts perpendicular to the long axis of the bone (lever). When the line of muscular force is at 90 degrees to the bone on which it attaches, all of the muscular force is rotary force; therefore, 100% of the force is contributing to the movement. That is, all of the force is being used to rotate the lever about its axis. The closer the angle of pull to 90 degrees, the greater the rotary component. At all other degrees of the angle of pull, one of the other two components of force is operating in addition to the rotary component. The same rotary component is continuing, although with less force, to rotate the lever about its axis. The horizontal or **nonrotary component** is either a **stabilizing component** or a **dislocating component** depending on whether the angle of pull is less than or greater than 90 degrees. If the angle is less than 90 degrees, the force is a stabilizing force because its pull directs the bone toward the joint axis. This increases the compressive forces within the joint and overall joint stability. If the angle is greater than 90 degrees, the force is dislocating because its pull directs the bone away from the joint axis (Fig. 2.22). Angles of pull greater than 90 degrees tend to decrease joint compressive forces and increase the distractive forces, thereby putting more stress on the joint's ligamentous structures. There is quite a bit of variance in both cases, depending on the actual joint structure.

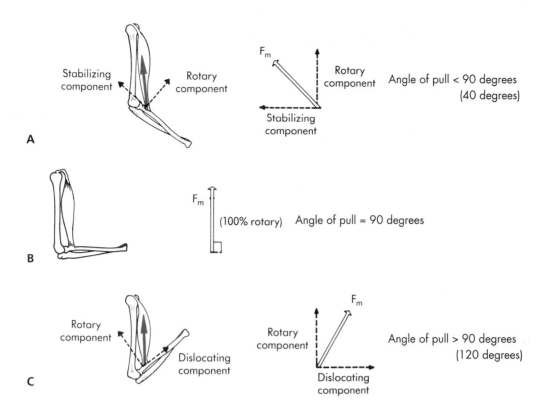

FIG. 2.22 • **A** to **C,** Components of force due to the angle of pull.

In some activities, it is desirable to have a person begin a movement when the angle of pull is at 90 degrees. Many boys and girls are unable to do a chin-up (pull-up) unless they start with the elbow in a position to allow the elbow flexor muscle group to approximate a 90-degree angle with the forearm.

This angle makes the chin-up easier because of the more advantageous angle of pull. The application of this fact can compensate for a lack of sufficient strength. In its range of motion, a muscle pulls a lever through a range characteristic of itself, but it is most effective when approaching and going beyond 90 degrees. An increase in strength is the only solution for muscles that operate at disadvantageous angles of pull and require a greater force to operate efficiently.

Uniarticular, biarticular, and multiarticular muscles

Uniarticular muscles are those that cross and act directly only on the joint that they cross. The brachialis of the elbow is an example in that it can only pull the humerus and ulna closer to each other upon concentric contraction. When the humerus is relatively stabilized, as in an elbow curl, the brachialis contracts to flex the elbow and pulls the ulna closer to the humerus. However, when the ulna is relatively stabilized, as in a pull-up, the brachialis indirectly causes motion at the shoulder even though it does not cross it. In this example the brachialis contracts and pulls the humerus closer to the ulna as an elbow flexor. Correspondingly, the shoulder has to move from flexion to extension for the pull-up to be accomplished.

Biarticular muscles are those that cross and act directly on two different joints. Depending on a variety of factors, a biarticular muscle may contract to cause, control, or prevent motion at either one or both of its joints. Biarticular muscles have two advantages over uniarticular muscles. They can cause, control, and/or prevent motion at more than one joint, and they may be able to maintain a relatively constant length due to "shortening" at one joint and "lengthening" at another joint. The muscle does not actually shorten at one joint and lengthen at the other; rather, the concentric shortening of the muscle to move one joint is offset by motion of the other joint, which moves its attachment of the muscle farther away. This maintenance of a relatively constant length results in the muscle's ability to continue to exert force. In the pull-up example, the biceps brachii acts as a flexor at the elbow. In the initial stage of the pull-up, the biceps brachii is in a relatively lengthened state at the elbow due to its extended position and in a relatively shortened state at the shoulder due to its flexed position. To accomplish the pull-up, the biceps brachii contracts concentrically to flex the elbow, so it effectively "shortens" at the elbow. Simultaneously, the shoulder is extending during the pull-up, which effectively "lengthens" the biceps brachii at the shoulder.

The biarticular muscles of the hip and knee provide excellent examples of two different patterns of action. **Concurrent** movement patterns allow the involved biarticular muscle to maintain a relatively consistent length because of the same action (extension) at both its joints. An example occurs within the rectus femoris (and also the hamstrings) when both the knee and the hip extend at the same time, as in standing from a squatting position as shown in Fig. 2.23. If only the knee were to extend, the rectus femoris would shorten and its ability to exert force similar to the other quadriceps muscles would decrease, but its relative length and subsequent force production capability are maintained due to its relative lengthening at the hip joint during extension.

A   **B**

FIG 2.23 ● Example of concurrent movement pattern. When moving from a squatted position to a standing position, the concurrent movement pattern of extension at the hip and extension at the knee allow the biarticular agonist muscles (hamstrings and rectus femoris, respectively) to maintain a relatively consistent length.

Due to opposite actions occurring simultaneously at both joints of a biarticular muscle, **countercurrent** movement patterns result in substantial shortening of the biarticular muscle. Substantial lengthening of its biarticular antagonist also occurs. This may be observed in the rectus femoris when kicking a ball. During the forward movement phase of the lower extremity, the rectus femoris is concentrically contracted to both flex the hip and extend the knee. These two movements, when combined, result in decreased force production capability in the rectus femoris and increased passive tension or stretch on the hamstring muscles at both the knee and the hip as the kick nears completion. Countercurrent movement patterns result in active insufficiency in the contracting agonist muscles and passive insufficiency in the antagonist muscles. See Fig. 2.24, *B*.

Multiarticular muscles act on three or more joints due to the line of pull between their origin and insertion crossing multiple joints. The principles discussed relative to biarticular muscles also apply to multiarticular muscles.

Active and passive insufficiency

As a muscle shortens, its ability to exert force diminishes, as discussed earlier. When the muscle becomes shortened to the point where it cannot generate or maintain active tension, **active insufficiency** is reached. As a result, the muscle cannot shorten any further. If the opposing muscle becomes stretched to the point where it can no longer lengthen and allow movement, **passive insufficiency** is reached. These principles are most easily observed in either biarticular or multiarticular muscles when the full range of motion is attempted in all the joints crossed by the muscle.

An example is when the rectus femoris contracts concentrically to both flex the hip and extend the knee. It may completely perform either action one at a time, as shown in Fig. 2.24, *A*, but is actively insufficient to obtain full range at both joints simultaneously, as shown in Fig. 2.24, *B*. Likewise, the hamstrings will not usually stretch enough to allow both maximal hip flexion and maximal knee extension; hence, they are passively insufficient. It is virtually impossible to actively extend the knee fully when beginning with the hip fully flexed, or vice versa.

A   **B**

FIG. 2.24 • Active and passive insufficiency. **A,** The rectus femoris is easily able to actively flex the hip or extend the knee through their respective full ranges of motion individually without fully stretching the hamstrings; **B,** However, when one tries to actively flex the hip and simultaneously extend the knee (countercurrent movement pattern), active insufficiency is reached in the rectus femoris and passive insufficiency is reached in the hamstrings, resulting in the inability to reach full range of motion in both joints.

REVIEW EXERCISES

1. Muscle nomenclature chart

Complete the chart by writing in the distinctive characteristics for which each of the muscles is named, such as shape, size, number of divisions, fiber direction, location, and/or action. Some muscles have more than one. Refer to Chapters 4, 5, 6, 7, 9, 10, 11, and 12 if needed.

Muscle name	Distinctive characteristic(s) for which it is named
Adductor magnus	
Biceps brachii	
Biceps femoris	
Brachialis	
Brachioradialis	
Coracobrachialis	
Deltoid	
Extensor carpi radialis brevis	
Extensor carpi ulnaris	
Extensor digiti minimi	
Extensor digitorum	

Extensor hallucis longus	
Extensor indicis	
Extensor pollicis brevis	
External oblique	
Fibularis brevis	
Flexor carpi radialis	
Flexor digitorum longus	
Flexor digitorum profundus	
Flexor digitorum superficialis	
Flexor pollicus longus	
Gastrocnemius	
Gluteus maximus	
Gluteus medius	
Iliacus	
Iliocostalis thoracis	
Infraspinatus	
Latissimus dorsi	
Levator scapulae	
Longissimus lumborum	
Obturator externus	
Palmaris longus	
Pectoralis minor	
Peroneus tertius	
Plantaris	
Pronator quadratus	
Pronator teres	
Psoas major	
Rectus abdominis	
Rectus femoris	
Rhomboid	
Semimembranous	
Semitendinosus	
Serratus anterior	
Spinalis cervicis	
Sternocleidomastoid	
Subclavius	
Subscapularis	
Supinator	
Supraspinatus	
Tensor fasciae latae	
Teres major	
Tibialis posterior	
Transversus abdominis	
Trapezius	
Triceps brachii	
Vastus intermedius	
Vastus lateralis	
Vastus medialis	

2. Muscle shape and fiber arrangement chart

For each muscle listed, determine first whether it should be classified as parallel or pennate. Complete the chart by writing in *flat, fusiform, strap, radiate,* or *sphincter* under those you classify as parallel. Write in *unipennate, bipennate,* or *multipennate* for those you classify as pennate.

Muscle	Parallel	Pennate
Adductor longus		
Adductor magnus		
Brachioradialis		
Extensor digitorum		
Flexor carpi ulnaris		
Flexor digitorum longus		
Gastrocnemius		
Gluteus maximus		
Iliopsoas		
Infraspinatus		
Latissimus dorsi		
Levator scapulae		
Palmaris longus		
Pronator quadratus		
Pronator teres		
Rhomboid		
Serratus anterior		
Subscapularis		
Triceps brachii		
Vastus intermedius		
Vastus medialis		

3. Choose a particular sport skill and determine the types of muscle contractions occurring in various major muscle groups throughout the body at different phases of the skill.

4. Muscle contraction typing chart

For each of the following exercises, write the type of contraction (isometric, concentric, or eccentric), if any, in the cell of the muscle group that is contracting. Place a dash in the cell if there is no contraction occurring. *Hint:* In some instances you may have more than one type of contraction in the same muscle groups throughout various portions of the exercises. If so, list them in the order of occurrence.

Exercise	Quadriceps	Hamstrings
a. Lie prone on a table with your knee in full extension.		
Maintain your knee in full extension.		
Very slowly flex your knee maximally.		
Maintain your knee in full flexion.		
From the fully flexed position, extend your knee fully as fast as possible but stop immediately before reaching maximal extension.		
From the fully flexed position, very slowly extend your knee fully.		
b. Begin sitting on the edge of the table with your knee in full extension.		
Maintain your knee in full extension.		
Very slowly flex your knee maximally.		
Maintain your knee in full flexion.		
Maintain your knee in approximately 90 degrees of flexion.		
From the fully flexed position, slowly extend your knee fully.		
c. Stand on one leg and move the other knee as directed.		
Maintain your knee in full extension.		
Very slowly flex your knee maximally.		
From the fully flexed position, slowly extend your knee fully.		
From the fully flexed position, extend your knee fully as fast as possible.		

5. With the wrist in neutral, extend the fingers maximally and attempt to maintain the position and then extend the wrist maximally. What happens to the fingers and why?

6. Maximally flex your fingers around a pencil with your wrist in neutral. Maintain the maximal finger flexion while you allow a partner to grasp your forearm with one hand and use his or her other hand to push your wrist into maximal flexion. Can you maintain control of the pencil? Explain.

7. You are walking in a straight line down the street when a stranger bumps into you. You stumble but "catch" your balance. Using the information from this chapter and other resources, explain what happened.

8. Drinking a glass of water is a normal daily activity in which the mind and body are involved in the controlled task. Explain how the movements happen once you become thirsty, in terms of the nerve roots, muscle contractions, and angle of pull.

LABORATORY EXERCISES

1. Observe on a fellow student some of the muscles shown in Figs. 2.1 and 2.2.

2. With a partner, choose a diarthrodial joint on the body and carry out each of the following exercises:
   a. Familiarize yourself with all of the joint's various movements and list them.
   b. Determine which muscles or muscle groups are responsible for each movement you listed in 2a.
   c. For the muscles or muscle groups you listed for each movement in 2b, determine the type of contraction occurring.
   d. Determine how to change the parameters of gravity and/or resistance so that the opposite muscles contract to control the same movements in 2c. Name the type of contraction occurring.
   e. Determine how to change the parameters of movement, gravity, and/or resistance so that the same muscles listed in 2c contract differently to control the opposite movement.

3. Utilizing a reflex hammer or the flexed knuckle of your long finger PIP joint, compare the patellar reflex among several subjects.

4. Request a partner to stand with eyes closed while you position his or her arms in an odd position at the shoulders, elbows, and wrists. Ask your partner to describe the exact position of each joint while keeping the eyes closed. Then have your partner begin in the anatomical position, close the eyes, and subsequently reassume the position in which you had previously placed him or her. Explain the neuromechanisms involved in your partner's being able both to sense the joint position in which you placed him or her and then to reassume the same position.

5. Stand up straight on one leg on a flat surface with the other knee flexed slightly and not in contact with anything. Look straight ahead and attempt to maintain your balance in this position for up to 5 minutes. What do you notice happening in terms of the muscles in your lower leg? Try this again with the knee of the leg you are standing on slightly flexed. What differences do you notice? Try it again standing on a piece of thick foam. Try it in the original position with your eyes closed. Elaborate on the differences among the various attempts.

6. Hold a heavy book in your hand with your forearm supinated and your elbow flexed approximately 90 degrees while standing. Have a partner suddenly place another heavy book atop the one you are holding. What is the immediate result regarding the angle of flexion in your elbow? Explain why this result occurs.

7. Sit up very straight on a table with the knees flexed 90 degrees and the feet hanging free. Maintain this position while flexing the right hip and attempting to cross your legs to place the right leg across the left knee. Is this difficult? What tends to happen to the low back and trunk? How can you modify this activity to make it easier?

8. Determine your one-repetition maximum for a biceps curl beginning in full extension and ending in full flexion. Carry out each of the following exercises with adequate periods for recovery in between:
   a. Begin with your elbow flexed 45 degrees, then have a partner hand you a weight slightly heavier than your one-repetition maximum (about 5 pounds). Attempt to lift this weight through the remaining range of flexion. Can you reach full flexion? Explain.
   b. Begin with your elbow in 90 degrees of flexion. Have your partner hand you a slightly heavier weight than in 8a. Attempt to hold the elbow flexed in this position for 10 seconds. Can you do this? Explain.
   c. Begin with your elbow in full flexion. Have your partner hand you an even slightly heavier weight than in 8b. Attempt to slowly lower the weight under control until you reach full extension. Can you do this? Explain.

References

Bernier MR: Perturbation and agility training in the rehabilitation for soccer athletes, *Athletic Therapy Today* 8(3):20–22, 2003.

Blackburn T, Guskiewicz KM, Petschauer MA, Prentice WE: Balance and joint stability: the relative contributions of proprioception and muscular strength, *Journal of Sport Rehabilitation* 9(4):315–328, 2000.

Carter AM, Kinzey SJ, Chitwood LF, Cole JL: Proprioceptive neuromuscular facilitation decreases muscle activity during the stretch reflex in selected posterior thigh muscles, *Journal of Sport Rehabilitation* 9(4):269–278, 2000.

Chimera N, Swanik K, Swanik C: Effects of plyometric training on muscle activation strategies and performance in female athletes. *Journal of Athletic Training* 39(1):24, 2004.

Dover G, Powers ME: Reliability of joint position sense and forcereproduction measures during internal and external rotation of the shoulder, *Journal of Athletic Training* 38(4):304–310, 2003.

Hall SJ: *Basic biomechanics,* ed 6, New York, 2012, McGraw-Hill.

Hamill J, Knutzen KM: *Biomechanical basis of human movement,* ed 3, Baltimore, 2008, Lippincott Williams & Wilkins.

Hamilton N, Weimar W, Luttgens K: *Kinesiology: scientific basis of human motion,* ed 12, New York, 2012, McGraw-Hill.

Knight KL, Ingersoll CD, Bartholomew J: Isotonic contractions might be more effective than isokinetic contractions in developing muscle strength, *Journal of Sport Rehabilitation* 10(2):124–131, 2001.

Kreighbaum E, Barthels KM: *Biomechanics: a qualitative approach for studying human movement,* ed 4, Boston, 1996, Allyn & Bacon.

Lindsay DT: *Functional human anatomy,* St. Louis, 1996, Mosby.

Logan GA, McKinney WC: *Anatomic kinesiology,* ed 3, Dubuque, IA, 1982, Brown.

Mader SS: *Biology,* ed 9, New York, 2007, McGraw-Hill.

McArdle WD, Katch FI, Katch VI: *Exercise physiology: nutrition, energy, and human performance,* ed 7, Baltimore, 2009, Lippincott Williams & Wilkins.

McCrady BJ, Amato HK: Functional strength and proprioception testing of the lower extremity, *Athletic Therapy Today* 9(5):60–61, 2005.

Myers JB, Guskiewicz KM, Schneider, RA, Prentice WE: Proprioception and neuromuscular control of the shoulder after muscle fatigue, *Journal of Athletic Training* 34(4):362–367, 1999.

National Strength and Conditioning Association; Baechle TR, Earle RW: *Essentials of strength training and conditioning,* ed 2, Champaign, IL, 2000, Human Kinetics.

Neumann DA: *Kinesiology of the musculoskeletal system: foundations for physical rehabilitation,* ed 2, St. Louis, 2010, Mosby.

Norkin CC, Levangie PK: *Joint structure and function—a comprehensive analysis,* ed 5, Philadelphia, 2011, Davis.

Northrip JW, Logan GA, McKinney WC: *Analysis of sport motion,* ed 3, Dubuque, IA, 1983, Brown.

Olmsted-Kramer LC, Hertel J: Preventing recurrent lateral ankle sprains: an evidence-based approach, *Athletic Therapy Today* 9(2):19–22, 2004.

Powers ME, Buckley BD, Kaminski TW, Hubbard TJ, Ortiz C: Six weeks of strength and proprioception training does not affect muscle fatigue and static balance in functional ankle stability, *Journal of Sport Rehabilitation* 13(3):201–227, 2004.

Powers SK, Howley ET: *Exercise physiology: theory and application of fitness and performance,* ed 8, New York, 2012, McGraw-Hill.

Rasch PJ: *Kinesiology and applied anatomy,* ed 7, Philadelphia, 1989, Lea & Febiger.

Raven PH, Johnson GB, Losos JB, Mason KA, Singer SR: *Biology,* ed 8, New York, 2008, McGraw-Hill.

Riemann BL, Lephart SM: The sensorimotor system, part I: the physiological basis of functional joint stability, *Journal of Athletic Training* 37(1):71–79, 2002.

Riemann BL, Lephart SM: The sensorimotor system, part II: the role of proprioception in motor control and functional joint stability, *Journal of Athletic Training* 37(1):80–84, 2002.

Riemann BL, Myers JB, Lephart SM: Sensorimotor system measurement techniques, *Journal of Athletic Training* 37(1):85–98, 2002.

Riemann BL, Tray NC, Lephart SM: Unilateral multiaxial coordination training and ankle kinesthesia, muscle strength, and postural control, *Journal of Sport Rehabilitation* 12(1):13–30, 2003.

Ross S, Guskiewicz K, Prentice W, Schneider R, Yu B: Comparison of biomechanical factors between the kicking and stance limbs, *Journal of Sport Rehabilitation* 13(2):135–150, 2004.

Saladin, KS: *Anatomy & physiology: the unity of form and function,* ed 5, New York, 2010, McGraw-Hill.

Sandrey MA: Using eccentric exercise in the treatment of lower extremity tendinopathies, *Athletic Therapy Today* 9(1):58–59, 2004.

Seeley RR, Stephens TD, Tate P: *Anatomy & physiology,* ed 7, New York, 2006, McGraw-Hill.

Shier D, Butler J, Lewis R: *Hole's essentials of human anatomy and physiology,* ed 10, New York, 2009, McGraw-Hill.

Shier D, Butler J, Lewis R: *Hole's human anatomy and physiology,* ed 12, New York, 2010, McGraw-Hill.

Van De Graaff KM: *Human anatomy,* ed 6, New York, 2002, McGraw-Hill.

Van De Graaf KM, Fox SI, LaFleur KM: *Synopsis of human anatomy & physiology,* Dubuque, IA, 1997, Brown.

Yaggie J, Armstrong WJ: Effects on lower extremity fatigue on indices of balance, *Journal of Sport Rehabilitation* 10(2):124–131, 2004.

 For additional resources and a list of related websites, visit **www.mhhe.com/floyd19e**.

Worksheet Exercises

For in- or out-of-class assignments, or for testing, utilize this tear-out worksheet.

Anterior muscular system worksheet

On the anterior muscular system worksheet, label the major superficial muscles on the right and the deeper muscles on the left.

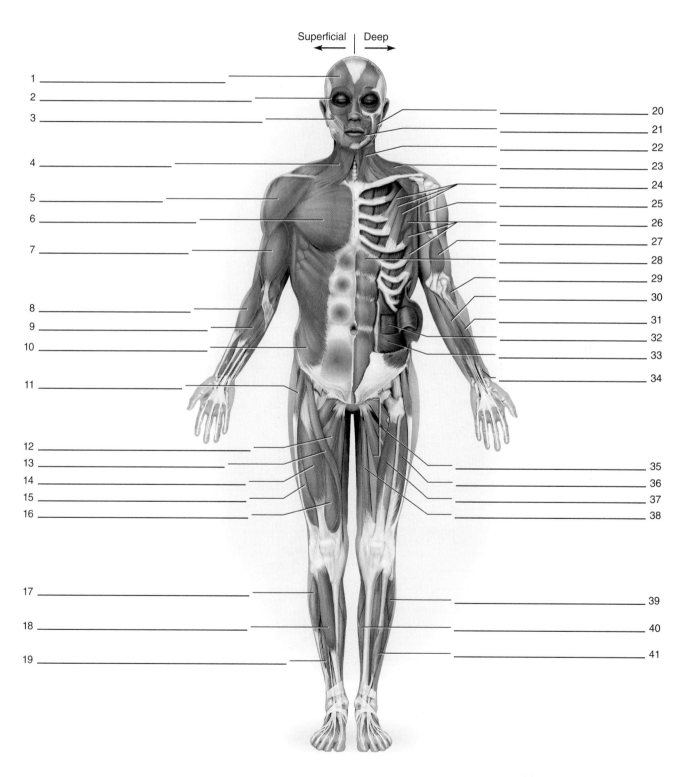

Superficial | Deep

1
2
3
4
5
6
7
8
9
10
11
12
13
14
15
16
17
18
19

20
21
22
23
24
25
26
27
28
29
30
31
32
33
34
35
36
37
38
39
40
41

Worksheet Exercises

For in- or out-of-class assignments, or for testing, utilize this tear-out worksheet.

Posterior muscular system worksheet

On the posterior muscular system worksheet, label the major superficial muscles on the right and the deeper muscles on the left.

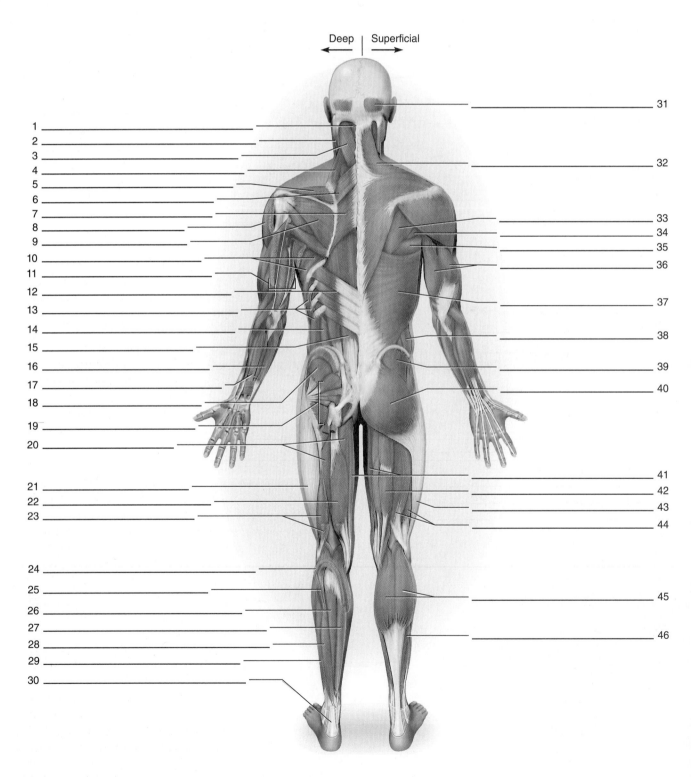

Deep | Superficial

BASIC BIOMECHANICAL FACTORS AND CONCEPTS

Objectives

- To know and understand how knowledge of levers can help improve physical performance

- To know and understand how the musculoskeletal system functions as a series of simple machines

- To know and understand how knowledge of torque and lever arm lengths can help improve physical performance

- To know and understand how knowledge of Newton's laws of motion can help improve physical performance

- To know and understand how knowledge of balance, equilibrium, and stability can help improve physical performance

- To know and understand how knowledge of force and momentum can help improve physical performance

- To know and understand the basic effects of mechanical loading on body tissues

Online Learning Center Resources

Visit *Manual of Structural Kinesiology*'s **Online Learning Center** at **www.mhhe.com/floyd19e** for additional information and study material for this chapter, including:

- *Self-grading quizzes*
- *Anatomy flashcards*
- *Animations*
- *Related websites*

In Chapter 1 we defined kinesiology, very simply, as the study of muscles, bones, and joints as they are involved in the science of movement. From this general definition we can go into greater depth in exploring the science of body movement, which primarily includes anatomy, physiology, and mechanics. For a true understanding of movement, a vast amount of knowledge is needed in all three areas. The focus of this text is primarily structural and functional anatomy. We have only very minimally touched on some physiology in the first two chapters. A much greater study of physiology as it relates to movement should be addressed in an exercise physiology course, for which there are many excellent texts and resources. Likewise, the study of mechanics as it relates to the functional and anatomical analysis of biological systems, known as **biomechanics**, should be addressed to a greater degree in a separate course. Human movement is quite complex. In order to make recommendations for its improvement, we need to study movements from a biomechanical perspective, both qualitatively and quantitatively. This chapter introduces some basic biomechanical factors and concepts, with the understanding that many readers will subsequently study these in more depth in a dedicated course utilizing much more thorough resources.

Many students in kinesiology classes have some knowledge, from a college or high school physics course, of the laws that affect motion. These principles and others are discussed briefly in this chapter, which should prepare you as you begin to apply them to motion in the human body. The more you can put these principles and concepts into practical application, the easier it will be to understand them.

Mechanics, the study of physical actions of forces, can be subdivided into **statics** and **dynamics**. Statics involves the study of systems that are in a constant state of motion, whether at rest with no motion or moving at a constant velocity without acceleration. In statics all forces acting on the body are in balance, resulting in the body being in equilibrium. Dynamics involves the study of systems in motion with acceleration. A system in acceleration is unbalanced due to unequal forces acting on the body. Additional components of biomechanical study include **kinematics** and **kinetics**. Kinematics is concerned with the description of motion and includes consideration of time, displacement, velocity, acceleration, and space factors of a system's motion. Kinetics is the study of forces associated with the motion of a body.

Types of machines found in the body

As discussed in Chapter 2, we utilize muscles to apply force to the bones on which they attach to cause, control, or prevent movement in the joints they cross. As is often the case, we utilize bones such as those in the hand to either hold, push, or pull on an object while using a series of bones and joints throughout the body to apply force via the muscles to affect the position of the object. In doing so we are using a series of simple machines to accomplish the tasks. Machines are used to increase or multiply the applied force in performing a task or to provide a **mechanical advantage**. The mechanical advantage provided by machines enables us to apply a relatively small force, or effort, to move a much greater resistance or to move one point of an object a relatively small distance to result in a relatively large amount of movement of another point of the same object. We can determine mechanical advantage by dividing the load by the effort. The mechanical aspect of each component should be considered with respect to the component's machinelike function.

Another way of thinking about machines is to note that they convert smaller amounts of force exerted over a longer distance to larger amounts of force exerted over a shorter distance. This may be turned around so that a larger amount of force exerted over a shorter distance is converted to a smaller amount of force over a greater distance. Machines function in four ways:

1. To balance multiple forces
2. To enhance force in an attempt to reduce the total force needed to overcome a resistance

3. To enhance range of motion and speed of movement so that resistance can be moved farther or faster than the applied force
4. To alter the resulting direction of the applied force

Simple machines are the lever, wheel and axle, pulley, inclined plane, screw, and wedge. The arrangement of the musculoskeletal system provides three types of machines in producing movement: levers, wheel/axles, and pulleys. Each of these involves a balancing of rotational forces about an axis. The lever is the most common form of simple machine found in the human body.

Levers

It may be difficult for a person to visualize his or her body as a system of levers, but this is actually the case. Human movement occurs through the organized use of a system of levers. While the anatomical levers of the body cannot be changed, when the system is properly understood they can be used more efficiently to maximize the muscular efforts of the body.

A lever is defined as a rigid bar that turns about an **axis of rotation**, or fulcrum. The axis is the point of rotation about which the lever moves. The lever rotates about the axis as a result of **force** (sometimes referred to as effort, E) being applied to it to cause its movement against a **resistance** (sometimes referred to as load or weight). In the body, the bones represent the bars, the joints are the axes, and the muscles contract to apply the force. The amount of resistance can vary from maximal to minimal. In fact, the bones themselves or the weight of the body segment may be the only resistance applied. All lever systems have each of these three components in one of three possible arrangements.

The arrangement or location of three points in relation to one another determines the type of lever and the application for which it is best suited. These points are the axis, the point of force application (usually the muscle insertion), and the point of resistance application (sometimes the center of gravity of the lever and sometimes the location of an external resistance). When the axis (A) is placed anywhere between the force (F) and the resistance (R), a first-class lever is produced (Fig. 3.1). In second-class levers, the resistance is somewhere between the axis and the force (Fig. 3.2). When the force is placed somewhere between the axis and the resistance, a third-class lever is created (Fig. 3.3). Table 3.1 provides a summary of the three classes of levers and the characteristics of each.

The mechanical advantage of levers may be determined using the following equations:

$$\text{Mechanical advantage} = \frac{\text{resistance}}{\text{force}}$$

or

$$\text{Mechanical advantage} = \frac{\text{length of force arm}}{\text{length of resistance arm}}$$

First-class levers

Typical examples of a first-class lever are the crowbar, the seesaw, pliers, oars, and the triceps in overhead elbow extension. In the body an example is when the triceps applies the force to the olecranon (F) in extending the nonsupported forearm (R) at the elbow (A). Other examples are when the agonist and antagonist muscle groups on either side of a joint axis are contracting simultaneously, with the agonist producing the force and the antagonist supplying the resistance. A

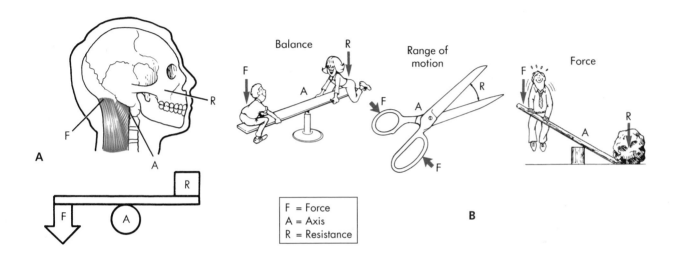

FIG. 3.1 • **A** and **B,** First-class levers.

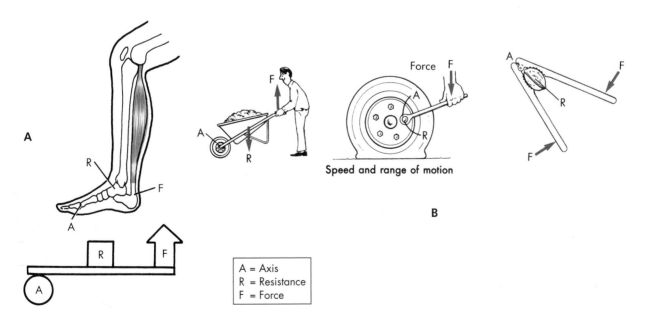

FIG. 3.2 • **A** and **B,** Second-class levers.

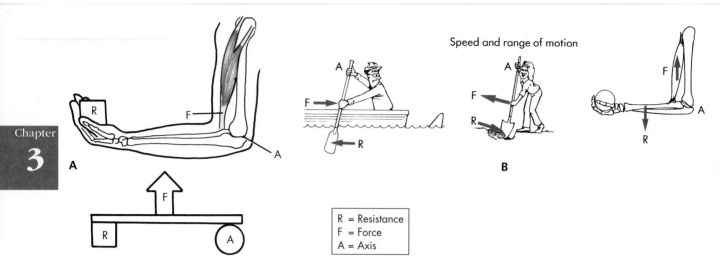

R = Resistance
F = Force
A = Axis

FIG. 3.3 • **A** and **B,** Third-class levers. Note that the paddle and shovel function as third-class levers only when the top hand does not apply force but serves as a fixed axis of rotation. If the top hand applied force and the lower hand acted as the axis, then these would represent first-class levers.

TABLE 3.1 • Classification of levers and characteristics of each

Class	Illustration	Arrangement	Arm movement	Direction of force vs. resistance	Functional design	Relationship to axis	Mechanical advantage	Practical example	Human example
1st	F R First class A	**F–A–R** Axis between force and resistance	Resistance arm and force arm move in opposite directions	Resistance and force are applied in same direction	Balanced movements	Axis in middle	Equal to 1	Seesaw	Erector spinae extending the head on cervical spine
					Speed and range of motion	Axis near force	Less than 1	Scissors	Triceps brachii in extending the elbow
					Force motion	Axis near resistance	Greater than 1	Crow bar	
2nd	R F A Second class	**A–R–F** Resistance between axis and force	Resistance arm and force arm move in the same direction	Resistance and force are applied in opposite directions	Force motion (large resistance can be moved with relatively small force)	Axis near resistance	Always greater than 1	Wheel barrow, nut-cracker	Gastrocnemius and soleus in plantar flexing the foot to raise the body on the toes
3rd	F R A Third class	**A–F–R** Force between axis and resistance	Resistance arm and force arm move in the same direction	Resistance and force are applied in opposite directions	Speed and range of motion (requires large force to move a relatively small resistance)	Axis near force	Always less than 1	Shovel-ing dirt, catapult	Biceps brachii and bra-chialis in flexing the elbow

first-class lever (see Fig. 3.1) is designed basically to produce balanced movements when the axis is midway between the force and the resistance (e.g., a seesaw). When the axis is close to the force, the lever produces speed and range of motion (e.g., the triceps in elbow extension). When the axis is close to the resistance, the lever produces force motion (e.g., a crowbar).

In applying the principle of levers to the body, it is important to remember that the force is applied where the muscle inserts in the bone, not in the belly of the muscle. For example, in elbow extension with the shoulder fully flexed and the arm beside the ear, the triceps applies the force to the olecranon of the ulna behind the axis of the elbow joint. As the applied force exceeds the amount of forearm resistance, the elbow extends.

The type of lever may be changed for a given joint and muscle depending on whether the body segment is in contact with a surface such as a floor or wall. For example, we have demonstrated that the triceps in elbow extension is a first-class lever with the hand free in space and the arm pushed away from the body. If the hand is placed in contact with the floor, as in performing a push-up to push the body away from the floor, the same muscle action at this joint now changes the lever to second class, because the axis is at the hand and the resistance is the body weight at the elbow joint.

Second-class levers

A second-class lever (see Fig. 3.2) is designed to produce force movements, since a large resistance can be moved by a relatively small force. Examples of second-class levers include a bottle opener, a wheelbarrow, and a nutcracker. We have just noted the example of the triceps extending the elbow in a push-up. A similar example of a second-class lever in the body is plantar flexion of the ankle to raise the body on the toes. The ball (A) of the foot serves as the axis of rotation as the ankle plantar flexors apply force to the calcaneus (F) to lift the resistance of the body at the tibiofibular articulation (R) with the talus. Opening the mouth against resistance provides another example of a second-class lever. There are relatively few other examples of second-class levers in the body.

Third-class levers

Third-class levers (see Fig. 3.3), with the force being applied between the axis and the resistance,

are designed to produce speed and range of motion. Most of the levers in the human body are of this type, which requires a great deal of force to move even a small resistance. Examples include a catapult, a screen door operated by a short spring, and the application of lifting force to a shovel handle with the lower hand while the upper hand on the shovel handle serves as the axis of rotation. The biceps brachii is a typical example in the body. Using the elbow joint (A) as the axis, the biceps brachii applies force at its insertion on the radial tuberosity (F) to rotate the forearm up, with its center of gravity (R) serving as the point of resistance application.

The brachialis is an example of true third-class leverage. It pulls on the ulna just below the elbow, and, since the ulna cannot rotate, the pull is direct and true. The biceps brachii, on the other hand, supinates the forearm as it flexes, so the third-class leverage applies to flexion only.

Other examples include the hamstrings contracting to flex the leg at the knee in a standing position and the iliopsoas being used to flex the thigh at the hip.

Factors in use of anatomical levers

Our anatomical leverage system can be used to gain a mechanical advantage that will improve simple or complex physical movements. Some individuals unconsciously develop habits of using human levers properly, but frequently this is not the case.

Torque and length of lever arms

To understand the leverage system, the concept of torque must be understood. **Torque**, or moment of force, is the turning effect of an eccentric force. **Eccentric force** is a force that is applied off center or in a direction not in line with the center of rotation of an object with a fixed axis. In objects without a fixed axis, it is an applied force that is not in line with the object's center of gravity; for rotation to occur, an eccentric force must be applied. In the human body, the contracting muscle applies an eccentric force (not to be confused with eccentric contraction) to the bone on which it attaches and causes the bone to rotate about an axis at the joint. The amount of torque can be determined by multiplying the **force magnitude** (amount of force) by the **force arm**. The perpendicular distance between the location of force application and the axis is known as the force arm, moment arm, or torque arm. The force arm may be best understood as the shortest distance from the axis of rotation to the line of action of the force. The greater the distance of the force arm, the more torque produced by the force. A frequent practical application of torque and levers occurs when we purposely increase the force arm length in order to increase the torque so that we can more easily move a relatively large resistance. This is commonly referred to as increasing our leverage.

It is also important to note the **resistance arm**, which may be defined as the distance between the axis and the point of resistance application. In discussing the application of levers, it is necessary to understand the length relationship between the two lever arms. There is an inverse relationship between force and the force arm, just as there is between resistance and the resistance arm. The longer the force arm, the less force required to move the lever if the resistance and resistance arm remain constant, as shown graphically in Fig. 3.4. In addition, if the force and force arm remain constant, a greater resistance may be moved by shortening the resistance arm. Because the muscular

force is applied internally, in musculoskeletal discussions the force arm may also be referred to as the internal moment arm; and because the load is applied externally, the resistance arm may be referred to as the external moment arm.

Also, there is a proportional relationship between the force components and the resistance components. That is, for movement to occur when either of the resistance components increases, there must be an increase in one or both of the force components. See Figs. 3.5, 3.6, and 3.7 to see how these relationships apply to

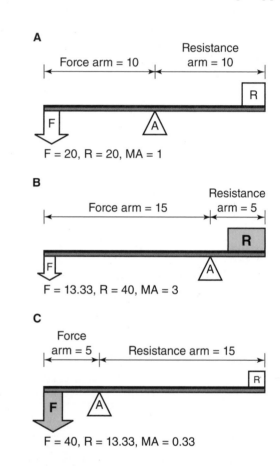

FIG. 3.5 • First-class levers. **A,** If the force arm and resistance arm are equal in length, a force equal to the resistance is required to balance it; **B,** As the force arm becomes longer, a decreasing amount of force is required to move a relatively larger resistance; **C,** As the force arm becomes shorter, an increasing amount of force is required to move a relatively smaller resistance, but the speed and range of motion that the resistance can be moved are increased. *Forces (moments) are calculated to balance the lever system. The effort and resistance forces sum to zero. If any of the components are moved in relation to one another, then either a greater force or a greater resistance will be required.*

Relationships among Force, Force Arm, and Resistance Arm with Constant Resistance of 20 kilograms

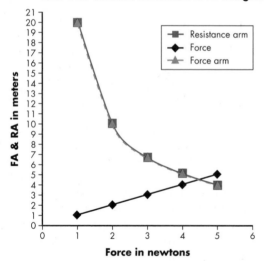

FIG. 3.4 • Relationships among forces, force arms, and resistance arms. (The graph assumes a constant resistance of 20 kilograms, and as a result the graphical representations of the resistance arm and force arm lie directly over each other.) With the resistance held constant at 20 kilograms and a resistance arm of 1 meter, the product of the (force) × (force arm) must equal 20 newtons. Thus there is an inverse relationship between the force and the force arm. As the force increases in newtons, the force arm length decreases in meters, and vice versa.

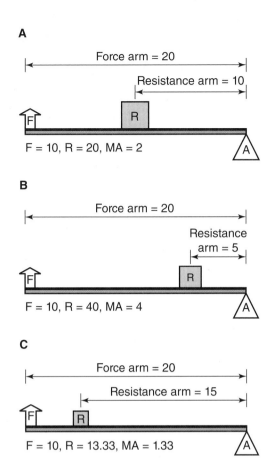

FIG. 3.6 • Second-class levers have a positive mechanical advantage due to the force arm always being longer than the resistance arm and are well suited for moving larger resistances with smaller forces. **A,** Placing the resistance halfway between the axis and the point of force application provides a mechanical advantage of 2; **B,** Moving the resistance closer to the axis increases the mechanical advantage but decreases the distance the resistance is moved; **C,** The closer the resistance is positioned to the point of force application, the less the mechanical advantage but the greater the distance the resistance is moved. *Forces (moments) are calculated to balance the lever system. The effort and resistance forces sum to zero. If any of the components are moved in relation to one another, then either a greater force or a greater resistance will be required.*

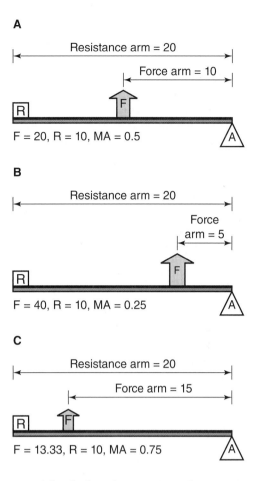

FIG. 3.7 • Third-class levers. **A,** A force greater than the resistance, regardless of the point of force application, is required due to the resistance arm always being longer; **B,** Moving the point of force application closer to the axis increases the range of motion and speed but requires more force; **C,** Moving the point of force application closer to the resistance decreases the force needed but also decreases the speed and range of motion. *Forces (moments) are calculated to balance the lever system. The effort and resistance forces sum to zero. If any of the components are moved in relation to one another, then either a greater force or a greater resistance will be required.*

first-, second-, and third-class levers, respectively. Even slight variations in the location of the force and the resistance are important in determining the mechanical advantage (MA) and effective force of the muscle. This point can be illustrated with the simple formula shown in Fig. 3.8, using the biceps brachii muscle in each example.

In Example A, the only way to move the insertion of the biceps brachii is with surgery, so this is not practical. In some orthopaedic conditions, the attachments of tendons are surgically relocated in an attempt to change the dynamic forces

the muscles exert on the joints. In Example B, we can and often do shorten the resistance arm to enhance our ability to move an object. When attempting a maximal weight in a biceps curl exercise, we may flex our wrist to move the weight just a little closer, which shortens the resistance arm. Example C is straightforward in that we obviously can reduce the force needed by reducing the resistance.

The system of leverage in the human body is built for speed and range of motion at the expense of force. Short force arms and long resistance arms require great muscular strength to produce movement. In the forearm, the attachments of the biceps and triceps muscles clearly illustrate this point, since the force arm of the biceps is 1 to 2 inches and that of the triceps is less than 1 inch. Many similar examples are found all over the body. From a practical point of view, this means that the muscular system should be strong to supply the necessary force for body movements, especially in strenuous sports activities.

When we speak of human leverage in relation to sport skills, we are generally referring to several levers. For example, throwing a ball involves levers at the shoulder, elbow, and wrist joints as well as from the ground up through the lower extremities and the trunk. In fact, it can be said that there is one long lever from the feet to the hand.

The longer the lever, the more effective it is in imparting velocity. A tennis player can hit a tennis ball harder (deliver more force to it) with a straight-arm drive than with a bent elbow, because the lever (including the racket) is longer and moves faster.

Fig. 3.9 indicates that a longer lever (Z1) travels faster than a shorter lever (S1) in traveling the same number of degrees. In sports activities in which it is possible to increase the length of a lever with a racket or bat, the same principle applies.

In baseball, hockey, golf, field hockey, and other sports, long levers produce more linear force and thus better performance. However, to be able to fully execute the movement in as short a time as possible, it is sometimes desirable to have a short lever arm. For example, a baseball catcher attempting to throw a runner out at second base does not have to throw the ball so that it travels as fast as when the pitcher is attempting to throw a strike. In the catcher's case, it is more important to initiate and complete the throw as soon as possible than to deliver as much velocity to the ball as possible. The pitcher, when attempting to throw a ball at 90-plus miles per hour, will

Lever equation for a child performing biceps curls
Lever equation

$$F \times FA = R \times RA$$
$$\text{(Force)} \times \text{(Force arm)} = \text{(Resistance)} \times \text{(Resistance arm)}$$

Initial Example

$$F \times 0.1 = 45 \text{ newtons} \times 0.25 \text{ meter}$$
$$F \times 0.1 = 11.25 \text{ newton-meters}$$
$$F = 112.5 \text{ newtons}$$

Example A – Lengthening the force arm

Increase FA by moving the insertion distally 0.05 meter:

$$F \times 0.15 = 45 \text{ newtons} \times 0.25 \text{ meter}$$
$$F \times 0.15 = 11.25 \text{ newton-meters}$$
$$F = 75 \text{ newtons}$$

An increase in the insertion from the axis by 0.05 meter results in a substantial reduction in the force necessary to move the resistance.

Example B – Shortening the resistance arm

Reduce RA by moving the point of resistance application proximally 0.05 meter:

$$F \times 0.1 = 45 \text{ newtons} \times 0.2 \text{ meter}$$
$$F \times 0.1 = 9 \text{ newton-meters}$$
$$F = 90 \text{ newtons}$$

A decrease in the resistance application from the axis by 0.05 meter results in a considerable reduction in the force necessary to move the resistance.

Example C – Reducing the resistance

Reduce R by reducing the resistance 1 newton:

$$F \times 0.1 = 44 \text{ newtons} \times 0.25 \text{ meter}$$
$$F \times 0.1 = 11 \text{ newton-meters}$$
$$F = 110 \text{ newtons}$$

Decreasing the amount of resistance can decrease the amount of force needed to move the lever.

FIG. 3.8 • Torque calculations with examples of modifications in the force arm, resistance arm, and resistance.

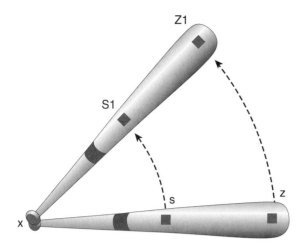

FIG. 3.9 • Length of levers. The end of a longer lever (Z1) travels faster than a shorter lever (S1) when moved over the same number of degrees in the same amount of time. The farther the resistance is from the axis, the farther it is moved and the more force is delivered to it, resulting in the resistance moving at a greater velocity. The same principle applies in sports in which it is possible to increase the lever length with a racket or bat.

utilize his body as a much longer lever system throughout a greater range of motion to impart velocity to the ball.

Wheels and axles

Wheels and axles are used primarily to enhance range of motion and speed of movement in the musculoskeletal system. A wheel and an axle essentially function as a form of lever. When either the wheel or the axle turns, the other must turn as well. Both complete one turn at the same time. The centers of the wheel and the axle both correspond to the fulcrum. Both the radius of the wheel and the radius of the axle correspond to the force arms. If the radius of the wheel is greater than the radius of the axle, then the wheel has a mechanical advantage over the axle due to the longer force arm. That is, a relatively smaller force may be applied to the wheel to move a relatively greater resistance applied to the axle. Very simply, if the radius of the wheel is five times the radius of the axle, then the wheel has a 5 to 1 mechanical advantage over the axle, as shown in Fig. 3.10. The mechanical advantage of a wheel and axle for this scenario may be calculated by considering the radius of the wheel over the radius of the axle. This application enables

the wheel and axle to act as a second-class lever to gain force motion.

$$\text{Mechanical advantage} = \frac{\text{radius of the wheel}}{\text{radius of the axle}}$$

In this case the mechanical advantage is always more than 1. An application of this example is using the outer portion of an automobile steering wheel to turn the steering mechanism. Before the development of power steering, steering wheels had a much larger diameter than today in order to give the driver more of a mechanical advantage. An example in the body of applying force to the wheel occurs when we attempt to manually force a person's shoulder into internal rotation while she or he holds it in external rotation isometrically. The humerus acts as the axle, and the person's hand and wrist are located near the outside of the wheel when the elbow is flexed approximately 90 degrees. If we unsuccessfully attempt to break the force of the contraction of the external rotators by pushing internally at the midforearm, we can increase our leverage or mechanical advantage and our likelihood of success by applying the force nearer to the hand and wrist.

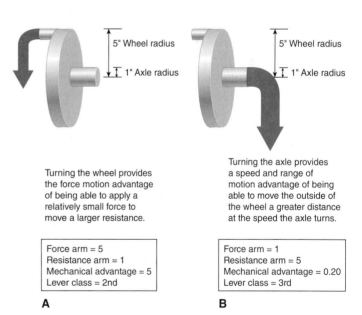

Turning the wheel provides the force motion advantage of being able to apply a relatively small force to move a larger resistance.

Turning the axle provides a speed and range of motion advantage of being able to move the outside of the wheel a greater distance at the speed the axle turns.

Force arm = 5
Resistance arm = 1
Mechanical advantage = 5
Lever class = 2nd

Force arm = 1
Resistance arm = 5
Mechanical advantage = 0.20
Lever class = 3rd

A   **B**

FIG. 3.10 • Wheel and axle. **A,** Mechanical advantage is gained by applying force to the outside of the wheel to more easily move a large resistance; **B,** The mechanical advantage in applying force to the axle is always less than 1 and requires a relatively large force, but the advantage is being able to move the larger wheel a relatively greater distance at a relatively greater speed than the axle.

If the application of force is reversed so that it is applied to the axle, then the mechanical advantage results from the wheel's turning a greater distance at greater speed. Using the same example, if the wheel radius is five times greater than the radius of the axle, the outside of the wheel will turn at a speed five times that of the axle. Additionally, the distance the outside of the wheel turns will be five times that of the outside of the axle. This application enables the wheel and axle to act as a third-class lever to gain speed and range of motion. The mechanical advantage of a wheel and axle for this scenario may be calculated by considering the radius of the axle over the radius of the wheel.

$$\text{Mechanical advantage} = \frac{\text{radius of the axle}}{\text{radius of the wheel}}$$

In this case the mechanical advantage is always less than 1. This is the principle utilized in the drivetrain of an automobile to turn the axle, which subsequently turns the tire one revolution for every turn of the axle. We utilize the powerful engine of the automobile to supply the force to increase the speed of the tire and subsequently carry us great distances. An example of the muscles applying force to the axle to result in greater range of motion and speed may again be seen in the upper extremity, in the case of the internal rotators attaching to the humerus. With the humerus acting as the axle and the hand and wrist located at the outside of the wheel (when the elbow is flexed approximately 90 degrees), the internal rotators apply force to the humerus. With the internal rotators concentrically internally rotating the humerus a relatively small amount, the hand and wrist travel a great distance. Using the wheel and axle in this manner enables us to significantly increase the speed at which we can throw objects.

Pulleys

Single pulleys have a fixed axle and function to change the effective direction of force application. Single pulleys have a mechanical advantage of 1, as shown in Fig. 3.11, *A*. Numerous weight machines utilize pulleys to alter the direction of the resistive force. Pulleys may be movable and can be combined to form compound pulleys to further increase the mechanical advantage. Every additional rope connected to movable pulleys increases the mechanical advantage by 1, as shown in Fig. 3.11, *B*.

In the human body, an excellent example is provided by the lateral malleolus, acting as a pulley around which the tendon of the peroneus longus runs. As this muscle contracts, it pulls toward its belly, which is toward the knee. Due to its use of the lateral malleolus as a pulley (Fig. 3.12), the force is transmitted to the plantar aspect of the foot, resulting in downward and outward movement of the foot. Other examples in the human body include pulleys on the volar aspect of the phalanges to redirect the force of the flexor tendons.

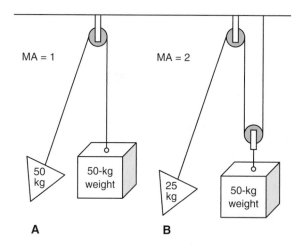

FIG. 3.11 • **A,** Single pulley; **B,** Compound movable pulley.

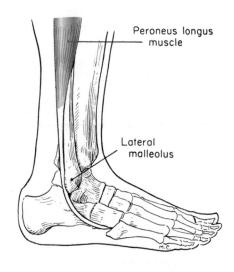

FIG. 3.12 • Pulley. The lateral malleolus serves as a pulley for the peroneus longus tendon.

Laws of motion and physical activities

Motion is fundamental in physical education and sports activity. Body motion is generally produced, or at least started, by some action of the muscular system. Motion cannot occur without a force, and the muscular system is the source of force in the human body. Thus, development of the muscular system is indispensable to movement.

Basically, there are two types of motion: **linear motion** and **angular motion**. Linear motion, also referred to as translatory motion, is motion along a line. If the motion is along a straight line, it is **rectilinear** motion, whereas motion along a curved line is known as **curvilinear** motion. Angular motion, also known as rotary motion, involves rotation around an axis. In the human body, the axis of rotation is provided by the various joints. In a sense, these two types of motion are related, since angular motion of the joints can produce the linear motion of walking. In many sports activities, the cumulative angular motion of the joints of the body imparts linear motion to a thrown object (ball, shot) or to an object struck with an instrument (bat, racket). It is also important to consider the **center of rotation**, which is the point or line around which all other points in the body move. In a door hinge, the axis of rotation is fixed and all points of the door have equal arcs of rotation around the center of the hinge. But in joints of the body the axis is not usually fixed, due to their accessory motion as discussed in Chapter 1. As a result, the location of the exact center of rotation changes with changes in the joint angle. So we have to consider the **instantaneous center of rotation**, which is the center of rotation at a specific instant in time during movement. See Fig. 3.13.

Quantity measurements—scalars versus vectors

In order to discuss measures of motion, we assign them a quantity of measurement. These mathematical quantities used to describe motion can be divided into either scalars or vectors. **Scalar** quantities are described by a magnitude (or numerical value) alone, such as speed as in miles per hour or meters per second. Other scalar quantities are length, area, volume, mass, time, density, temperature, pressure, energy, work, and power. **Vector** quantities are described by both a magnitude and a direction, such as velocity as in miles per hour in an eastward direction. Other vector quantities are acceleration, direction, displacement, force, drag, momentum, lift, weight, and thrust.

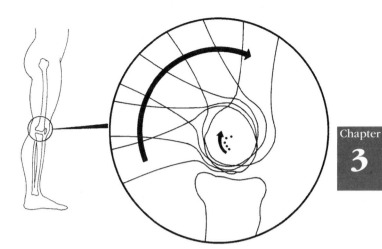

FIG. 3.13 ● The path of the instantaneous center of rotation for the knee during extension.

Displacement is a change in the position or location of an object from its original point of reference, whereas **distance**, or the path of movement, is the actual sum length it is measured to have traveled. Thus an object may have traveled a distance of 10 meters along a linear path in two or more directions but be displaced from its original reference point by only 6 meters. Fig. 3.14 provides an example. **Angular displacement** is the change in location of a rotating body. **Linear displacement** is the distance a system moves in a straight line.

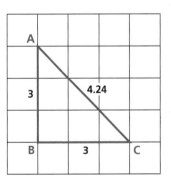

FIG. 3.14 ● Displacement. If the path of movement is from A to B and then from B to C, the distance covered is AB + BC, but the displacement is the distance from A to C, or AC. If each cell is 1 square meter, then AB is 3 meters and BC is 3 meters, so the distance covered would be 6 meters. Using the Pythagorean Theorem (in a right triangle the square of the measure of the hypotenuse is equal to the sum of the squares of the measures of the legs, or $a^2 + b^2 = c^2$), we can then determine the displacement (AC) to be 4.24 meters with $AB^2 + BC^2 = AC^2$.

We are sometimes concerned about the time it takes for displacement to occur. **Speed** is how fast an object is moving, or the distance an object travels in a specific amount of time. **Velocity**, the rate at which an object changes its position, includes the direction and describes the rate of displacement.

A brief review of Newton's laws of motion will indicate the many applications of these laws to physical education activities and sports. Newton's laws explain all the characteristics of motion, and they are fundamental to understanding human movement.

Law of inertia

A body in motion tends to remain in motion at the same speed in a straight line unless acted on by a force; a body at rest tends to remain at rest unless acted on by a force.

Inertia can be described as the resistance to action or change. In terms of human movement, *inertia* refers to resistance to acceleration or deceleration. Inertia is the tendency for the current state of motion to be maintained, whether the body segment is moving at a particular velocity or is motionless.

Muscles produce the force necessary to start motion, stop motion, accelerate motion, decelerate motion, or change the direction of motion. Put another way, inertia is the reluctance to change status; only force can do so. The greater the mass of an object, the greater its inertia. Therefore, the greater the mass, the more force needed to significantly change an object's inertia. Numerous examples of this law are found in physical education activities. A sprinter in the starting blocks must apply considerable force to overcome resting inertia. A runner on an indoor track must apply considerable force to overcome moving inertia and stop before hitting the wall. Fig. 3.15 provides an example of how a skier in motion remains in motion even though airborne after skiing off a hill. We routinely experience inertial forces when our upper body tends to move forward if we are driving a car at the speed limit and then suddenly have to slow down. Balls and other objects that are thrown or struck require force to stop them. Starting, stopping, and changing direction—a part of many physical activities—provide many examples of the law of inertia applied to body motion.

Because force is required to change inertia, it is obvious that any activity that is carried out at a steady pace in a consistent direction will conserve energy and that any irregularly paced or directed

FIG. 3.15 ● An example of Newton's first law of motion. The skier continues airborne in space because of the previously established inertia.

activity will be very costly to energy reserves. This explains in part why activities such as handball and basketball are so much more fatiguing than jogging and dancing.

Law of acceleration

A change in the acceleration of a body occurs in the same direction as the force that caused it. The change in acceleration is directly proportional to the force causing it and inversely proportional to the mass of the body.

Acceleration may be defined as the rate of change in velocity. To attain speed in moving the body, a strong muscular force is generally necessary. **Mass**, the amount of matter in a body, affects the speed and acceleration in physical movements. A much greater force is required from the muscles to accelerate an 80-kilogram man than to accelerate a 58-kilogram man to the same running speed. Also, it is possible to accelerate a baseball faster than a shot because of the difference in mass. The force required to run at half speed is less than the force required to run at top speed. To impart speed to a ball or an object, it is necessary to rapidly accelerate the part of the body holding the object. Football, basketball, track, and field hockey are a few sports that demand speed and acceleration.

Law of reaction

For every action there is an opposite and equal reaction.

As we place force on a supporting surface by walking over it, the surface provides an equal resistance back in the opposite direction to the

FIG. 3.16 • An example of Newton's third law of motion. To accelerate forward, a walker must push backward. Note that the front part of the footprint in the sand is more deeply depressed than the rear part.

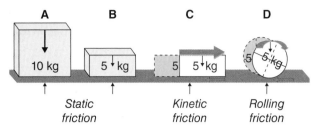

FIG. 3.17 • Friction. **A,** Static friction; **B,** Static friction also, but less than in *A* because there is less mass (weight); **C,** Kinetic friction is always less than static friction; **D,** Rolling friction is always less than kinetic friction.

soles of our feet. Our feet push down and back, while the surface pushes up and forward. The force of the surface reacting to the force we place on it is referred to as **ground reaction force**. We provide the action force, while the surface provides the reaction force. It is easier to run on a hard track than on a sandy beach because of the difference in the ground reaction forces of the two surfaces. The track resists the runner's propulsion force, and the reaction drives the runner ahead. The sand dissipates the runner's force, and the reaction force is correspondingly reduced, with an apparent loss in forward force and speed (Fig. 3.16). A sprinter applies a force in excess of 1335 Newtons on the starting blocks, which resist with an equal force. When a body is in flight, as in jumping, movement of one part of the body produces a reaction in another part because there is no resistive surface to supply a reaction force.

Friction

Friction is the force that results from the resistance between the surfaces of two objects moving on each other. Depending on the activity involved, we may desire increased or decreased friction. In running, we depend on friction forces between our feet and the ground so that we may exert force against the ground and propel ourselves forward. When friction is reduced due to a slick ground or shoe surface, we are more likely to slip. In skating, we desire decreased friction so that we may slide across the ice with less resistance. Friction may be further characterized as either static or kinetic. See Fig. 3.17, *A, B,* and *C*. **Static friction** is the amount of friction between two objects that have not yet

begun to move, whereas **kinetic friction** is the friction between two objects that are sliding along each other. Static friction is always greater than kinetic friction. As a result, it is always more difficult to initiate dragging an object across a surface than it is to continue dragging it. Static friction may be increased by increasing the normal or perpendicular forces pressing the two objects together, as by adding more weight to one object sitting on another object. To determine the amount of friction forces, we must consider both the forces pressing the two objects together and the **coefficient of friction**, which depends on the hardness and roughness of the surface textures. The coefficient of friction is the ratio of the force needed to over come the friction to the force holding the surfaces together. **Rolling friction** (Fig. 3.17, *D*) is the resistance to an object rolling across a surface, such as a ball rolling across a court or a tire rolling across the ground. Rolling friction is always much less than static or kinetic friction.

Balance, equilibrium, and stability

Balance is the ability to control equilibrium, either static or dynamic. In relation to human movement, **equilibrium** refers to a state of zero acceleration, where there is no change in the speed or direction of the body. Equilibrium may be either static or dynamic. If the body is at rest or completely motionless, it is in **static equilibrium**. **Dynamic equilibrium** occurs when all the applied and inertial forces acting on the moving body are in balance, resulting in movement with unchanging speed or direction. For us to control equilibrium and hence achieve balance, we need to maximize **stability**. Stability is the resistance to a change in the body's acceleration or, more appropriately, the resistance to a disturbance of the body's equilibrium. Stability may be enhanced by determining the body's **center of gravity** and changing it appropriately. The center of gravity is the point at which all of the body's mass and weight is equally balanced or equally distributed in all directions. Very generally, the center of gravity for humans is located in the vicinity of the umbilicus.

Balance is important for the resting body as well as for the moving body. Generally, balance is to be desired, but there are circumstances in which movement is improved when the body tends to be unbalanced. Following are certain general factors that apply toward enhancing equilibrium, maximizing stability, and ultimately achieving balance.

1. A person has balance when the center of gravity falls within the base of support (Fig. 3.18).
2. A person has balance in direct proportion to the size of the base. The larger the base of support, the more balance.
3. A person has balance depending on the weight (mass). The greater the weight, the more balance.
4. A person has balance depending on the height of the center of gravity. The lower the center of gravity, the more balance.

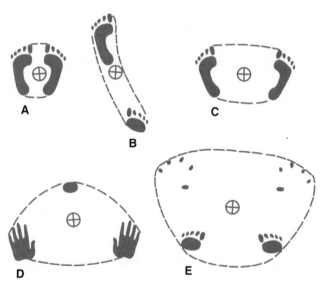

A   **B**   **C**   **D**   **E**

FIG. 3.18 ● Base of support. The base of support includes the part(s) of the body in contact with the supporting surface and the intervening area. **A, B,** and **C,** The weight is supported by the feet; **D,** The weight is supported by the forehead and hands during a headstand; **E,** The weight is supported by the hands and feet while the body is in a squat position. Circled crosses indicate the point of intersection of the line of gravity with the base of support.

5. A person has balance depending on where the center of gravity is in relation to the base of support. The balance is less if the center of gravity is near the edge of the base. However, when anticipating an oncoming force, stability may be improved by placing the center of gravity nearer the side of the base of support expected to receive the force.
6. In anticipation of an oncoming force, stability may be increased by enlarging the size of the base of support in the direction of the anticipated force.
7. Equilibrium may be enhanced by increasing the friction between the body and the surfaces it contacts.
8. Rotation about an axis aids balance. A moving bike is easier to balance than a stationary bike.
9. Kinesthetic physiological functions contribute to balance. The semicircular canals of the inner ear, vision, touch (pressure), and kinesthetic sense all provide balance information to the performer. Balance and its components of equilibrium and stability are essential in all movements. All are affected by the constant force of gravity, as well as by inertia. Walking has been described as an activity in which a person throws the body in and out of balance with each step. In rapid running movements in which moving inertia is high, the individual has to lower the center of gravity to maintain balance when stopping or changing direction. Conversely, in jumping activities, the individual attempts to raise the center of gravity as high as possible.

Force

Muscles are the main source of force that produces or changes movement of a body segment, the entire body, or an object thrown, struck, or stopped. As discussed previously, a variety of factors affect the ability of a muscle to exert force. We obviously need to understand these various factors. And we must utilize this knowledge in properly managing the factors to condition our muscles appropriately to achieve the desired response in dealing with both internal and external forces. As a result, we usually desire stronger muscles in order to be able to produce more force for both maximum and sustained exertion.

Forces either push or pull on an object in an attempt to affect motion or shape. Without forces acting on an object, there is no motion. Force is the product of mass times acceleration. The mass of a body segment or the entire body times the speed of acceleration determines the force. Obviously, in football this is very important, yet it is just as important in other activities that use only a part of the human body. In throwing a ball, the force applied to the ball is equal to the mass of the arm times the arm's speed of acceleration. Also, as previously discussed, leverage is important.

$$\text{Force} = \text{mass} \times \text{acceleration}$$
$$F = m \times a$$

The quality of motion, or, more scientifically stated, the **momentum**, which is equal to mass times velocity, is important in skill activities. The greater the momentum, the greater the resistance to change in the inertia or state of motion. In other words, a larger person with greater mass moving at the same velocity as a smaller person will have more momentum. On the other hand, a person with less mass moving at a higher velocity may have more momentum than a person with greater mass moving at a lower velocity. Momentum may be altered by **impulse**, which is the product of force and time.

It is not necessary to apply maximal force and thereby increase the momentum of a ball or an object being struck in all situations. In skillful performance, regulation of the amount of force is necessary. Judgment as to the amount of force required to throw a softball a given distance, hit a golf ball 200 yards, or hit a tennis ball across the net and into the court is important.

In activities involving movement of various joints, as in throwing a ball or putting a shot, there should be a summation of forces from the beginning of movement in the lower segment of the body to the twisting of the trunk and movement at the shoulder, elbow, and wrist joints. The speed at which a golf club strikes the ball is the result of a summation of forces of the lower extremities, trunk, shoulders, arms, and wrists. Shot-putting and discus and javelin throwing are other good examples that show that a summation of forces is essential.

Mechanical loading basics

As we utilize the musculoskeletal system to exert force on the body to move and to interact with the ground and other objects or people, significant mechanical loads are generated and absorbed by the tissues of the body. The forces causing these loads may be internal or external. Only muscles can actively generate internal force, but tension in tendons, connective tissues, ligaments, and joint capsules may passively generate internal forces.

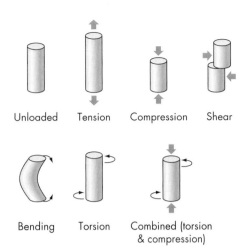

FIG. 3.19 • Mechanical loading forces.

External forces are produced from outside the body and originate from gravity, inertia, or direct contact. All tissues, in varying degrees, resist changes in their shape. Obviously, tissue deformation may result from external forces, but we also have the ability to generate internal forces large enough to fracture bones, dislocate joints, and disrupt muscles and connective tissues. To prevent injury or damage from tissue deformation, we must use the body to absorb energy from both internal and external forces. Along this line, it is to our advantage to absorb such force over larger aspects of our body rather than smaller ones, and to spread the absorption rate over a longer period of time. Additionally, the stronger and healthier we are, the more likely we are to be able to withstand excessive mechanical loading and the resultant excessive tissue deformation. Tension (stretching or strain), compression, shear, bending, and torsion (twisting) are all forces that act individually or in combination to provide mechanical loading that may result in excessive tissue deformation. Fig. 3.19 illustrates the mechanical forces that act on tissues of the body.

Functional application

In the performance of various sport skills, many applications of the laws of leverage, motion, and balance may be found. A skill common to many activities is throwing. The object thrown may be some type of ball, but it is frequently an object of another size or shape, such as a rock, beanbag, Frisbee, discus, or javelin. A brief analysis of some of the basic mechanical principles involved in the skill of throwing will help indicate the importance of understanding the applications of these principles. Many activities involve these and often other mechanical principles. Motion is basic to throwing when the angular motion (Fig. 2.21) of the levers (bones) of the body (trunk, shoulder, elbow, and wrist) is used to give linear motion to the ball when it is released.

Newton's laws of motion apply in throwing because the individual's inertia and the ball's inertia (see p. 82) must be overcome by the application of force. The muscles of the body provide the force to move the body parts and the ball held in the hand. The **law of acceleration** (Newton's second law) comes into operation with the muscular force necessary to accelerate the arm, wrist, and hand. The greater the force (mass times acceleration) a person can produce, the faster the arm will move and, thus, the greater the speed that will be imparted to the ball. The reaction of the feet against the surface on which the person stands illustrates the application of the **law of reaction**.

The leverage factor is very important in throwing a ball or an object. For all practical purposes, the body from the feet to the fingers can be considered one long lever. The longer the lever, either from natural body length or from the movement of the body to the extended backward position (as in throwing a softball, with extension of the shoulder and elbow joints), the greater will be the arc through which it accelerates and thus the greater will be the speed imparted to the thrown object.

In certain circumstances, when the ball is to be thrown only a short distance, as in baseball when it is thrown by the catcher to the bases, the short lever is advantageous because it takes less total time to release the ball.

Balance, or equilibrium, is a factor in throwing when the body is rotated to the rear in the beginning of the throw. This motion moves the body nearly out of balance to the rear, and balance then changes again in the body with the forward movement. Balance is again established with the followthrough, when the feet are spread and the knees and trunk are flexed to lower the center of gravity.

Summary

The preceding discussion has been a brief overview of some of the factors affecting motion. Analysis of human motion in light of the laws of physics poses a problem: How comprehensive is the analysis to be? It can become very complex, particularly when body motion is combined with the manipulation of an object in the hand involved in throwing, kicking, striking, or catching.

These factors become involved when we attempt an analysis of the activities common to our physical education programs—football, baseball, basketball, track and field, field hockey, and swimming, to mention a few. However, to have a complete view of which factors control human movement, we must have a working knowledge of both the physiological and the biomechanical principles of kinesiology.

It is beyond the scope of this book to make a detailed analysis of other activities. Some sources that consider these problems in detail are listed in the references.

REVIEW AND LABORATORY EXERCISES

1. Lever component identification chart

Determine and list two practical examples of levers (in the body or in daily living) for each lever class. Do not use examples already discussed in this chapter. For each example identify the force, axis, and resistance. Also explain the advantage of using each lever—that is, whether it is to achieve balance, force, motion, speed, or range of motion.

Lever class	Example	Force	Axis	Resistance	Advantage provided
1st					
1st					
2nd					
2nd					
3rd					
3rd					

2. Anatomical levers can improve physical performance. Explain how this occurs using the information you have learned in relation to throwing.
3. If your biceps muscle inserts to your forearm 2 inches below your elbow, the distance from the elbow to the palm of your hand is 18 inches, and you lift a 20-pound weight, how much force must your muscle exert to achieve elbow flexion?
4. If the weight of an object is 50 kilograms and your mechanical advantage is 4, how much force would you need to exert to lift the object with a lever system?
5. For the lever system component calculation chart, arrange the lever components as listed for each task a. through j. Determine the lever class and calculate the values for the force arm (FA), the resistance arm (RA), the force, and the mechanical advantage (MA). You may want to draw all the various arrangements of the components on a separate sheet of paper.

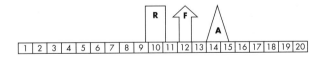

Each vertical line on the lever bar represents the points at which the components are to be arranged, with the left endpoint representing 0 and the right endpoint representing 20.

Lever system component calculation chart

Lever components			Variables				
Force applied at	Axis placed at	25 Newtons of resistance placed at	Lever class	FA	RA	Force	MA
a.	0	2	20				
b.	0	9	15				
c.	3	17	13				
d.	8	4	19				
e.	12	0	18				
f.	19	9	3				
g.	16	2	7				
h.	13	20	4				
i.	8	17	1				
j.	20	4	11				

6. What class of lever is an automobile steering wheel?
7. List two different wheel and axles in which the force is applied to the wheel. From your observation, estimate which of the two has the greater mechanical advantage.

8. List two different wheel and axles in which the force is applied to the axle. From your observation, estimate which of the two has the greater mechanical advantage.

9. When attempting to remove a screw, is it easier when using a screwdriver with a larger grip on the handle? Why?

10. If a pulley setup has five supporting ropes, what is the MA of the setup?

11. What amount of force is needed to lift an object in a pulley system if the weight of the object being lifted is 200 kg and the number of supporting ropes is four?

12. Identify a practical example of Newton's law of inertia. Explain how the example illustrates the law.

13. Identify a practical example of Newton's law of acceleration. Explain how the example illustrates the law.

14. Identify a practical example of Newton's law of reaction. Explain how the example illustrates the law.

15. If a baseball player hit a triple and ran around the bases to third base, what would be his displacement? *Hint:* The distance from each base to the next is 90 feet.

16. Select a sporting activity and explain how the presence of too much friction becomes a problem in the activity.

17. Select a sporting activity and explain how the presence of too little friction becomes a problem in the activity.

18. Using the mechanical loading basic forces of compression, torsion, and shear, describe each force by using examples from soccer or volleyball.

19. **Laws of motion task comparison chart**

	For this chart, assume that you possess the skill, strength, etc. to be able to perform each of the paired tasks. Circle the task that would be easier to perform based on Newton's laws of motion and explain why.	
	Paired tasks	Explanation
a.	Throw a baseball 60 mph OR Throw a shot-put 60 mph.	
b.	Kick a bowling ball 40 yards OR Kick a soccer ball 40 yards.	
c.	Bat a whiffle ball over a 320-yard fence OR Bat a baseball over a 320-yard fence.	
d.	Catch a shot-put that was thrown at 60 mph OR Catch a softball that was thrown at 60 mph.	
e.	Tackle a 240-pound running back sprinting toward you full speed OR Tackle a 200-pound running back sprinting toward you full speed.	
f.	Run a 40-yard dash in 4.5 seconds on a wet field OR Run a 40-yard dash in 4.5 seconds on a dry field.	

20. Develop special projects and class reports by individuals or small groups of students on the mechanical analysis of all the skills involved in the following:
    a. Basketball
    b. Baseball
    c. Dancing
    d. Diving
    e. Football
    f. Field hockey
    g. Golf
    h. Gymnastics
    i. Soccer
    j. Swimming
    k. Tennis
    l. Wrestling

21. Develop term projects and special class reports by individuals or small groups of students on the following factors in motion:
    a. Acceleration
    b. Aerodynamics
    c. Angular displacement
    d. Balance
    e. Base of support
    f. Buoyancy
    g. Center of gravity
    h. Drag
    i. Equilibrium
    j. Force
    k. Friction
    l. Gravity
    m. Hydrodynamics
    n. Impulse
    o. Inertia
    p. Instantaneous center of rotation
    q. Leverage
    r. Lift
    s. Linear displacement
    t. Mass
    u. Momentum
    v. Motion
    w. Projectiles
    x. Rebound angle
    y. Restitution
    z. Speed
    aa. Spin
    bb. Stability
    cc. Thrust
    dd. Torque
    ee. Velocity
    ff. Vector composition
    gg. Vector resolution
    hh. Weight
    ii. Work

22. Develop demonstrations, term projects, or special reports by individuals or small groups of students on the following activities:
    a. Lifting
    b. Throwing
    c. Standing
    d. Walking
    e. Running
    f. Jumping
    g. Falling
    h. Sitting
    i. Pushing and pulling
    j. Striking

Chapter

3

References

Adrian MJ, Cooper JM: *The biomechanics of human movement,* Indianapolis, IN, 1989, Benchmark.

American Academy of Orthopaedic Surgeons; Schenck RC, ed.: *Athletic training and sports medicine,* ed 3, Rosemont, IL, 1999, American Academy of Orthopaedic Surgeons.

Barham JN: *Mechanical kinesiology,* St. Louis, 1978, Mosby.

Broer MR: *An introduction to kinesiology,* Englewood Cliffs, NJ, 1968, Prentice-Hall.

Broer MR, Zernicke RF: *Efficiency of human movement,* ed 3, Philadelphia, 1979, Saunders.

Bunn JW: *Scientific principles of coaching,* ed 2, Englewood Cliffs, NJ, 1972, Prentice-Hall.

Cooper JM, Adrian M, Glassow RB: *Kinesiology,* ed 5, St. Louis, 1982, Mosby.

Donatelli R, Wolf SL: *The biomechanics of the foot and ankle,* Philadelphia, 1990, Davis.

Hall SJ: *Basic biomechanics,* ed 6, New York, 2012, McGraw-Hill.

Hamill J, Knutzen KM: *Biomechanical basis of human movement,* ed 3, Baltimore, 2008, Lippincott Williams & Wilkins.

Hamilton N, Weimar W, Luttgens K: *Kinesiology: scientific basis of human motion,* ed 12, New York, 2012, McGraw-Hill.

Hinson M: *Kinesiology,* ed 4, New York, 1981, McGraw-Hill.

Kegerreis S, Jenkins WL, Malone TR: Throwing injuries, *Sports Injury Management* 2:4, 1989.

Kelley DL: *Kinesiology: fundamentals of motion description,* Englewood Cliffs, NJ, 1971, Prentice-Hall.

Kreighbaum E, Barthels KM: *Biomechanics: a qualitative approach for studying human movement,* ed 4, New York, 1996, Allyn & Bacon.

Logan GA, McKinney WC: *Anatomic kinesiology,* ed 3, New York, 1982, McGraw-Hill.

McCreary EK, Kendall FP, Rodgers MM, Provance PG, Romani WA: *Muscles: testing and function with posture and pain,* ed 5, Philadelphia, 2005, Lippincott Williams & Wilkins.

McGinnis PM: *Biomechanics of sport and exercise,* ed 2, Champaign, IL, 2005, Human Kinetics.

Neumann, DA: *Kinesiology of the musculoskeletal system: foundations for physical rehabilitation,* ed 2, St. Louis, 2010, Mosby.

Nordin M, Frankel VH: *Basic biomechanics of the musculoskeletal system,* ed 3, Philadelphia, 2001, Lippincott Williams & Wilkins.

Norkin CC, Levangie PK: *Joint structure and function—a comprehensive analysis,* ed 5, Philadelphia, 2011, Davis.

Northrip JW, Logan GA, McKinney WC: *Analysis of sport motion: anatomic and biomechanic perspectives,* ed 3, New York, 1983, McGraw-Hill.

Piscopo J, Baley J: *Kinesiology: the science of movement,* New York, 1981, Wiley.

Prentice WE: *Principles of athletic training: a competency based approach,* ed 15, New York, 2014, McGraw-Hill.

Rasch PJ: *Kinesiology and applied anatomy,* ed 7, Philadelphia, 1989, Lea & Febiger.

Scott MG: *Analysis of human motion,* ed 2, New York, 1963, Appleton-Century-Crofts.

Segedy A: Braces' joint effects spur research surge, *Biomechanics,* February 2005.

Weineck J: *Functional anatomy in sports,* ed 2, St. Louis, 1990, Mosby.

Whiting WC, Zermicke R: *Biomechanics of musculoskeletal injury,* ed 2, Champaign, IL, 2008, Human Kinetics.

Wirhed R: *Athletic ability and the anatomy of motion,* ed 3, St. Louis, 2006, Mosby Elsevier.

For additional resources and a list of related websites, visit **www.mhhe.com/floyd19e**.

THE SHOULDER GIRDLE

Objectives

- To identify on the skeleton important bony features of the shoulder girdle

- To label on a skeletal chart the important bony features of the shoulder girdle

- To draw on a skeletal chart the muscles of the shoulder girdle and indicate shoulder girdle movements using arrows

- To demonstrate, using a human subject, all the movements of the shoulder girdle and list their respective planes of movement and axes of rotation

- To palpate the muscles of the shoulder girdle on a human subject and list their antagonists

- To palpate the joints of the shoulder girdle on a human subject during each movement through the full range of motion

- To determine, through analysis, the shoulder girdle movements and the muscles involved in selected skills and exercises

Online Learning Center Resources

Visit *Manual of Structural Kinesiology*'s Online Learning Center at **www.mhhe.com/floyd19e** for additional information and study material for this chapter, including:

- *Self-grading quizzes*
- *Anatomy flashcards*
- *Animations*
- *Related websites*

The entire upper extremity depends upon the shoulder girdle to serve as a base from which to function. The only attachment of the upper extremity to the axial skeleton is via the scapula and its attachment through the clavicle at the sternoclavicular joint. To enhance understanding of how the shoulder joint and the rest of the upper extremity depend on the shoulder girdle, we will discuss it separately from the other structures.

Brief descriptions of the most important bones in the shoulder region will help you understand the skeletal structure and its relationship to the muscular system.

Bones

Two bones are primarily involved in movements of the shoulder girdle. They are the scapula and the clavicle, which generally move as a unit. Their only bony link to the axial skeleton is provided by the clavicle's articulation with the sternum. Key bony landmarks for studying the shoulder girdle are the manubrium, clavicle, coracoid process, acromion process, glenoid fossa, lateral border, inferior angle, medial border, superior angle, and spine of the scapula (Figs. 4.1, 4.2, 4.3, and 4.4).

Joints

When analyzing shoulder girdle (scapulothoracic) movements, it is important to realize that the scapula moves on the rib cage as a consequence of joint motion actually occurring at the sternoclavicular joint and to a lesser extent at the acromioclavicular joint (see Figs. 4.1 and 4.3).

Sternoclavicular (SC)

This is classified as a (multiaxial) anthrodial joint. In relation to the manubrium of the sternum, the clavicle moves anteriorly 15 degrees with protraction and moves posteriorly 15 degrees with retraction.

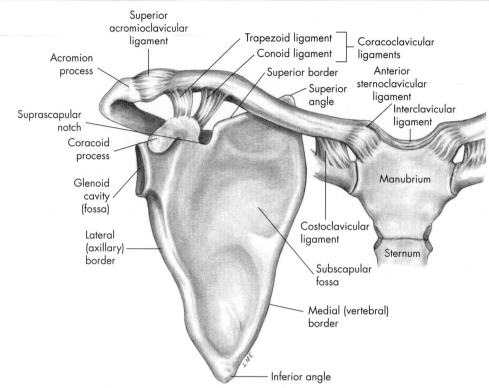

FIG. 4.1 ● Right shoulder girdle, anterior view.

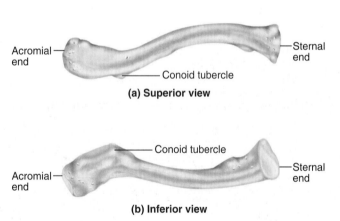

(a) Superior view

(b) Inferior view

FIG 4.2 ● Right clavicle. **A,** Superior view; **B,** Inferior view.

Some rotation of the clavicle along its axis during various movements of the shoulder girdle results in slight rotary gliding movement at the sternoclavicular joint. It is supported anteriorly by the anterior sternoclavicular ligament and posteriorly by the posterior sternoclavicular ligament. The costoclavicular and interclavicular ligaments also provide stability against superior displacement.

Acromioclavicular (AC)

This joint is classified as an arthrodial joint. It has a 20- to 30-degree total gliding and rotational motion accompanying other shoulder girdle and shoulder joint motions. In addition to the strong support provided by the coracoclavicular ligaments (conoid and trapezoid), the superior and inferior acromioclavicular ligaments provide stability to this often-injured joint. The coracoclavicular joint, classified as a syndesmotic-type joint, functions through its ligaments to greatly increase the stability of the acromioclavicular joint.

Scapulothoracic

This joint is not a true synovial joint, due to its not having regular synovial features and to the fact that its movement is totally dependent on the sternoclavicular and acromioclavicular joints. Even though scapula movement occurs as a result of motion at the SC and AC joints, the scapula can be described as having a total range of 25-degree abduction–adduction movement, 60-degree upward–downward rotation, and 55-degree elevation–depression. The scapulothoracic joint is supported dynamically by its muscles and lacks ligamentous support, since it has no synovial features.

There is not a typical articulation between the anterior scapula and the posterior rib cage. Between these two osseous structures is the serratus anterior muscle originating off the upper nine ribs laterally and running just behind the rib cage posteriorly to insert on the medial border of the scapula. Immediately posterior to the serratus anterior is the subscapularis muscle (see Chapter 5) on the anterior scapula.

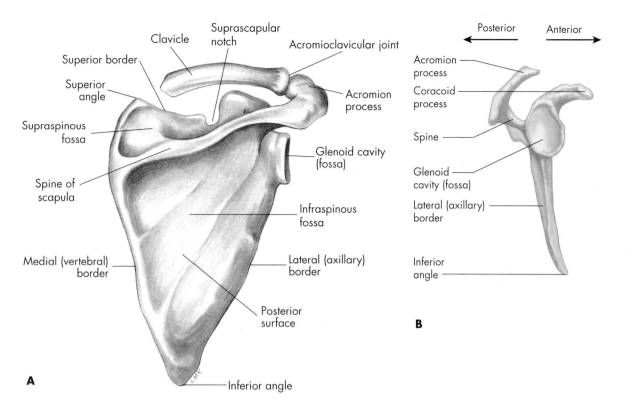

FIG. 4.3 • Right scapula. **A,** Posterior view; **B,** Lateral view.

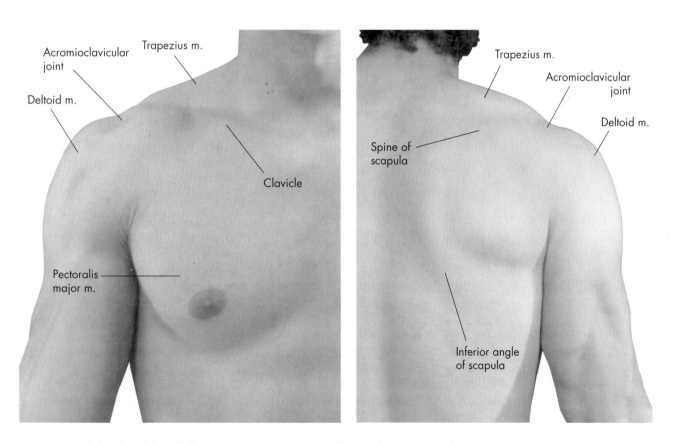

FIG. 4.4 • Right shoulder girdle surface anatomy, anterior and posterior views.

Movements FIGS. 4.5, 4.6

In analyzing shoulder girdle movements, it is often helpful to focus on a specific scapular bony landmark, such as the inferior angle (posteriorly), the glenoid fossa (laterally), or the acromion process (anteriorly). All of these movements have their pivotal point where the clavicle joins the sternum at the sternoclavicular joint.

Scapula
adduction

Scapula
elevation

Scapula
depression

Scapula
upward
rotation

Scapula
downward
rotation

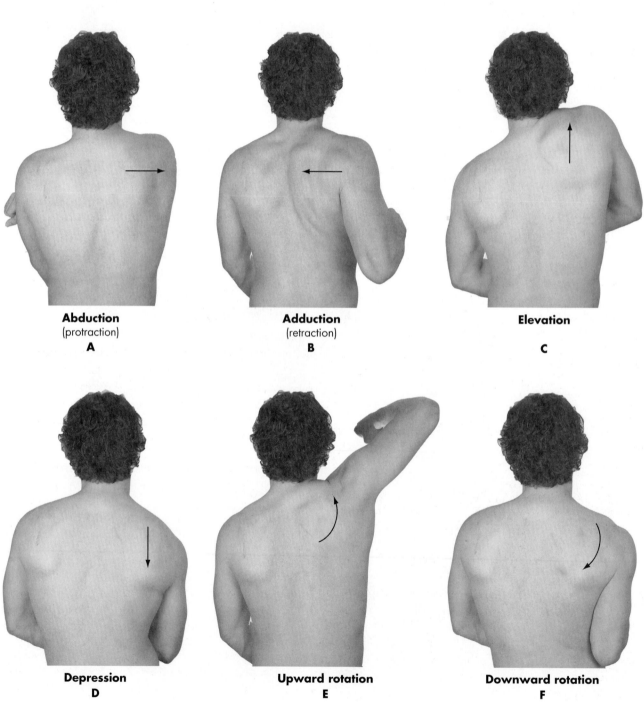

FIG. 4.5 • Movements of the shoulder girdle.

Movements of the shoulder girdle can be described as movements of the scapula, but it is important to remember that wherever the scapula goes, the clavicle follows. Figs. 4.5 and 4.6 show movements of the shoulder girdle.

Abduction (protraction): Movement of the scapula laterally away from the spinal column, as in reaching for an object in front of the body

Adduction (retraction): Movement of the scapula medially toward the spinal column, as in pinching the shoulder blades together

Elevation: Upward or superior movement of the scapula, as in shrugging the shoulders

Depression: Downward or inferior movement of the scapula, as in returning to a normal position from a shoulder shrug

Upward rotation: Turning the glenoid fossa upward and moving the inferior angle superiorly and laterally away from the spinal column to assist in raising the arm out to the side

Downward rotation: Returning the inferior angle medially and inferiorly toward the spinal column and the glenoid fossa to its normal position, as in bringing the arm down to the side. (Once the scapula has returned to its anatomical position, further downward rotation actually results in the superior angle moving slightly superomedial.)

To accomplish some of the previously listed shoulder girdle movements, the scapula must rotate or tilt on its axis. While these are not primary movements of the shoulder girdle, they are necessary for the scapula to move normally throughout its range of motion during shoulder girdle movements.

Lateral tilt (outward tilt): Consequential movement during abduction in which the scapula rotates about its vertical axis, resulting in posterior movement of the medial border and anterior movement of the lateral border

Medial tilt (inward tilt): Return from lateral tilt; consequential movement during extreme adduction in which the scapula rotates about its vertical axis, resulting in anterior movement of the medial border and posterior movement of the lateral border

Anterior tilt (upward tilt): Consequential rotational movement of the scapula about the frontal axis occurring during hyperextension of the glenohumeral joint, resulting in the superior border moving anteroinferiorly and the inferior angle moving posterosuperiorly

Posterior tilt (downward tilt): Consequential rotational movement of the scapula about the frontal axis occurring during hyperflexion of the glenohumeral joint, resulting in the superior border moving posteroinferiorly and the inferior angle moving anterosuperiorly

Synergy with the muscles of the glenohumeral joint

The shoulder joint and shoulder girdle work together in carrying out upper-extremity activities. It is critical to understand that movement of the shoulder girdle is not dependent on the shoulder joint and its muscles. However, the muscles of the shoulder girdle are essential in providing a scapula-stabilizing effect, so that the muscles of the shoulder joint will have a stable base from which to exert force for powerful movement involving the humerus. Consequently, the shoulder girdle muscles contract to maintain the scapula in a relatively static position during many shoulder joint actions.

As the shoulder joint goes through more extreme ranges of motion, the scapular muscles contract to move the shoulder girdle so that its glenoid fossa will be in a more appropriate position

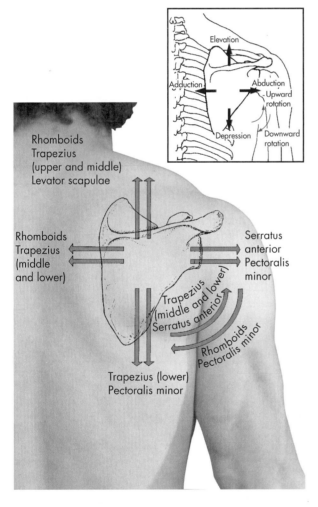

FIG. 4.6 ● Actions of the scapular muscles. Posterior view with actions.

from which the humerus can move. Without the accompanying movement of the scapula, we can raise our humerus only to around 90 to 120 degrees of total shoulder abduction and flexion. This works through the action of the appropriate muscles of both joints working in synergy to accomplish the desired action of the entire upper extremity. For example, if we want to raise our hand out to the side laterally as high as possible, the serratus anterior and trapezius (middle and lower fibers) muscles upwardly rotate the scapula as the supraspinatus and deltoid initiate glenohumeral abduction. This synergy between the scapula and shoulder joint muscles enhances movement of the entire upper extremity. Further discussion of the interaction and teamwork between these joints is provided at the beginning of Chapter 5, with Table 5.1 listing the shoulder girdle movements that usually accompany shoulder joint movements. Additional discussion of scapulohumeral rhythm is provided in Chapter 5 under Joint.

Muscles

There are five muscles primarily involved in shoulder girdle movements, as shown in Fig. 4.7: the pectoralis minor, serratus anterior, trapezius, rhomboid, and levator scapulae. To avoid confusion, it is helpful to group muscles of the shoulder girdle separately from those of the shoulder joint. The subclavius muscle is also included in this group, but it is not regarded as a primary mover in any actions of the shoulder girdle. All five shoulder girdle muscles have their origin on the axial skeleton, with their insertion located on the scapula and/or the clavicle. Shoulder girdle muscles do not attach to the humerus, nor do they cause actions of the shoulder joint. The pectoralis minor and subclavius are located anteriorly in relation to the trunk. The serratus anterior is located anteriorly to the scapula but posteriorly and laterally to the trunk. Located posteriorly to the trunk and cervical spine are the trapezius, rhomboid, and levator scapulae.

The shoulder girdle muscles are essential in providing dynamic stability of the scapula so that it can serve as a relative base of support for shoulder joint activities such as throwing, batting, and blocking.

The scapula muscles also play a role in spinal posture. Typically, due to poor posture and the way we use our muscles through life, we tend to develop a forward shoulder posture that results in the scapula protractors and depressors becoming stronger and tighter and the retractors becoming weaker. This leads to further depression and protraction, or a forward shoulder posture, which also contributes to increased kyphosis (increased posterior convexity of the thoracic spine) and a forward head with increased lordosis (increased posterior concavity of the cervical spine). See Chapter 12. This, in turn, places more stress on the posterior spinal muscles and also places the glenohumeral joint in a less functional and more compromised position. To avoid this, we should routinely practice good posture, beginning with an appropriate lumbar lordotic curve, and keep our shoulder girdle directly over our pelvis instead of forward. This will make it easier to maintain our head and cervical spine over the trunk in a correctly balanced position. An additional benefit of good scapular and spinal posture is easier inspiration due to less weight and mass over the rib cage and thoracic cavity.

Scapular winging is relatively rare but can affect normal functional activity of the upper extremity. Most commonly it affects the serratus anterior, leading to medial winging when pushing forward or raising the arm. The serratus anterior weakness or paralysis is typically due to injury of the long thoracic nerve, which can have a variety of causes. Much less commonly the trapezius and/or rhomboid may be affected, leading to lateral winging.

Shoulder girdle muscles—location and action

Anterior
  Pectoralis minor—abduction, downward rotation, and depression
  Subclavius—depression and abduction
Posterior and laterally
  Serratus anterior—abduction and upward rotation
Posterior
  Trapezius
    Upper fibers—elevation and extension and rotation of the head at the neck
    Middle fibers—elevation, adduction, and upward rotation
    Lower fibers—adduction, depression, and upward rotation
  Rhomboid—adduction, downward rotation, and elevation
  Levator scapulae—elevation

It is important to understand that muscles may not necessarily be active throughout the absolute full range of motion for which they are noted as being agonists.

Table 4.1 provides a detailed breakdown of the muscles responsible for primary movements of the shoulder girdle.

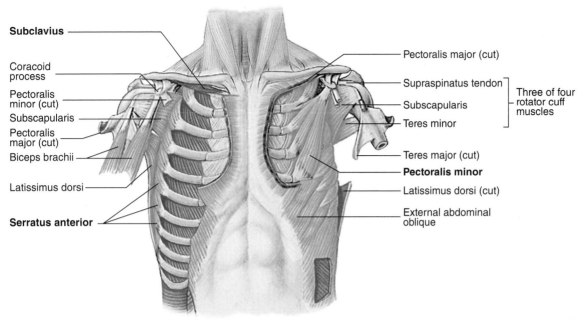

A

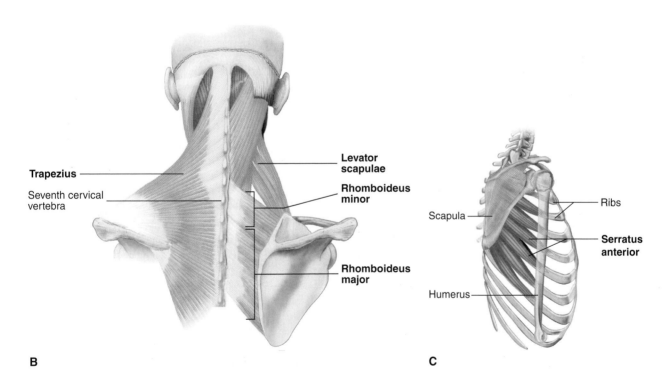

B

C

FIG. 4.7 • Muscles acting on the scapula. **A,** Anterior view: The pectoralis major is removed on both sides; **B,** Posterior view: The trapezius is removed on the right to reveal the deeper muscles; **C,** Lateral view: The serratus anterior.

TABLE 4.1 • Agonist muscles of the shoulder girdle

	Muscle	Origin	Insertion	Action	Plane of motion	Palpation	Innervation
Anterior muscle	Pectoralis minor	Anterior surfaces of the 3rd to 5th ribs	Coracoid process of the scapula	Abduction	Transverse	Difficult, but under the pectoralis major muscle and just inferior to the coracoid process during resisted depression; enhanced by placing the subject's hand behind the back and having him or her actively lift the hand away	Medial pectoral nerve (C8 and T1)
				Downward rotation	Frontal		
				Depression			
Posterior and lateral muscle	Serratus anterior	Surface of the upper 9 ribs at the side of the chest	Anterior aspect of the whole length of the medial border of the scapula	Abduction	Transverse	Frontal and lateral side of the chest below the 5th and 6th ribs just proximal to the origin during abduction; best accomplished with the glenohumeral joint flexed 90 degrees; in the same position palpate the upper fibers between the lateral borders of the pectoralis major and latissimus dorsi in the axilla	Long thoracic nerve (C5–C7)
				Upward rotation	Frontal		
Posterior muscles	Trapezius upper fibers	Base of skull, occipital protuberance, and posterior ligaments of neck	Posterior aspect of the lateral 3rd of the clavicle	Elevation	Frontal	Between occipital protuberance and C6 and laterally to acromion, particularly during elevation and extension of the head at the neck	Spinal accessory nerve and branches of C3 and C4
				Upward rotation			
				Extension of head at neck	Sagittal		
				Rotation of head at neck	Transverse		
	Trapezius middle fibers	Spinous process of 7th cervical and upper 3 thoracic vertebrae	Medial border of the acromion process and superior border of the scapular spine	Elevation	Frontal	From C7 to T3 and laterally to acromion process and scapular spine, particularly during adduction	
				Adduction	Transverse		
				Upward rotation	Frontal		
	Trapezius lower fibers	Spinous process of 4th to 12th thoracic vertebrae	Triangular space at the base of the scapular spine	Adduction	Transverse	From T4 to T12 and medial aspect of scapular spine, particularly during depression and adduction	
				Depression	Frontal		
				Upward rotation			
	Rhomboids	Spinous processes of the 7th cervical and first 5 thoracic vertebrae	Medial border of the scapula, inferior to the scapular spine	Adduction	Transverse	Difficult due to being deep to trapezius, but may be palpated through it during adduction; best accomplished with subject's ipsilateral hand behind the back to relax the trapezius and bring the rhomboid into action when the subject lifts the hand away from the back	Dorsal scapular nerve (C5)
				Downward rotation	Frontal		
				Elevation			
	Levator scapulae	Transverse processes of the upper 4 cervical vertebrae	Medial border of the scapula from the superior angle to the scapular spine	Elevation	Frontal	Difficult to palpate due to being deep to trapezius; best palpated at insertion just medial to the superior angle of scapula, particularly during slight elevation	Dorsal scapular nerve C5 and branches of C3 and C4

Note: The subclavius is not listed because it is not a prime mover in shoulder girdle movements.

96   www.mhhe.com/floyd19e

Nerves

The muscles of the shoulder girdle are innervated primarily from the nerves of the cervical plexus and brachial plexus, as illustrated in Figs. 4.8 and 4.9. The trapezius is innervated by the spinal accessory nerve and from branches of C3 and C4. In addition to supplying the trapezius, C3 and C4 also innervate the levator scapula. The levator scapula receives further innervation from the dorsal scapular nerve originating from C5. The dorsal scapular nerve also innervates the rhomboid. The long thoracic nerve originates from C5, C6, and C7 and innervates the serratus anterior. The medial pectoral nerve arises from C8 and T1 to innervate the pectoralis minor.

■ Roots: C5, C6, C7, C8, T1
□ Trunks: upper, middle, lower
□ Anterior divisions
■ Posterior divisions
■ Cords: posterior, lateral, medial
□ Branches: Axillary nerve
           Radial nerve
           Musculocutaneous nerve
           Median nerve
           Ulnar nerve

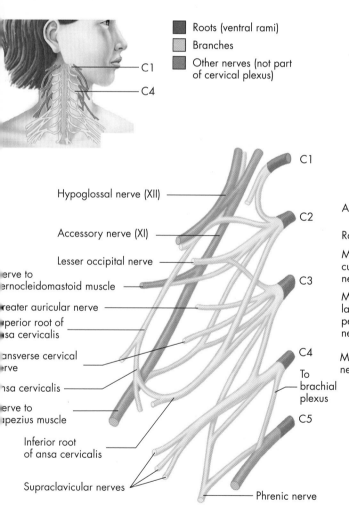

■ Roots (ventral rami)
□ Branches
■ Other nerves (not part of cervical plexus)

Hypoglossal nerve (XII)

Accessory nerve (XI)

Lesser occipital nerve

...erve to ...ernocleidomastoid muscle

...reater auricular nerve

...perior root of ...sa cervicalis

...ansverse cervical ...rve

...nsa cervicalis

...erve to ...apezius muscle

Inferior root of ansa cervicalis

Supraclavicular nerves

C1
C2
C3
C4
To brachial plexus
C5

Phrenic nerve

FIG. 4.8 ● Cervical plexus, anterior view. The roots of the plexus are formed by the ventral rami of the spinal nerves C1–C4.

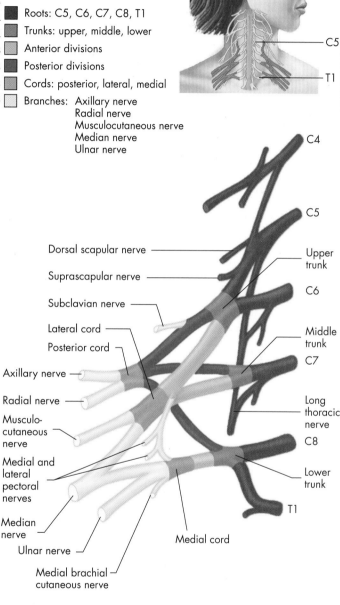

C5
T1

C4
C5
C6

Dorsal scapular nerve
Suprascapular nerve
Subclavian nerve
Lateral cord
Posterior cord
Axillary nerve
Radial nerve
Musculo-cutaneous nerve
Medial and lateral pectoral nerves
Median nerve
Ulnar nerve
Medial brachial cutaneous nerve

Upper trunk
Middle trunk
C7
Long thoracic nerve
C8
Lower trunk
T1
Medial cord

FIG. 4.9 ● Brachial plexus, anterior view. The roots of the plexus are formed by the ventral rami of the spinal nerves C5–T1 and join to form an upper, middle, and lower trunk. Each trunk divides into anterior and posterior divisions. The divisions join together to form the posterior, lateral, and medial cords from which the major brachial plexus nerves arise.

Trapezius muscle   FIG. 4.10

(tra-pe´zi-us)

Origin

Upper fibers: base of skull, occipital protuberance, and posterior ligaments of neck

Middle fibers: spinous processes of seventh cervical and upper three thoracic vertebrae

Lower fibers: spinous processes of fourth through twelfth thoracic vertebrae

Insertion

Upper fibers: posterior aspect of the lateral third of the clavicle

Middle fibers: medial border of the acromion process and upper border of the scapular spine

Lower fibers: triangular space at the base of the scapular spine

Action

Upper fibers: elevation of the scapula, upward rotation, and extension and rotation of the head at the neck

Middle fibers: elevation, upward rotation, and adduction (retraction) of the scapula

Lower fibers: depression, adduction (retraction), and upward rotation of the scapula

Palpation

Upper fibers: between occipital protuberance and C6 and laterally to acromion, particularly during elevation and extension of the head at the neck

Middle fibers: from C7 to T3 and laterally to acromion process and scapular spine, particularly during adduction

Lower fibers: from T4 to T12 and medial aspect of scapular spine, particularly during depression and adduction

Scapula elevation

Cervical extension

Cervical rotation unilaterally

Scapula upward rotation

Scapula adduction

Scapula depression

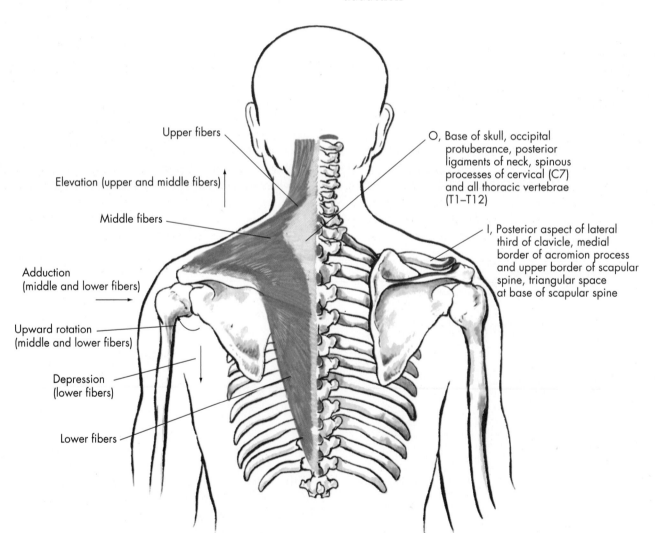

Upper fibers

Elevation (upper and middle fibers)

Middle fibers

Adduction (middle and lower fibers)

Upward rotation (middle and lower fibers)

Depression (lower fibers)

Lower fibers

O, Base of skull, occipital protuberance, posterior ligaments of neck, spinous processes of cervical (C7) and all thoracic vertebrae (T1–T12)

I, Posterior aspect of lateral third of clavicle, medial border of acromion process and upper border of scapular spine, triangular space at base of scapular spine

FIG. 4.10 • Trapezius muscle, posterior view. O, Origin; I, Insertion.

Innervation

Spinal accessory nerve (cranial nerve XI) and branches of C3, C4

Application, strengthening, and flexibility

The upper fibers are a thin and relatively weak part of the muscle. They provide some elevation of the clavicle. Due to their origin on the base of the skull, they assist in extension of the head.

The middle fibers are stronger and thicker and provide strong elevation, upward rotation, and adduction (retraction) of the scapula. Rarely is this portion of the muscle weak, because it is so active in positioning the shoulder for function and posture. As a result, it is often a source of tenderness and discomfort due to chronic tension.

The lower fibers assist in adduction (retraction) and rotate the scapula upward. This portion is typically weak, particularly in individuals whose activities demand a significant amount of scapula abduction.

When all the parts of the trapezius are working together, they tend to pull upward and adduct at the same time. This may be seen in lifting the handles of a wheelbarrow. Typical action of the trapezius muscle is fixation of the scapula for deltoid action. Continuous action in upward rotation of the scapula permits the arms to be raised over the head. The muscle is always used in preventing the glenoid fossae from being pulled down during the lifting of objects with the arms. It is also typically seen in action during the holding of an object overhead. Holding the arm at the side horizontally shows typical fixation of the scapula by the trapezius muscle, while the deltoid muscle holds the arm in that position. The muscle is used strenuously when lifting with the hands, as in picking up a heavy wheelbarrow. The trapezius must prevent the scapula from being pulled downward. Carrying objects on the tip of the shoulder also calls this muscle into play. Strengthening of the upper and middle fibers can be accomplished through shoulder-shrugging exercises. The middle and lower fibers can be strengthened through bent-over rowing and shoulder joint horizontal abduction exercises from a prone position. The lower fibers can be emphasized with a chestproud shoulder retraction exercise attempting to place the elbows in the back pants pockets with depression. Parallel dips or body dips are also helpful for emphasizing the lower trapezius. See Appendix 3 for more commonly used exercises to address the trapezius and other muscles in this chapter.

To stretch the trapezius, each portion needs to be specifically addressed. The upper fibers may be stretched by using one hand to pull the head and neck forward into flexion or slight lateral flexion to the opposite side while the ipsilateral hand is hooked under a table edge to maintain the scapula in depression. The middle fibers are stretched to some extent with the procedure used for the upper fibers, but they may be stretched further by using a partner to passively pull the scapula into full protraction. The lower fibers are perhaps best stretched with the subject in a side-lying position while a partner grasps the lateral border and inferior angle of the scapula and moves it passively into maximal elevation and protraction.

Levator scapulae muscle   FIG. 4.11
(le-va´tor scap´u-lae)

Origin

Transverse processes of the upper four cervical vertebrae

Scapula elevation

Insertion

Medial border of the scapula from the superior angle to the scapular spine

Action

Elevates the medial margin of the scapula
Weak downward rotation
Weak adduction

Palpation

Difficult to palpate due to being deep to trapezius; best palpated at insertion just medial to the superior angle of the scapula, particularly during slight elevation

Scapula downward rotation

Scapula adduction

Innervation

Dorsal scapular nerve C5 and branches of C3 and C4

Application, strengthening, and flexibility

Shrugging the shoulders calls the levator scapulae muscle into play, along with the upper trapezius muscle. Fixation of the scapula by the pectoralis minor muscle allows the levator scapulae muscles on both sides to extend the neck or to flex laterally if used on one side only.

The levator scapulae is perhaps best stretched by rotating the head approximately 45 degrees contralaterally and flexing the cervical spine actively while maintaining the scapula in a relaxed, depressed position.

Like the trapezius, the levator scapulae is a very common site for tightness, tenderness, and discomfort secondary to chronic tension and from carrying items with straps over the shoulder.

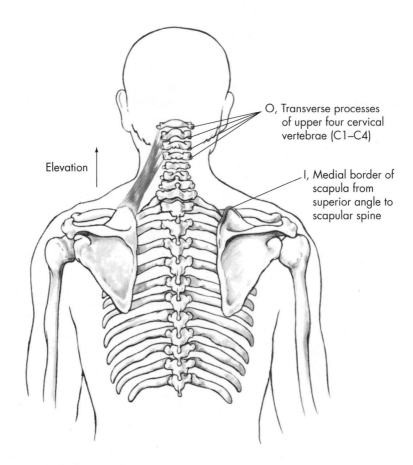

Elevation

O, Transverse processes of upper four cervical vertebrae (C1–C4)

I, Medial border of scapula from superior angle to scapular spine

FIG. 4.11 • Levator scapulae muscle, posterior view. O, Origin; I, Insertion.

Rhomboid muscles—major and minor   FIG. 4.12

(rom´boyd)

Origin

Spinous processes of the seventh cervical and first five thoracic vertebrae

Insertion

Medial border of the scapula, below the spine of the scapula

Action

The rhomboid major and minor muscles work together. Adduction (retraction): draw the scapula toward the spinal column
Downward rotation: from the upward rotated position; draw the scapula into downward rotation
Elevation: slight upward movement accompanying adduction

Palpation

Difficult to palpate due to being deep to the trapezius, but may be palpated through the relaxed trapezius during adduction. This is best accomplished by placing the subject's ipsilateral hand behind the back (glenohumeral internal rotation and scapula downward rotation), which relaxes the trapezius and brings the rhomboid into action when the subject lifts the hand away from the back.

Innervation

Dorsal scapular nerve (C5)

Application, strengthening, and flexibility

The rhomboid muscles fix the scapula in adduction (retraction) when the muscles of the shoulder joint adduct or extend the arm. These muscles are used powerfully in chinning. As one hangs from the horizontal bar, suspended by the hands, the scapula tends to be pulled away from the top of the chest. When the chinning movement begins, it is the rhomboid muscles that rotate the medial border of the scapula down and back toward the spinal column. Note their favorable position to do this. Related to this, the rhomboids work in a similar manner to prevent scapula winging.

The trapezius and rhomboid muscles working together produce adduction with slight elevation of the scapula. To prevent this elevation, the latissimus dorsi muscle is called into play.

Chin-ups, dips, and bent-over rowing are excellent exercises for developing strength in this muscle. The rhomboids may be stretched by passively moving the scapula into full protraction while maintaining depression. Upward rotation may assist in this stretch as well.

Scapula adduction

Scapula downward rotation

Scapula elevation

Chapter 4

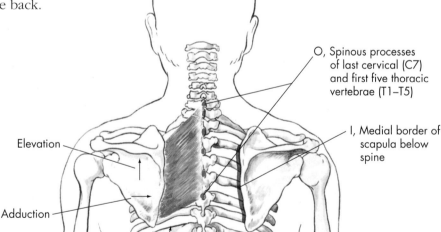

Elevation

Adduction

Downward rotation

O, Spinous processes of last cervical (C7) and first five thoracic vertebrae (T1–T5)

I, Medial border of scapula below spine

FIG. 4.12 • Rhomboid muscles (major and minor), posterior view. O, Origin; I, Insertion.

Serratus anterior muscle   FIG. 4.13

(ser-a´tus an-tir´e-or)

Origin

Surface of the upper nine ribs at the side of the chest

Insertion

Anterior aspect of the whole length of the medial
border of the scapula

Action

Abduction (protraction): draws the medial border of
the scapula away from the vertebrae

Upward rotation: longer, lower fibers tend to draw the
inferior angle of the scapula farther away from the
vertebrae, thus rotating the scapula upward slightly

Palpation

Front and lateral side of the chest below the fifth and
sixth ribs just proximal to their origin during abduction, which is best accomplished from a supine
position with the glenohumeral joint in 90 degrees
of flexion. The upper fibers may be palpated in the
same position between the lateral borders of the
pectoralis major and latissimus dorsi in the axilla.

Innervation

Long thoracic nerve (C5–C7)

Application, strengthening, and flexibility

The serratus anterior muscle is used commonly
in movements drawing the scapula forward with
slight upward rotation, such as throwing a baseball, punching in boxing, shooting and guarding
in basketball, and tackling in football. It works
along with the pectoralis major muscle in typical
action, such as throwing a baseball.

The serratus anterior muscle is used strongly
in doing push-ups, especially in the last 5 to 10
degrees of motion. The bench press and overhead
press are good exercises for this muscle. A winged
scapula condition usually results from weakness
of the rhomboid and/or the serratus anterior. Serratus anterior weakness may result from an injury
to the long thoracic nerve.

The serratus anterior can be stretched by
standing, facing a corner and placing each hand
at shoulder level on the two walls. As you lean
in and attempt to place your nose in the corner,
both scapulae are pushed into an adducted position, which stretches the serratus anterior.

Chapter
4

Scapula
upward
rotation

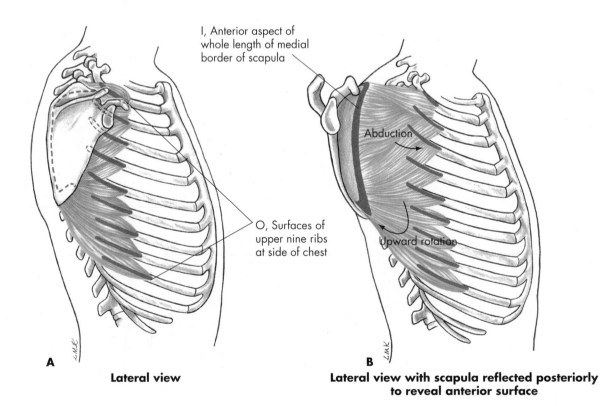

FIG. 4.13 • Serratus anterior muscle, lateral view. O, Origin; I, Insertion.

Pectoralis minor muscle   FIG. 4.14

(pek-to-ra´lis mi´nor)

Origin
Anterior surfaces of the third to fifth ribs

Insertion
Coracoid process of the scapula

Action
Abduction (protraction): draws the scapula forward and tends to tilt the lower border away from the ribs

Downward rotation: as it abducts, it draws the scapula downward

Depression: when the scapula is rotated upward, it assists in depression

Palpation
Difficult to palpate, but can be palpated under the pectoralis major muscle and just inferior to the coracoid process during resisted depression. This may be enhanced by placing the subject's hand behind the back and having him or her actively lift the hand away, which causes downward rotation.

Innervation
Medial pectoral nerve (C8–T1)

Application, strengthening, and flexibility

The pectoralis minor muscle is used, along with the serratus anterior muscle, in true abduction (protraction) without rotation. This is seen particularly in movements such as push-ups in which true abduction of the scapula is necessary. Therefore, the serratus anterior draws the scapula forward with a tendency toward upward rotation, the pectoralis minor pulls forward with a tendency toward downward rotation, and the two pulling together give true abduction. These muscles will be seen working together in most movements of pushing with the hands.

The pectoralis minor is most used in depressing and rotating the scapula downward from an upwardly rotated position, as in pushing the body upward on dip bars or in body dips.

The pectoralis minor is often tight due to being overused in activities involving abduction, which may lead to forward and rounded shoulders. As a result stretching may be indicated, which can be accomplished with a wall push-up in the corner as used for stretching the serratus anterior. Additionally, lying supine with a rolled towel directly under the thoracic spine while a partner pushes each scapula into retraction places this muscle on stretch.

Scapula abduction

Scapula downward rotation

Scapula depression

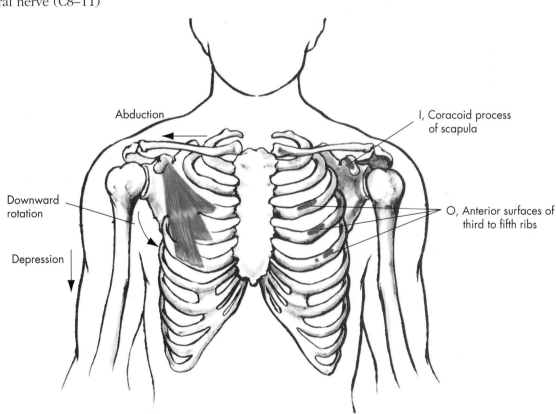

Abduction

Downward rotation

Depression

I, Coracoid process of scapula

O, Anterior surfaces of third to fifth ribs

FIG. 4.14 ● Pectoralis minor muscle, anterior view. O, Origin; I, Insertion.

Subclavius muscle   FIG. 4.15

(sub-klá ve-us)

Origin

Superior aspect of first rib at its junction with its costal cartilage

Insertion

Inferior groove in the midportion of the clavicle

Action

Stabilization and protection of the sternoclavicular joint
Depression
Abduction (protraction)

Palpation

Difficult to distinguish from the pectoralis major, but may be palpated just inferior to the middle third of the clavicle with the subject side-lying and in

Scapula abduction

a somewhat upwardly rotated position and the humerus supported in a partially passively flexed position. Slight active depression and abduction of the scapula may enhance palpation.

Innervation

Nerve fibers from C5 and C6

Application, strengthening, and flexibility

The subclavius pulls the clavicle anteriorly and inferiorly toward the sternum. In addition to assisting in abducting and depressing the clavicle and the shoulder girdle, it has a significant role in protecting and stabilizing the sternoclavicular joint during upper-extremity movements. It may be strengthened during activities in which there is active depression, such as dips, or active abduction, such as push-ups. Extreme elevation and retraction of the shoulder girdle provide a stretch to the subclavius.

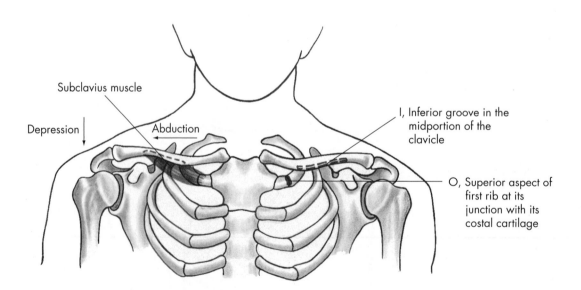

FIG. 4.15 ● Subclavius muscle, anterior view. O, Origin; I, Insertion.

REVIEW EXERCISES

1. List the planes in which each of the following shoulder girdle movements occurs. List the
   respective axis of rotation for each movement in
   each plane.
   a. Adduction
   b. Abduction
   c. Rotation upward
   d. Rotation downward
   e. Elevation
   f. Depression

2. **Muscle analysis chart** • Shoulder girdle

Complete the chart by listing the muscles primarily involved in each movement.	
Abduction	Adduction
Elevation	Depression
Upward rotation	Downward rotation

3. **Antagonistic muscle action chart** • Shoulder girdle

Complete the chart by listing the muscle(s) or parts of muscles that are antagonist in their actions to the muscles in the left column.	
Agonist	Antagonist
Serratus anterior	
Trapezius (upper fibers)	
Trapezius (middle fibers)	
Trapezius (lower fibers)	
Rhomboid	
Levator scapulae	
Pectoralis minor	

LABORATORY EXERCISES

1. Locate the following prominent skeletal features on a human skeleton and on a subject:
   a. Scapula
      1. Medial border
      2. Inferior angle
      3. Superior angle
      4. Coracoid process
      5. Spine of scapula
      6. Glenoid cavity
      7. Acromion process
      8. Supraspinatus fossa
      9. Infraspinatus fossa
   b. Clavicle
      1. Sternal end
      2. Acromial end
   c. Joints
      1. Sternoclavicular joint
      2. Acromioclavicular joint
2. Describe how and where you palpate the following muscles on a human subject:
   a. Serratus anterior
   b. Trapezius
   c. Rhomboid major and minor
   d. Levator scapulae
   e. Pectoralis minor
      Note: "How" means palpating during active contraction and possibly resisting a primary movement of the muscle. Some muscles have several primary movements, such as the trapezius with rotation upward and adduction. "Where" refers to the location on the body where the muscle can be felt.
3. Palpate the sternoclavicular and acromioclavicular joint movements and the muscles primarily involved while demonstrating the following shoulder girdle movements:
   a. Adduction
   b. Abduction
   c. Rotation upward
   d. Rotation downward
   e. Elevation
   f. Depression

4. **Shoulder girdle movement analysis chart**

After analyzing each exercise in the chart, break each into two primary movement phases, such as a lifting phase and a lowering phase. For each phase, determine the shoulder girdle movements occurring, and then list the shoulder girdle muscles primarily responsible for causing/controlling those movements. Beside each muscle in each movement, indicate the type of contraction as follows: I—isometric; C—concentric; E—eccentric.

| Exercise | Initial movement (lifting) phase | | Secondary movement (lowering) phase | |
	Movement(s)	Agonist(s)—(contraction type)	Movement(s)	Agonist(s)—(contraction type)
Push-up				
Chin-up				
Bench press				
Dip				
Lat pull				
Overhead press				
Prone row				
Barbell shrugs				

5. Shoulder girdle sport skill analysis chart

Analyze each skill in the chart and list the movements of the right and left shoulder girdle in each phase of the skill. You may prefer to list the initial position the shoulder girdle is in for the stance phase. After each movement, list the shoulder girdle muscle(s) primarily responsible for causing/controlling that movement. Beside each muscle in each movement, indicate the type of contraction as follows: I—isometric; C—concentric; E—eccentric. It may be desirable to review the concepts for analysis in Chapter 8 for the various phases.

Exercise		Stance phase	Preparatory phase	Movement phase	Follow-through phase
Baseball pitch	(R)				
	(L)				
Volleyball serve	(R)				
	(L)				
Tennis serve	(R)				
	(L)				
Softball pitch	(R)				
	(L)				
Tennis backhand	(R)				
	(L)				
Batting	(R)				
	(L)				
Bowling	(R)				
	(L)				
Basketball free throw	(R)				
	(L)				

References

Andrews JR, Zarins B, Wilk KE: *Injuries in baseball,* Philadelphia, 1998, Lippincott-Raven.

DePalma MJ, Johnson EW: Detecting and treating shoulder impingement syndrome: the role of scapulothoracic dyskinesis, *The Physician and Sportsmedicine* 31(7), 2003.

Field D: *Anatomy: palpation and surface markings,* ed 3, Oxford, 2001, Butterworth-Heinemann.

Hislop HJ, Montgomery J: *Daniels and Worthingham's muscle testing: techniques of manual examination,* ed 8, Philadelphia, 2007, Saunders.

Johnson RJ: Acromioclavicular joint injuries: identifying and treating "separated shoulder" and other conditions, Harmon K, Rubin A, eds: *The Physician and Sportsmedicine* 29(11), 2001.

Loftice JW, Fleisig GS, Wilk KE, Reinold MM, Chmielewski T, Escamilla RF, Andrews JR, eds: *Conditioning program for baseball pitchers,* Birmingham, AL, 2004, American Sports Medicine Institute.

McMurtrie H, Rikel JK: *The coloring review guide to human anatomy,* New York, 1991, McGraw-Hill.

Muscolino JE: *The muscular system manual: the skeletal muscles of the human body,* ed 3, St. Louis, 2010, Elsevier Mosby.

Neumann DA: *Kinesiology of the musculoskeletal system: foundations for physical rehabilitation,* ed 2, St. Louis, 2010, Mosby.

Norkin CC, Levangie PK: *Joint structure and function—a comprehensive analysis,* ed 5, Philadelphia, 2011, Davis.

Rasch PJ: *Kinesiology and applied anatomy,* ed 7, Philadelphia, 1989, Lea & Febiger.

Seeley RR, Stephens TD, Tate P: *Anatomy & physiology,* ed 8, Dubuque, IA, 2008, McGraw-Hill.

Smith LK, Weiss EL, Lehmkuhl LD: *Brunnstrom's clinical kinesiology,* ed 5, Philadelphia, 1996, Davis.

Sobush DC, et al: The Lennie test for measuring scapula position in healthy young adult females: a reliability and validity study, *Journal of Orthopedic and Sports Physical Therapy* 23:39, January 1996.

Soderburg GL: *Kinesiology—application to pathological motion,* Baltimore, 1986, Williams & Wilkins.

Van De Graaff KM: *Human anatomy,* ed 6, Dubuque, IA, 2002, McGraw-Hill.

Wilk KE, Reinold MM, Andrews JR, eds: *The athlete's shoulder,* ed 2, Philadelphia, 2009, Churchill Livingstone Elsevier.

Williams CC: Posterior sternoclavicular joint dislocation emergencies series, Howe WB, ed.: *The Physician and Sportsmedicine* 27(2), 1999.

For additional resources and a list of related websites, visit **www.mhhe.com/floyd19e**.

Worksheet Exercises

For in- or out-of-class assignments, or for testing, utilize this tear-out worksheet.

Worksheet 1

Using crayons or colored markers, draw and label on the worksheet the following muscles. Indicate the origin and insertion of each muscle with an "O" and an "I," respectively, and draw in the origin and insertion on the contralateral side of the skeleton.

- a. Trapezius
- b. Rhomboid major and minor
- c. Serratus anterior
- d. Levator scapulae
- e. Pectoralis minor

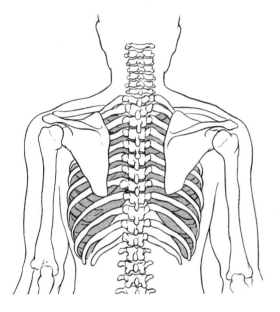

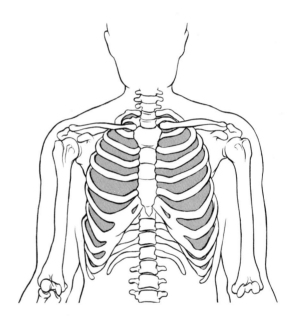

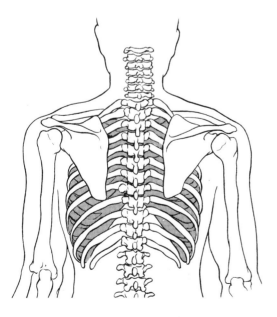

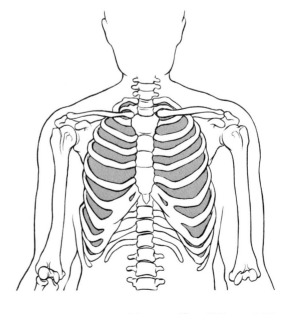

Worksheet Exercises

For in- or out-of-class assignments, or for testing, utilize this tear-out worksheet.

Worksheet 2

Label each of lines 1 through 6 on the drawing with the letter, from the following list, that corresponds to the movements of the shoulder girdle indicated by the arrow.

a. Adduction (retraction)
b. Abduction (protraction)
c. Rotation upward
d. Rotation downward
e. Elevation
f. Depression

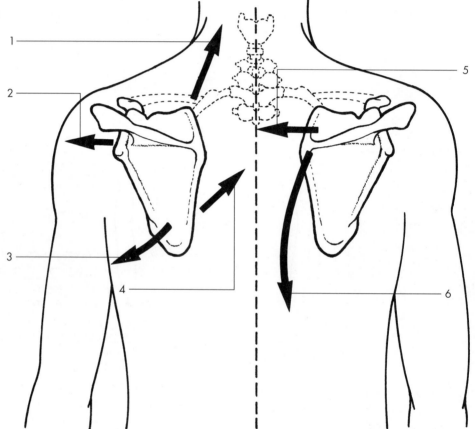

On the lines below, which correspond to the numbers of the arrows above, list the muscle(s) or parts of muscles primarily responsible for causing each movement.

1. _____

2. _____

3. _____

4. _____

5. _____

6. _____

THE SHOULDER JOINT

Objectives

- To identify on a human skeleton or human subject selected bony structures of the shoulder joint

- To label on a skeletal chart selected bony structures of the shoulder joint

- To draw on a skeletal chart the muscles of the shoulder joint and indicate, using arrows, shoulder joint movements

- To demonstrate with a fellow student all the movements of the shoulder joints and list their respective planes and axes of rotation

- To learn and understand how movements of the scapula accompany movements of the humerus in achieving movement of the entire shoulder complex

- To determine and list the muscles of the shoulder joint and their antagonists

- To organize and list the muscles that produce the movements of the shoulder girdle and the shoulder joint

- To determine, through analysis, the shoulder joint movements and muscles involved in selected skills and exercises

Online Learning Center Resources

Visit *Manual of Structural Kinesiology*'s **Online Learning Center** at **www.mhhe.com/floyd19e** for additional information and study material for this chapter, including:

- *Self-grading quizzes*
- *Anatomy flashcards*
- *Animations*
- *Related websites*

The only attachment of the shoulder joint to the axial skeleton is via the scapula and its attachment through the clavicle at the sternoclavicular joint. Movements of the shoulder joint are many and varied. It is unusual to have movement of the humerus without scapula movement. When the humerus is flexed above shoulder level, the scapula is elevated, rotated upward, and abducted. With glenohumeral abduction above shoulder level, the scapula is elevated and rotated upward. Adduction of the humerus results in depression and rotation downward, whereas extension of the humerus results in depression, rotation downward, and adduction of the scapula. The scapula abducts with humeral internal rotation and horizontal adduction. The scapula adducts with external rotation and horizontal abduction of the humerus. For a summary of these movements and the muscles primarily responsible for them, refer to Table 5.1.

Because the shoulder joint has such a wide range of motion in so many different planes, it also has a significant amount of laxity, which often results in instability problems such as rotator cuff impingement, subluxations, and dislocations. The price of mobility is reduced stability. The concept that the more mobile a joint is, the less stable it is and that the more stable it is, the less mobile it is applies generally throughout the body, but particularly in the shoulder joint. See the section on stability and mobility of diarthrodial joints in Chapter 1.

TABLE 5.1 • Pairing of shoulder girdle and shoulder joint movements. When the muscles of the shoulder joint (second column) perform the actions in the first column through any substantial range of motion, the muscles of the shoulder girdle (fourth column) work in concert by performing the actions in the third column.

Shoulder joint actions	Shoulder joint agonists	Shoulder girdle actions	Shoulder girdle agonists
Abduction	Supraspinatus, deltoid, upper pectoralis major	Upward rotation/elevation	Serratus anterior, middle and lower trapezius, levator scapulae, rhomboids
Adduction	Latissimus dorsi, teres major, lower pectoralis major	Downward rotation	Pectoralis minor, rhomboids
Flexion	Anterior deltoid, upper pectoralis major, coracobrachialis	Elevation/upward rotation	Levator scapulae, serratus anterior, upper and middle trapezius, rhomboids
Extension	Latissimus dorsi, teres major, lower pectoralis major, posterior deltoid	Depression/downward rotation	Pectoralis minor, lower trapezius
Internal rotation	Latissimus dorsi, teres major, pectoralis major, subscapularis	Abduction (protraction)	Serratus anterior, pectoralis minor
External rotation	Infraspinatus, teres minor	Adduction (retraction)	Middle and lower trapezius, rhomboids
Horizontal abduction	Middle and posterior deltoid, infraspinatus, teres minor	Adduction (retraction)	Middle and lower trapezius, rhomboids
Horizontal adduction	Pectoralis major, anterior deltoid, coracobrachialis	Abduction (protraction)	Serratus anterior, pectoralis minor
Diagonal abduction (overhand activities)	Posterior deltoid, infraspinatus, teres minor	Adduction (retraction)/upward rotation/elevation	Trapezius, rhomboids, serratus anterior, levator scapulae
Diagonal adduction (overhand activities)	Pectoralis major, anterior deltoid, coracobrachialis	Abduction (protraction)/depression/downward rotation	Serratus anterior, pectoralis minor

Bones

The scapula, clavicle, and humerus serve as attachments for most of the muscles of the shoulder joint. Learning the specific location and importance of certain bony landmarks is critical to understanding the functions of the shoulder complex. Some of these scapular landmarks are the supraspinous fossa, infraspinous fossa, subscapular fossa, spine of the scapula, glenoid cavity, coracoid process, acromion process, and inferior angle. Humeral landmarks are the head, greater tubercle, lesser tubercle, intertubercular groove, and deltoid tuberosity (Figs. 5.1 and 5.2, and review Figs. 4.1 and 4.3).

Joint

The shoulder joint, specifically known as the glenohumeral joint, is a multiaxial ball-and-socket joint classified as enarthrodial (see Fig. 5.1). As such, it moves in all planes and is the most movable joint in the body. It is similar to the hip in its joint classification; however, the socket provided by the glenoid fossa is much shallower and relatively small in comparison to the rather large humeral head. Its stability is enhanced slightly by the glenoid labrum (see Fig. 5.5), a cartilaginous ring that surrounds the glenoid fossa just inside its periphery. It is further stabilized by the glenohumeral ligaments, especially anteriorly and inferiorly. The anterior glenohumeral ligaments become taut as external rotation, extension, abduction, and horizontal abduction occur, whereas the very thin posterior capsular ligaments become taut in internal rotation, flexion, and horizontal adduction. In recent years the importance of the inferior glenohumeral ligament in providing both anterior and posterior stability has come to light (Figs. 5.3 and 5.4). It is,

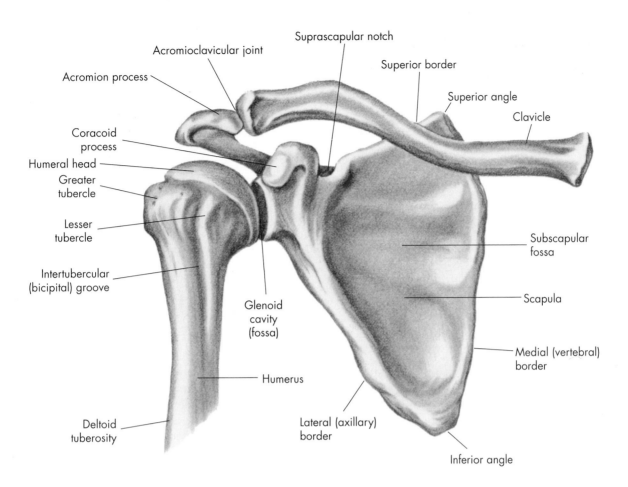

FIG. 5.1 • Right glenohumeral joint, anterior view.

however, important to note that, due to the wide range of motion involved in the glenohumeral joint, the ligaments are quite lax until the extreme ranges of motion are reached. This relative lack of static stability provided by the ligaments emphasizes the need for optimal dynamic stability to be provided by muscles such as the rotator cuff group. Stability is sacrificed to gain mobility.

Movement of the humerus from the side position is common in throwing, tackling, and striking activities. Flexion and extension of the shoulder joint are performed frequently when supporting body weight in a hanging position or in a movement from a prone position on the ground.

Determining the exact range of each movement for the glenohumeral joint is difficult because of the accompanying shoulder girdle movement. However, if the shoulder girdle is prevented from moving, then the glenohumeral joint movements are generally thought to be in the following ranges: 90 to 100 degrees of abduction, 0 degrees adduction (prevented by the trunk) or 75 degrees anterior to the trunk, 40 to 60 degrees of extension, 90 to 100 degrees of flexion, 70 to 90 degrees of internal and external rotation, 45 degrees of horizontal abduction, and 135 degrees of horizontal adduction. If the shoulder girdle is free to move, then the total range of the combined joints is 170 to 180 degrees of abduction, 170 to 180 degrees of flexion, and 140 to 150 degrees of horizontal adduction.

As discussed in Chapter 4 and emphasized in Table 5.1, the glenohumeral joint is paired with the shoulder girdle to accomplish the total shoulder range of motion. As an example, the 170 to 180 degrees of total abduction includes approximately 60 degrees of scapula upward rotation, 25 degrees of scapula elevation, and 95 degrees of glenohumeral abduction. These respective actions do not necessarily happen in a totally sequential fashion, but this synergistic relationship is often referred to as scapulohumeral rhythm. While the exact number of degrees in one segment compared to another may vary within and between individuals, the generally accepted ratio is 2 to 1; that is, for every 2 degrees of glenohumeral motion, there is 1 degree of scapula motion.

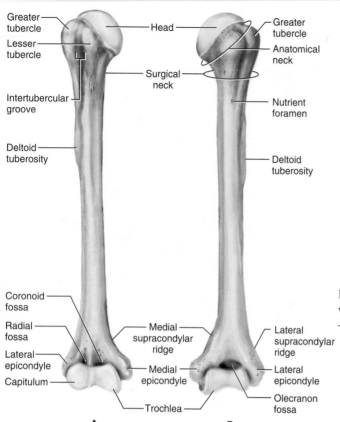

Greater tubercle — Head — Greater tubercle

Lesser tubercle — Anatomical neck

Surgical neck

Intertubercular groove

Nutrient foramen

Deltoid tuberosity — Deltoid tuberosity

Coronoid fossa

Radial fossa — Medial supracondylar ridge — Lateral supracondylar ridge

Lateral epicondyle — Medial epicondyle — Lateral epicondyle

Capitulum — Olecranon fossa

Trochlea

A   **B**

FIG. 5.2 • The right humerus. **A,** Anterior view; **B,** Posterior view.

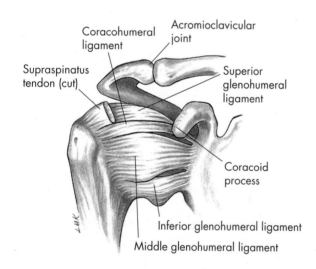

Coracohumeral ligament — Acromioclavicular joint

Supraspinatus tendon (cut) — Superior glenohumeral ligament

Coracoid process

Inferior glenohumeral ligament

Middle glenohumeral ligament

FIG. 5.3 • Glenohumeral ligaments, anterior view.

The shoulder joint is frequently injured because of its anatomical design. A number of factors contribute to its injury rate, including the shallowness of the glenoid fossa, the laxity of the ligamentous structures necessary to accommodate

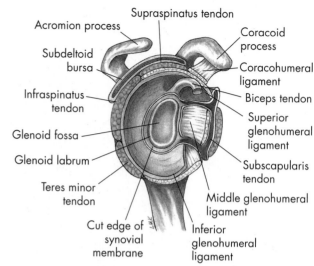

Acromion process — Supraspinatus tendon — Coracoid process

Subdeltoid bursa — Coracohumeral ligament

Infraspinatus tendon — Biceps tendon

Glenoid fossa — Superior glenohumeral ligament

Glenoid labrum — Subscapularis tendon

Teres minor tendon — Middle glenohumeral ligament

Cut edge of synovial membrane — Inferior glenohumeral ligament

FIG. 5.4 • Right glenohumeral joint, lateral view with humerus removed.

its wide range of motion, and the lack of strength and endurance in the muscles, which are essential in providing dynamic stability to the joint. As a result, anterior or anteroinferior glenohumeral subluxations and dislocations are quite common with physical activity. Although posterior dislocations are fairly rare, shoulder problems due to posterior instability are somewhat commonplace.

Another frequent injury is to the rotator cuff. The subscapularis, supraspinatus, infraspinatus, and teres minor muscles make up the rotator cuff. They are small muscles whose tendons cross the front, top, and rear of the head of the humerus to attach on the lesser and greater tubercles, respectively. Their point of insertion enables them to rotate the humerus, an essential movement in this freely movable joint. Most important, however, is the vital role that the rotator cuff muscles play in maintaining the humeral head in correct approximation within the glenoid fossa while the more powerful muscles of the joint move the humerus through its wide range of motion.

In recent years the phenomenon of glenohumeral internal rotation deficit, or GIRD, has received attention. GIRD represents a difference in internal rotation range of motion between an individual's throwing and nonthrowing shoulders. Studies have demonstrated that overhead athletes who had a GIRD of greater than 20% had a higher risk of injury than those who did not. Appropriate stretching exercises may be used to regain the amount of internal rotation necessary to improve performance and reduce the likelihood of injury.

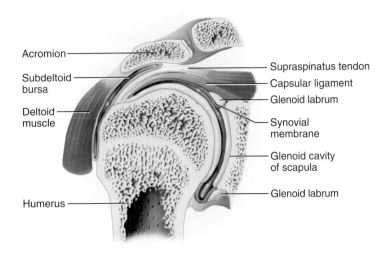

FIG. 5.5 • The right glenohumeral joint, frontal section.

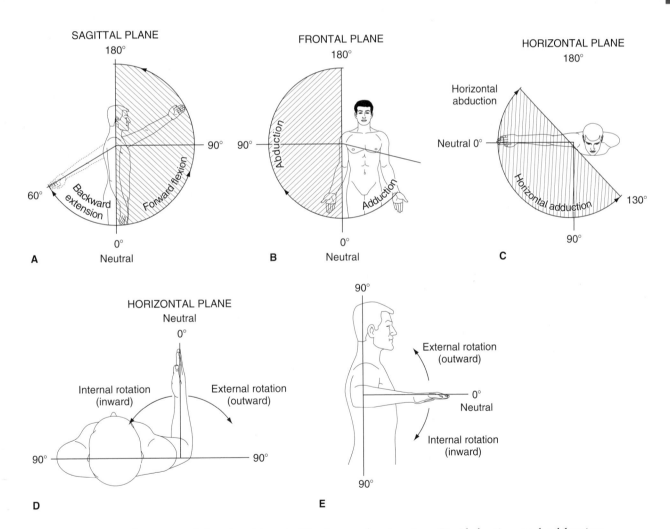

FIG. 5.6 • Range of motion of the shoulder. **A,** Flexion and extension; **B,** Abduction and adduction; **C,** Horizontal abduction and adduction; **D,** Internal and external rotation with the arm at the side of the body; **E,** Internal and external rotation with the arm abducted to 90 degrees.

Movements FIGS. 5.6, 5.7

Flexion: Movement of the humerus straight anteriorly from any point in the sagittal plane

Extension: Movement of the humerus straight posteriorly from any point in the sagittal plane, sometimes referred to as hyperextension

Abduction: Upward lateral movement of the humerus in the frontal plane out to the side, away from the body

Adduction: Downward movement of the humerus in the frontal plane medially toward the body from abduction

External rotation: Movement of the humerus laterally in the transverse plane around its long axis away from the midline

Internal rotation: Movement of the humerus in the transverse plane medially around its long axis toward the midline

Horizontal abduction (extension): Movement of the humerus in a horizontal or transverse plane away from the chest

Horizontal adduction (flexion): Movement of the humerus in a horizontal or transverse plane toward and across the chest

Diagonal abduction: Movement of the humerus in a diagonal plane away from the midline of the body

Diagonal adduction: Movement of the humerus in a diagonal plane toward the midline of the body

Flexion

A

Extension

B

Abduction

C

Adduction

D

FIG. 5.7 • Movements of the shoulder joint.

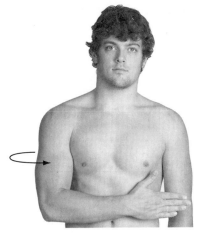

Internal rotation

E

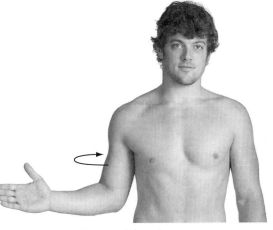

External rotation

F

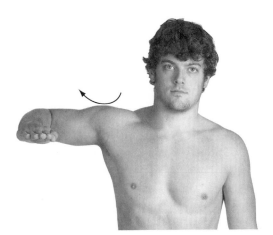

Horizontal abduction

G

Horizontal adduction

H

Shoulder
horizontal
adduction

Diagonal abduction

I

Diagonal adduction

J

FIG. 5.7 (continued) • Movements of the shoulder joint.

Muscles

In attempting to learn and understand the muscles of the glenohumeral joint, it may be helpful to group them according to their location and function. Muscles that originate on the scapula and clavicle and insert on the humerus may be thought of as muscles intrinsic to the glenohumeral joint, whereas muscles originating on the trunk and inserting on the humerus are considered extrinsic to the joint. The intrinsic muscles include the deltoid, the coracobrachialis, the teres major, and the rotator cuff group, which is composed of the subscapularis, the supraspinatus, the infraspinatus, and the teres minor. Extrinsic glenohumeral muscles are the latissimus dorsi and the pectoralis major. It may also be helpful to organize the muscles according to their general location. The pectoralis major, coracobrachialis, and subscapularis are anterior muscles. The deltoid and supraspinatus are located superiorly. The latissimus dorsi, teres major, infraspinatus, and teres minor are located posteriorly. Table 5.1 (p. 114) lists the glenohumeral joint movements and the muscles primarily responsible for them, and Table 5.2 provides the action of each muscle.

The biceps brachii and triceps brachii (long head) are also involved in glenohumeral movements. Primarily, the biceps brachii assists in flexing and horizontally adducting the shoulder, whereas the long head of the triceps brachii assists in extension and horizontal abduction. Further discussion of these muscles appears in Chapter 6.

Chapter 5

TABLE 5.2 • **Agonist muscles of the glenohumeral joint**

	Muscle	Origin	Insertion	Action	Plane of motion	Palpation	Innervation
Anterior muscles	Pectoralis major upper fibers	Medial half of anterior surface of clavicle	Flat tendon 2 or 3 inches wide to lateral lip of intertubercular groove of humerus	Internal rotation	Transverse	From medial end of the clavicle to the intertubercular groove of the humerus, during flexion and adduction from the anatomical position	Lateral pectoral nerve (C5, C6, C7)
				Horizontal adduction			
				Diagonal adduction	Diagonal		
				Flexion	Sagittal		
				Abduction	Frontal		
	Pectoralis major lower fibers	Anterior surface of costal cartilages of first six ribs, and adjoining portion of sternum	Flat tendon 2 or 3 inches wide to lateral lip of intertubercular groove of humerus	Internal rotation	Transverse	From the lower ribs and sternum to the intertubercular groove of the humerus, during resisted extension from a flexed position	Medial pectoral nerve (C8, T1)
				Horizontal adduction			
				Diagonal adduction	Diagonal		
				Extension from flexed position	Sagittal		
				Adduction	Frontal		
	Sub-scapularis	Entire anterior surface of subscapular fossa	Lesser tubercle of humerus	Internal rotation	Transverse	Mostly inaccessible, lateral portion may be palpated on supine subject (arm in slight flexion and adduction with elbow lying across abdomen); pull medial border laterally with one hand while palpating between the scapula and rib cage with other hand (subject actively internally rotates)	Upper and lower subscapular nerve (C5, C6)
				Adduction	Frontal		
				Extension	Sagittal		
	Coraco-brachialis	Coracoid process of scapula	Middle of medial border of humeral shaft	Horizontal adduction	Transverse	The belly may be palpated high up on the medial arm just posterior to the short head of the biceps brachii and toward the coracoid process, particularly with resisted adduction	Musculotaneous nerve (C5, C6, C7)
				Diagonal adduction	Diagonal		

TABLE 5.2 (continued) • Agonist muscles of the glenohumeral joint

Superior muscles	Deltoid anterior fibers	Anterior lateral third of clavicle	Deltoid tuberosity on lateral humerus	Abduction	Frontal	From the clavicle toward the anterior humerus during resisted flexion or horizontal adduction	Axillary nerve (C5, C6)
				Flexion	Sagittal		
				Horizontal adduction	Transverse		
				Internal rotation			
				Diagonal adduction	Diagonal		
	Deltoid middle fibers	Lateral aspect of acromion	Deltoid tuberosity on lateral humerus	Abduction	Frontal	From the lateral border of the acromion down toward the deltoid tuberosity during resisted abduction	
				Horizontal abduction	Transverse		
	Deltoid posterior fibers	Inferior edge of spine of scapula	Deltoid tuberosity on lateral humerus	Abduction	Frontal	From the lower lip of the spine of the scapula toward the posterior humerus during resisted extension or horizontal abduction	
				Horizontal abduction	Transverse		
				External rotation			
				Diagonal abduction	Diagonal		
	Supra-spinatus	Medial two-thirds of supraspinous fossa	Superiorly on greater tubercle of humerus	Abduction	Frontal	Above the spine of the scapula in supraspinous fossa during initial abduction in the scapula plane; tendon may be palpated just off acromion on greater tubercle	Suprascapula nerve (C5)
Posterior muscles	Latissimus dorsi	Posterior crest of ilium, back of sacrum and spinous processes of lumbar and lower six thoracic vertebrae, slips from lower three ribs	Medial side of intertubercular groove of humerus, just anterior to the insertion of the teres major	Extension	Sagittal	Tendon may be palpated as it passes under the teres major at the posterior axillary wall, particularly during resisted extension and internal rotation. The muscle can be palpated in the upper lumbar/lower thoracic area during extension from a flexed position and throughout most of its length during resisted adduction from a slightly abducted position	Thoracodorsal (C6, C7, C8)
				Adduction	Frontal		
				Internal rotation	Transverse		
				Horizontal abduction			
	Teres major	Posteriorly on inferior third of lateral border of scapula and just superior to inferior angle	Medial lip of intertubercular groove of humerus, just posterior to the insertion of the latissimus dorsi	Extension	Sagittal	Just above the latissimus dorsi and below the teres minor on the posterior scapula surface, moving diagonally upward and laterally from the inferior angle of the scapula during resisted internal rotation	Lower subscapular nerve (C5, C6)
				Adduction	Frontal		
				Internal rotation	Transverse		
	Infra-spinatus	Infraspinous fossa just below spine of scapula	Posteriorly on greater tubercle of humerus	External rotation	Transverse	Just below the spine of the scapula passing upward and laterally to the humerus during resisted external rotation	Suprascapula nerve (C5, C6)
				Horizontal abduction			
				Extension	Sagittal		
				Diagonal abduction	Diagonal		
	Teres minor	Posteriorly on upper and middle aspect of lateral border of scapula	Posteriorly on greater tubercle of humerus	External rotation	Transverse	Just above the teres major on the posterior scapula surface, moving diagonally upward and laterally from the inferior angle of the scapula during resisted external rotation	Axillary nerve (C5, C6)
				Horizontal abduction			
				Extension	Sagittal		
				Diagonal abduction	Diagonal		

Note: The biceps brachii assists in flexion, horizontal adduction, and diagonal adduction, while the long head of the triceps brachii assists in extension, adduction, horizontal abduction, and diagonal abduction. Because they are covered in Chapter 6, neither is listed above.

Chapter

5

Shoulder joint muscles—location

Anterior
    Pectoralis major
    Coracobrachialis
    Subscapularis
Superior
    Deltoid
    Supraspinatus
Posterior
    Latissimus dorsi
    Teres major
    Infraspinatus
    Teres minor

Muscle identification

Figs. 5.8 and 5.9 identify the anterior and posterior muscles of the shoulder joint and shoulder girdle. Compare Fig. 5.8 with Fig. 5.10 and Fig. 5.9 with Fig. 5.11, and refer to Table 5.2 for a detailed breakdown of the agonist muscles for the glenohumeral joint.

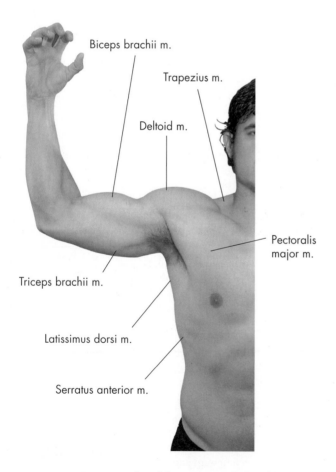

FIG. 5.8 • Anterior shoulder joint and shoulder girdle muscles.

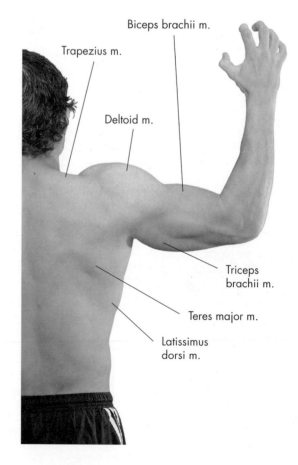

FIG. 5.9 • Posterior shoulder joint and shoulder girdle muscles.

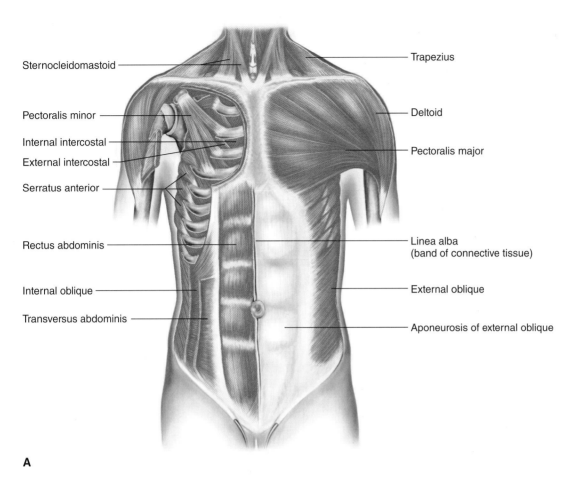

Sternocleidomastoid

Pectoralis minor

Internal intercostal

External intercostal

Serratus anterior

Rectus abdominis

Internal oblique

Transversus abdominis

Trapezius

Deltoid

Pectoralis major

Linea alba
(band of connective tissue)

External oblique

Aponeurosis of external oblique

A

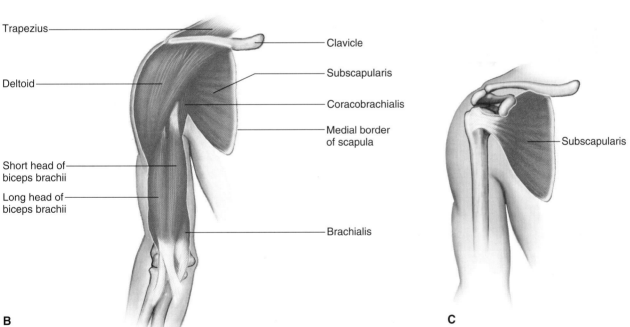

Trapezius

Deltoid

Short head of
biceps brachii

Long head of
biceps brachii

Clavicle

Subscapularis

Coracobrachialis

Medial border
of scapula

Brachialis

Subscapularis

B

C

FIG. 5.10 ● Anterior muscles of the shoulder. **A,** The right pectoralis major is removed to show the pectoralis minor and serratus anterior; **B,** Muscles of the anterior right shoulder and arm, with the rib cage removed; **C,** Subscapularis.

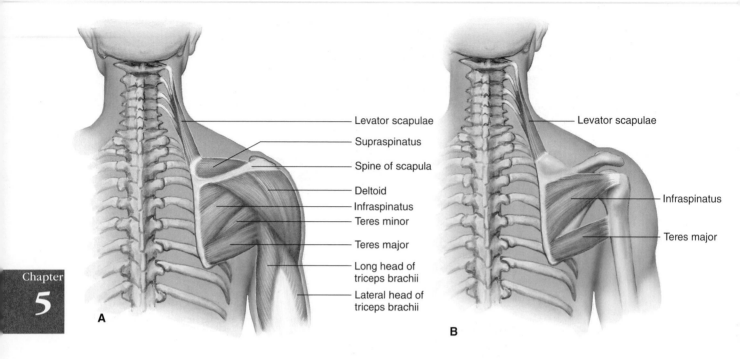

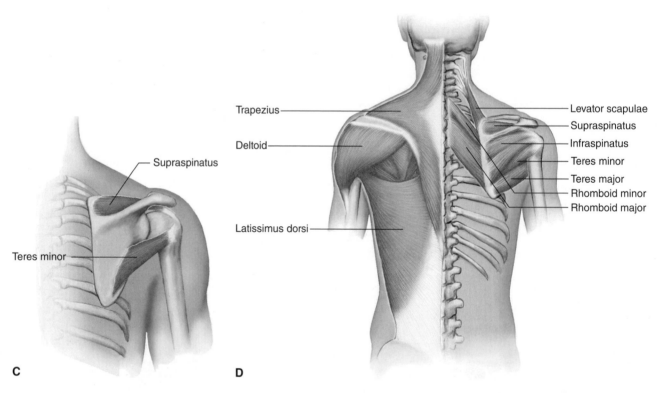

FIG. 5.11 • Posterior muscles of the shoulder. **A,** The right trapezius and deltoid are removed to show the underlying muscles; **B,** Levator scapulae, infraspinatus, and teres major; **C,** Supraspinatus and teres minor; **D,** Muscles of the posterior surface of the scapula and arm.

Nerves   FIGS. 5.12, 5.13

The muscles of the shoulder joint are all innervated by the nerves of the brachial plexus. The pectoralis major is innervated by the pectoral nerves. Specifically, the lateral pectoral nerve arising from C5, C6, and C7 innervates the clavicular head, while the medial pectoral nerve arising from C8 and T1 innervates the sternal head. The thoracodorsal nerve, arising from C6, C7, and C8, supplies the latissimus dorsi. The axillary nerve (Fig. 5.12), branching from C5 and C6, innervates the deltoid and teres minor. A lateral patch of skin over the deltoid region of the arm is provided sensation by the axillary nerve. Both the upper and lower subscapular nerves arising from C5 and C6 innervate the subscapularis, while only the lower subscapular nerve supplies the teres major. The supraspinatus and infraspinatus are innervated by the suprascapular nerve, which originates from C5 and C6. The musculocutaneous nerve, as seen in Fig. 5.13, branches from C5, C6, and C7 and innervates the coracobrachialis. It supplies sensation to the radial aspect of the forearm.

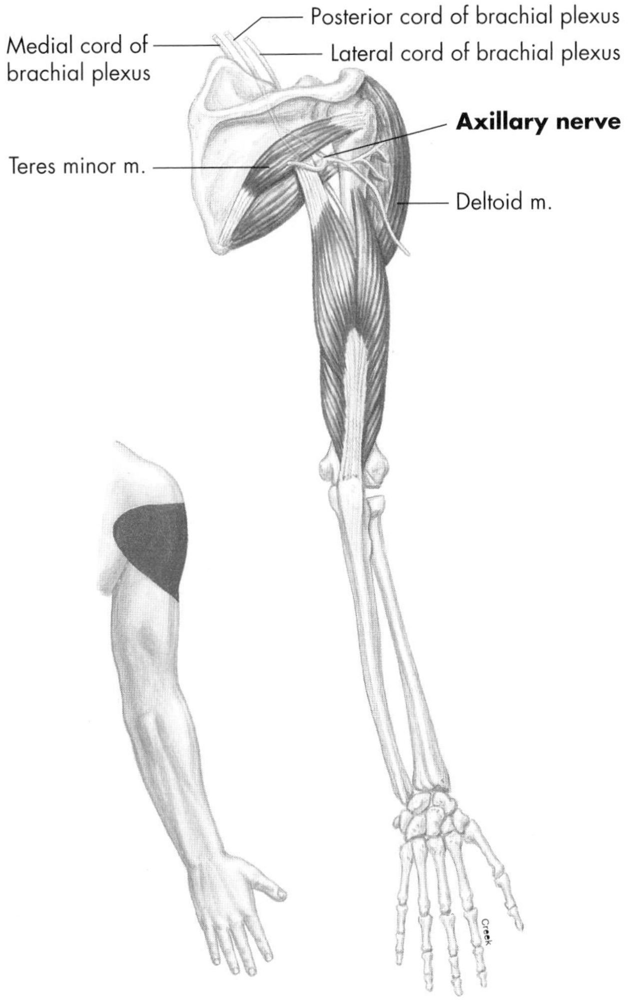

Chapter

5

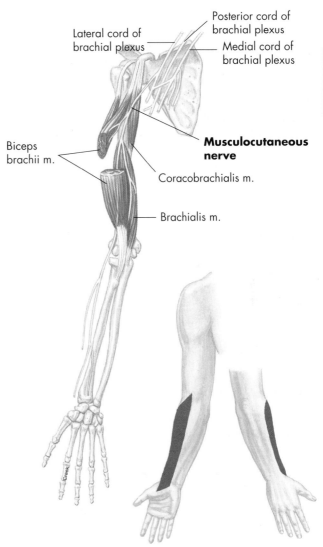

FIG. 5.12 • Muscular and cutaneous distribution of the axillary nerve.

FIG. 5.13 • Muscular and cutaneous distribution of the musculocutaneous nerve.

Deltoid muscle   FIG. 5.14

(del-toyd´)

Shoulder
abduction

Origin

Anterior fibers: anterior lateral third of the clavicle
Middle fibers: lateral aspect of the acromion
Posterior fibers: inferior edge of the spine of the scapula

Shoulder
flexion

Insertion

Deltoid tuberosity on the lateral humerus

Action

Anterior fibers: abduction, flexion, horizontal adduction, and internal rotation of the glenohumeral joint
Middle fibers: abduction of the glenohumeral joint
Posterior fibers: abduction, extension, horizontal abduction, and external rotation of the glenohumeral joint

Shoulder
horizontal
adduction

Palpation

Anterior fibers: from the clavicle toward the anterior humerus during resisted flexion or horizontal adduction
Middle fibers: from the lateral border of the acromion down toward the deltoid tuberosity during resisted abduction
Posterior fibers: from the lower lip of the spine of the scapula toward the posterior humerus during resisted extension or horizontal abduction

Shoulder
internal
rotation

Innervation

Axillary nerve (C5, C6)

Shoulder
extension

Shoulder
horizontal
abduction

Shoulder
external
rotation

Application, strengthening, and flexibility

The deltoid muscle is used in any lifting movement. The trapezius muscle stabilizes the scapula as the deltoid pulls on the humerus. The anterior fibers of the deltoid muscle flex and internally rotate the humerus. The posterior fibers extend and externally rotate the humerus. The anterior fibers also horizontally adduct the humerus, while the posterior fibers horizontally abduct it. Any movement of the humerus on the scapula will involve part or all of the deltoid muscle.

Lifting the humerus from the side to the position of abduction is a typical action of the deltoid. Sidearm dumbbell raises are excellent for strengthening the deltoid, especially the middle fibers. By abducting the arm in a slightly horizontally adducted (30 degrees) position, the anterior deltoid fibers can be emphasized. The posterior fibers can be strengthened better by abducting the arm in a slightly horizontally abducted (30 degrees) position. See Appendix 3 for more commonly used exercises for the deltoid and other muscles in this chapter.

Stretching the deltoid requires varying positions, depending on the fibers to be stretched. The anterior deltoid is stretched by taking the humerus into extreme horizontal abduction or by extreme extension and adduction. The middle deltoid is stretched by taking the humerus into extreme adduction behind the back. Extreme horizontal adduction stretches the posterior deltoid.

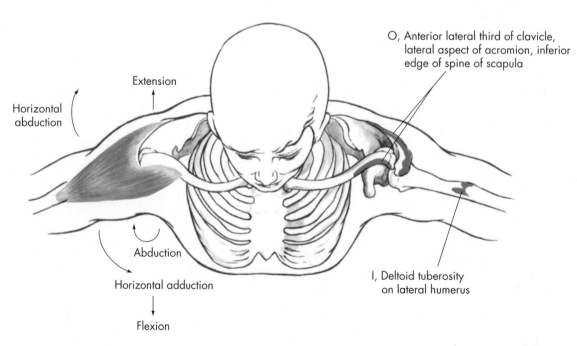

FIG. 5.14 ● Deltoid muscle, superior view. O, Origin; I, Insertion.

Pectoralis major muscle   FIG. 5.15

(pek-to-ra´lis ma´jor)

Origin

Upper fibers (clavicular head): medial half of the anterior surface of the clavicle

Lower fibers (sternal head): anterior surface of the costal cartilages of the first six ribs, and adjacent portion of the sternum

Insertion

Flat tendon 2 or 3 inches wide to the lateral lip of the intertubercular groove of the humerus

Action

Upper fibers (clavicular head): internal rotation, horizontal adduction, flexion up to about 60 degrees, abduction (once the arm is abducted 90 degrees, the upper fibers assist in further abduction), and adduction (with the arm below 90 degrees of abduction) of the glenohumeral joint

Lower fibers (sternal head): internal rotation, horizontal adduction, and adduction and extension of the glenohumeral joint from a flexed position to the anatomical position

Palpation

Upper fibers: from the medial end of the clavicle to the intertubercular groove of the humerus, during flexion and adduction from the anatomical position

Lower fibers: from the ribs and sternum to the intertubercular groove of the humerus, during resisted extension from a flexed position and resisted adduction from the anatomical position

Innervation

Upper fibers: lateral pectoral nerve (C5–C7)
Lower fibers: medial pectoral nerve (C8, T1)

Shoulder internal rotation

Shoulder horizontal adduction

Chapter
5

Shoulder flexion

Shoulder abduction

Shoulder adduction

Shoulder extension

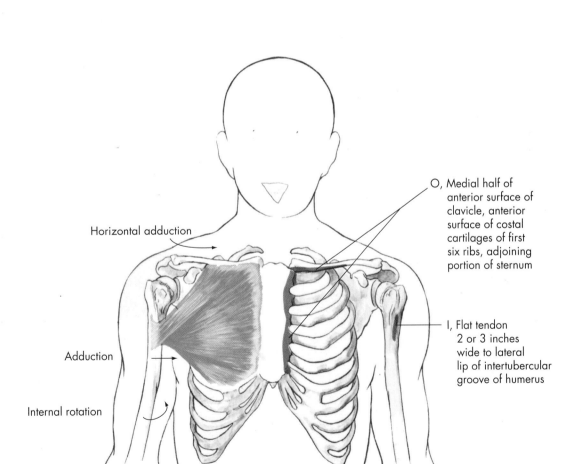

Horizontal adduction

O, Medial half of anterior surface of clavicle, anterior surface of costal cartilages of first six ribs, adjoining portion of sternum

Adduction

Internal rotation

I, Flat tendon 2 or 3 inches wide to lateral lip of intertubercular groove of humerus

FIG. 5.15 ● Pectoralis major muscle, anterior view. O, Origin; I, Insertion.

Application, strengthening, and flexibility

The anterior axillary fold is formed primarily by the pectoralis major (Fig. 5.16). It aids the serratus anterior muscle in drawing the scapula forward as it moves the humerus in flexion and internal rotation. Even though the pectoralis major is not attached to the scapula, it is effective in this scapula protraction because of its anterior pull on the humerus, which joins to the scapula at the glenohumeral joint. Typical action is shown in throwing a baseball. As the glenohumeral joint is flexed, the humerus is internally rotated and the scapula is drawn forward with upward rotation. It also works as a helper of the latissimus dorsi muscle when extending and adducting the humerus from a raised position.

The pectoralis major and the anterior deltoid work closely together. The pectoralis major is used powerfully in push-ups, pull-ups, throwing, and tennis serves. With a barbell, the subject takes a supine position on a bench with the arms at the side and moves the arms to a horizontally adducted position. This exercise, known as bench pressing, is widely used for pectoralis major development.

Due to the popularity of bench pressing and other weight-lifting exercises that emphasize the pectoralis major and its use in most sporting activities, it is often overdeveloped in comparison to its antagonists. As a result, stretching is often needed and can be done by passive external rotation. It is also stretched when the shoulder is horizontally abducted. Extending the shoulder fully provides stretching to the upper pectoralis major, while full abduction stretches the lower pectoralis major.

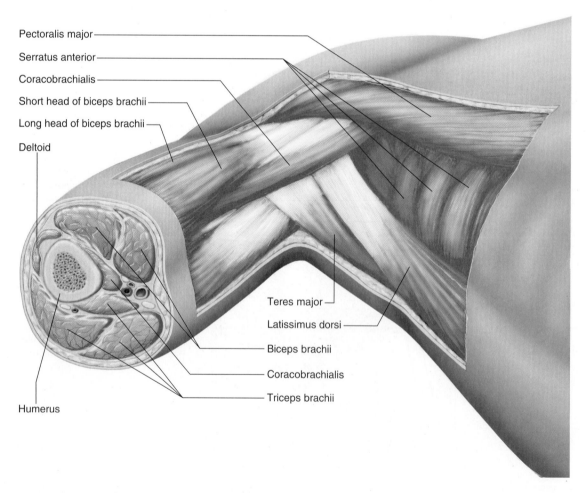

FIG. 5.16 • Cross section of right arm and relationship of glenohumeral muscles in axilla.

Latissimus dorsi muscle   FIG. 5.17
(lat-is´i-mus dor´si)

Origin
Posterior crest of the ilium, back of the sacrum and spinous processes of the lumbar and lower six

thoracic vertebrae (T6–T12); slips from the lower three ribs

Insertion
Medial lip of the intertubercular groove of the humerus, just anterior to the insertion of the teres major

Shoulder adduction

Shoulder extension

Shoulder internal rotation

Shoulder horizontal abduction

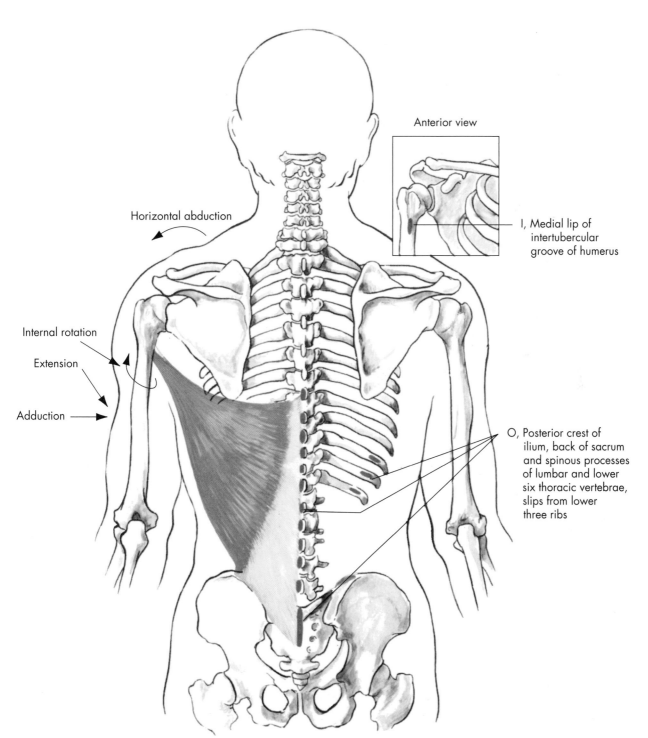

Anterior view

Horizontal abduction

Internal rotation

Extension

Adduction

I, Medial lip of intertubercular groove of humerus

O, Posterior crest of ilium, back of sacrum and spinous processes of lumbar and lower six thoracic vertebrae, slips from lower three ribs

FIG. 5.17 ● Latissimus dorsi muscle, posterior view. O, Origin; I, Insertion.

Action

Adduction of the glenohumeral joint
Extension of the glenohumeral joint
Internal rotation of the glenohumeral joint
Horizontal abduction of the glenohumeral joint

Palpation

The tendon may be palpated as it passes under
the teres major at the posterior wall of the axilla,
particularly during resisted extension and internal rotation. The muscle can be palpated in the
upper lumbar/lower thoracic area during extension
from a flexed position. The muscle may be palpated throughout most of its length during resisted
adduction from a slightly abducted position.

Innervation

Thoracodorsal nerve (C6–C8)

Application, strengthening, and flexibility

Latissimus dorsi means broadest muscle of the
back. This muscle, along with the teres major,
forms the posterior axillary fold (see Fig. 5.16).
It has a strong action in adduction, extension,
and internal rotation of the humerus. Due to the
upward rotation of the scapula that accompanies
glenohumeral abduction, the latissimus effectively
downwardly rotates the scapula by way of its
action in pulling the entire shoulder girdle downward in active glenohumeral adduction. It is one
of the most important extensor muscles of the
humerus and contracts powerfully in chinning.
The latissimus dorsi is assisted in all its actions by
the teres major and is sometimes referred to as the
swimmer's muscle because of its function in pulling the body forward in the water during internal
rotation, adduction, and extension. Development
of this muscle contributes significantly to what is
known as a "swimmer's build."

Exercises in which the arms are pulled down
bring the latissimus dorsi muscle into powerful
contraction. Chinning, rope climbing, and other
uprise movements on the horizontal bar are good
examples. In barbell exercises, the basic rowing
and pullover exercises are good for developing
the "lats." Pulling the bar of an overhead pulley
system down toward the shoulders, known as "lat
pulls," is a common exercise for this muscle.

The latissimus dorsi is stretched with the teres
major when the shoulder is externally rotated
while in a 90-degree abducted position. This
stretch may be accentuated further by abducting the shoulder fully while maintaining external
rotation and then laterally flexing and rotating the
trunk to the opposite side.

Teres major muscle   FIG. 5.18
(te´rez ma´jor)

Origin

Posteriorly on the inferior third of the lateral border of the scapula and just superior to the inferior angle

Insertion

Medial lip of the intertubercular groove of the humerus just posterior to the insertion of the latissimus dorsi

Action

Extension of the glenohumeral joint, particularly from the flexed position to the posteriorly extended position

Internal rotation of the glenohumeral joint

Adduction of the glenohumeral joint, particularly from the abducted position down to the side and toward the midline of the body

Palpation

Just above the latissimus dorsi and below the teres minor on the posterior scapula surface, moving diagonally upward and laterally from the inferior angle of the scapula during resisted internal rotation

Innervation

Lower subscapular nerve (C5, C6)

Application, strengthening, and flexibility

The teres major muscle is effective only when the rhomboid muscles stabilize the scapula or move the scapula in downward rotation. Otherwise, the scapula would move forward to meet the arm.

This muscle works effectively with the latissimus dorsi. It assists the latissimus dorsi, pectoralis major, and subscapularis in adducting, internally rotating, and extending the humerus. It is said to be the latissimus dorsi's "little helper." It may be strengthened by lat pulls, rope climbing, and internal rotation exercises against resistance.

Externally rotating the shoulder in a 90-degree abducted position stretches the teres major.

Shoulder extension

Shoulder internal rotation

Chapter
5

Shoulder adduction

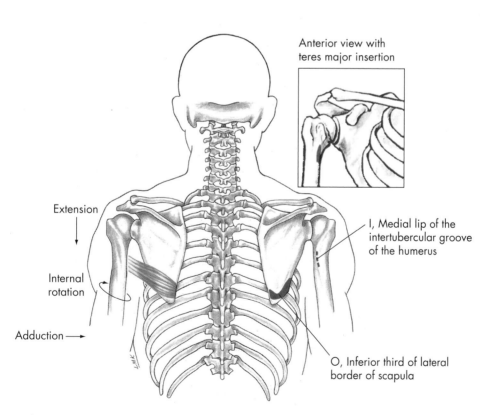

Anterior view with teres major insertion

Extension

Internal rotation

Adduction →

I, Medial lip of the intertubercular groove of the humerus

O, Inferior third of lateral border of scapula

FIG. 5.18 ● Teres major muscle, posterior view. O, Origin; I, Insertion.

Shoulder
flexion

Coracobrachialis muscle FIG. 5.19

(kor-a-ko-bra´ki-a´lis)

Origin
Coracoid process of the scapula

Insertion
Middle of the medial border of the humeral shaft

Action
Flexion of the glenohumeral joint
Adduction of the glenohumeral joint
Horizontal adduction of the glenohumeral joint

Palpation
The belly may be palpated high up on the medial
arm just posterior to the short head of the biceps
brachii and toward the coracoid process, particularly with resisted adduction.

Shoulder
adduction

Shoulder
horizontal
adduction

Innervation
Musculocutaneous nerve (C5–C7)

Application, strengthening, and flexibility
The coracobrachialis is not a powerful muscle,
but it does assist in flexion and adduction and is
most functional in moving the arm horizontally
toward and across the chest. It is best strengthened by horizontally adducting the arm against
resistance, as in bench pressing. It may also be
strengthened by performing lat pulls (defined on
p. 128).

The coracobrachialis is best stretched in
extreme horizontal abduction, although extreme
extension also stretches this muscle.

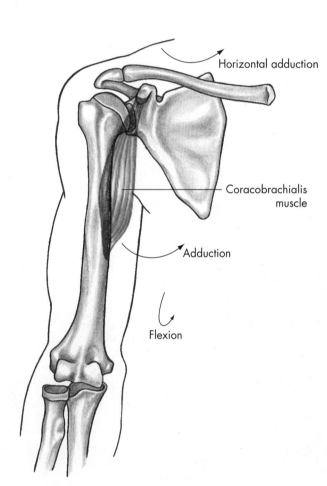

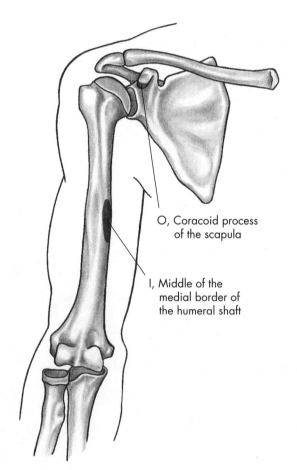

FIG. 5.19 • Coracobrachialis muscle, anterior view. O, Origin; I, Insertion.

Rotator cuff muscles

Figs. 5.20 and 5.21 illustrate the rotator cuscle group, which, as previously mentioned, is most important in maintaining the humeral head in its proper location within the glenoid cavity. The acronym **SITS** may be used in learning the names of the supraspinatus, infraspinatus, teres minor, and subscapularis. These muscles, which are not very large in comparison with the deltoid and pectoralis major, must possess not only adequate strength but also a significant amount of muscular endurance to ensure their proper functioning, particularly in repetitive overhead activities such as throwing, swimming, and pitching. Quite often when these types of activities are conducted with poor technique, muscle fatigue, or inadequate warm-up and conditioning, the rotator cuff muscle group—particularly the supraspinatus—fails to dynamically stabilize the humeral head in the glenoid cavity, leading to further rotator cuff problems such as tendinitis and rotator cuff impingement within the subacromial space.

Rotator cuff **impingement syndrome** occurs when the tendons of these muscles, particularly the supraspinatus and infraspinatus, become irritated and inflamed as they pass through the subacromial space between the acromion process of the scapula and the head of the humerus. Pain, weakness, and loss of movement at the shoulder can result. Loss of function of the rotator cuff muscles, due to injury or loss of strength and endurance, may cause the humerus to move superiorly, resulting in this impingement.

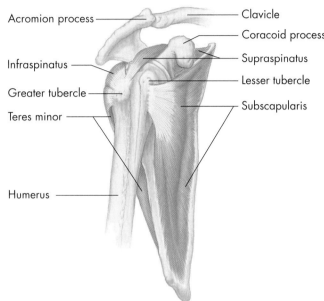

FIG. 5.20 ● Rotator cuff muscles, anterolateral view, right shoulder.

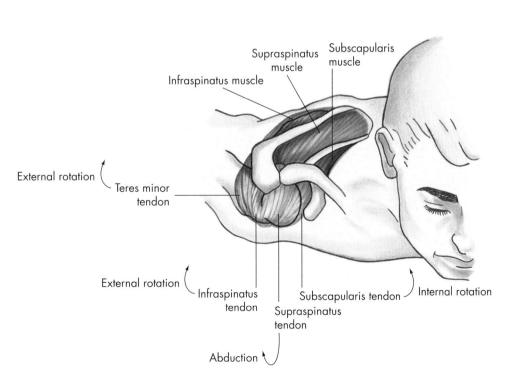

FIG. 5.21 ● Rotator cuff muscles, superior view, right shoulder.

Shoulder internal rotation

Shoulder abduction

Chapter 5

Shoulder external rotation

Shoulder extension

Shoulder adduction

Shoulder horizontal abduction

Subscapularis muscle   FIG. 5.22

(sub-skap-u-la´ris)

Shoulder
internal
rotation

Origin

Entire anterior surface of the subscapular fossa

Shoulder
adduction

Insertion

Lesser tubercle of the humerus

Action

Internal rotation of the glenohumeral joint
Adduction of the glenohumeral joint
Extension of the glenohumeral joint
Stabilization of the humeral head in the glenoid fossa

Chapter
5

Palpation

The subscapularis, latissimus dorsi, and teres major,
 in conjunction, form the posterior axillary fold.
 Most of the subscapularis is inaccessible on the
 anterior scapula behind the rib cage. The lateral
 portion may be palpated with the subject supine
 and arm in slight flexion and adduction so that the
 elbow is lying across the abdomen. Use one hand
 posteriorly to grasp the medial border and pull
 it laterally, while palpating between the scapula
 and rib cage with the other hand with the subject

Shoulder
extension

FIG. 5.22 ● Subscapularis muscle, anterior view.
O, Origin; I, Insertion.

actively internally rotating by pressing the forearm
against the chest.

Innervation

Upper and lower subscapular nerve (C5, C6)

Application, strengthening, and flexibility

The subscapularis muscle, another rotator cuff
muscle, holds the head of the humerus in the glenoid fossa from in front and below. It acts with
the latissimus dorsi and teres major muscles in its
typical movement but is less powerful in its action
because of its proximity to the joint. The muscle
also requires the help of the rhomboid in stabilizing
the scapula to make it effective in the movements
described. The subscapularis is relatively hidden
behind the rib cage in its location on the anterior
aspect of the scapula in the subscapular fossa. It
may be strengthened with exercises similar to those
used for the latissimus dorsi and teres major, such
as rope climbing and lat pulls. A specific exercise
for its development is done by internally rotating
the arm against resistance in the beside-the-body
position at 0 degrees of glenohumeral abduction.

External rotation with the arm adducted by the
side stretches the subscapularis.

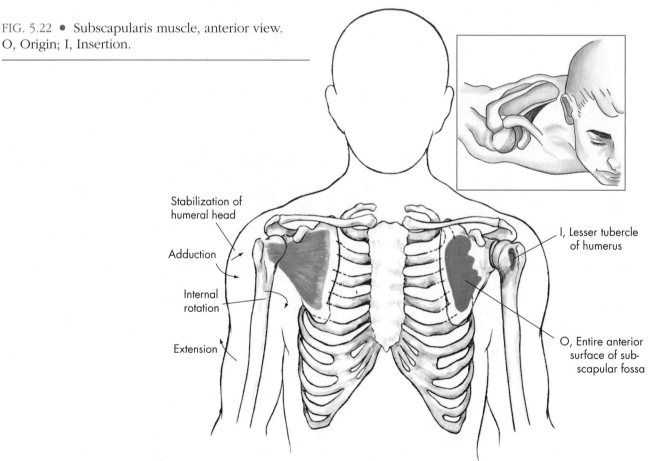

Stabilization of
humeral head

Adduction

Internal
rotation

Extension

I, Lesser tubercle
of humerus

O, Entire anterior
surface of sub-
scapular fossa

Supraspinatus muscle   FIG. 5.23

(su´pra-spi-na´tus)

Origin

Medial two-thirds of the supraspinous fossa

Insertion

Superiorly on the greater tubercle of the humerus

Action

Abduction of the glenohumeral joint
Stabilization of the humeral head in the glenoid fossa

Palpation

Anterior and superior to the spine of the scapula in the supraspinous fossa during initial abduction in the scapula plane. Also, the tendon may be palpated in a seated position just off the acromion on the greater tubercle.

Innervation

Suprascapular nerve (C5)

Application, strengthening, and flexibility

The supraspinatus muscle holds the head of the humerus in the glenoid fossa. In throwing movements, it provides important dynamic stability by maintaining the proper relationship between the humeral head and the glenoid fossa. In the cocking phase of throwing, there is a tendency for the humeral head to subluxate anteriorly. In the follow-through phase, the humeral head tends to move posteriorly.

The supraspinatus, along with the other rotator cuff muscles, must have excellent strength and endurance to prevent abnormal and excessive movement of the humeral head in the fossa.

The supraspinatus is the most often injured rotator cuff muscle. Acute severe injuries may occur with trauma to the shoulder. However, mild to moderate strains or tears often occur with athletic activity, particularly if the activity involves repetitious overhead movements, such as throwing or swimming.

Injury or weakness in the supraspinatus may be detected when the athlete attempts to substitute the scapula elevators and upward rotators to obtain humeral abduction. An inability to smoothly abduct the arm against resistance is indicative of possible rotator cuff injury.

The supraspinatus muscle may be called into play whenever the middle fibers of the deltoid muscle are used. A "full-can exercise" may be used to emphasize supraspinatus action. This is performed by placing the arm in thumbs-up position, followed by abducting the arm to 90 degrees in a 30- to 45-degree horizontally adducted position (scaption), as if one were holding a full can.

Adducting the arm behind the back with the shoulder internally rotated and extended stretches the supraspinatus.

Shoulder abduction

Chapter
5

FIG. 5.23 ● Supraspinatus muscle, posterior view. O, Origin; I, Insertion.

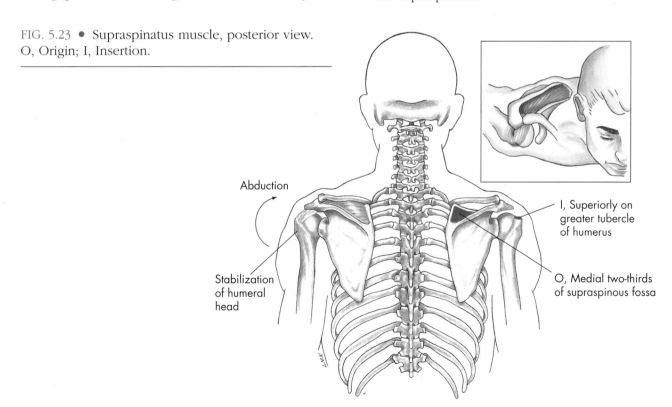

Abduction

Stabilization of humeral head

I, Superiorly on greater tubercle of humerus

O, Medial two-thirds of supraspinous fossa

Shoulder
external
rotation

Shoulder
horizontal
abduction

Chapter
5

Shoulder
extension

Infraspinatus muscle   FIG. 5.24

(in´fra-spi-na´tus)

Origin

Posterior surface of scapula below spine

Insertion

Greater tubercle on posterior side of the humerus

Action

External rotation of the glenohumeral joint
Horizontal abduction of the glenohumeral joint
Extension of the glenohumeral joint
Stabilization of the humeral head in the glenoid
   fossa

Palpation

Just below the spine of the scapula passing upward
   and laterally to the humerus during resisted external rotation

Innervation

Suprascapular nerve (C5, C6)

Application, strengthening, and flexibility

The infraspinatus and teres minor muscles are effective when the rhomboid muscles stabilize the scapula. When the humerus is rotated outward, the rhomboid muscles flatten the scapula to the back and fixate it so that the humerus may be rotated.

An appropriate amount of strength and endurance is critical in both the infraspinatus and teres minor as they are called upon eccentrically to slow down the arm from high velocity internal rotation activities such as baseball pitching and serving in tennis. The infraspinatus is vital to maintaining the posterior stability of the glenohumeral joint. It is the most powerful of the external rotators and is the second most commonly injured rotator cuff muscle.

Both the infraspinatus and the teres minor can best be strengthened by externally rotating the arm against resistance in the 15- to 20-degree abducted position and the 90-degree abducted position.

Stretching of the infraspinatus is accomplished with internal rotation and extreme horizontal adduction.

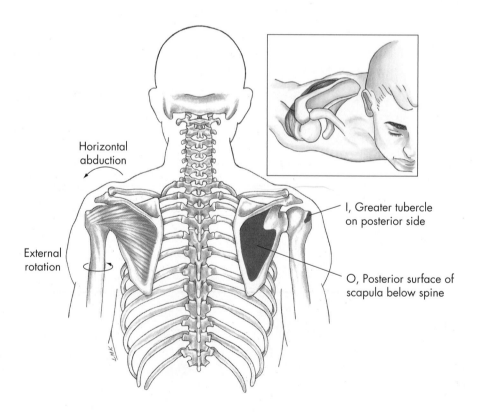

FIG. 5.24 ● Infraspinatus muscle, posterior view. O, Origin; I, Insertion.

Teres minor muscle   FIG. 5.25

(te´rez mi´nor)

Origin

Posteriorly on the upper and middle aspect of the lateral border of the scapula

Insertion

Posteriorly on the greater tubercle of the humerus

Action

External rotation of the glenohumeral joint
Horizontal abduction of the glenohumeral joint
Extension of the glenohumeral joint
Stabilization of the humeral head in the glenoid fossa

Palpation

Just above the teres major on the posterior scapula surface, moving diagonally upward and laterally from the inferior angle of the scapula during resisted external rotation

Innervation

Axillary nerve (C5, C6)

Application, strengthening, and flexibility

The teres minor functions very similarly to the infraspinatus in providing dynamic posterior stability to the glenohumeral joint. Both of these muscles perform the same actions together. The teres minor is strengthened with the same exercises that are used in strengthening the infraspinatus.

The teres minor is stretched similarly to the infraspinatus by internally rotating the shoulder while moving into extreme horizontal adduction.

Shoulder external rotation

Shoulder horizontal abduction

Chapter 5

Shoulder extension

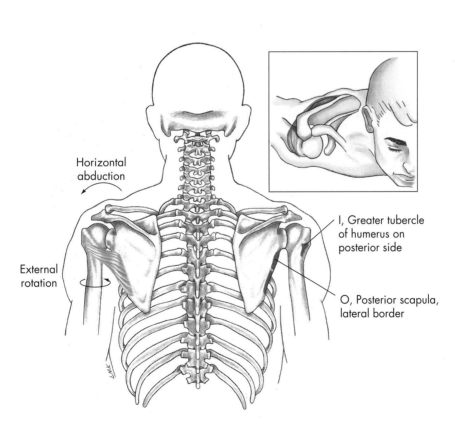

Horizontal abduction

External rotation

I, Greater tubercle of humerus on posterior side

O, Posterior scapula, lateral border

FIG. 5.25 ● Teres minor muscle, posterior view. O, Origin; I, Insertion.

REVIEW EXERCISES

1. List the planes in which each of the following glenohumeral joint movements occurs. List the respective axis of rotation for each movement in each plane.
   a. Abduction
   b. Adduction
   c. Flexion
   d. Extension
   e. Horizontal adduction
   f. Horizontal abduction
   g. External rotation
   h. Internal rotation
2. Why is it essential that both anterior and posterior muscles of the shoulder joint be properly developed? What are some activities or sports that would cause unequal development? equal development?
3. What practical application do the activities or sports in Question 2 support if each of the following is true?
   a. The rotator cuff muscles are not functioning properly due to fatigue or lack of appropriate strength and endurance.
   b. The scapula stabilizers are not functioning properly due to fatigue or lack of strength and endurance.
4. The movements of the scapula in relation to the humerus can be explained by discussing the movement of the shoulder complex in its entirety. How does the position of the scapula affect shoulder joint abduction? How does the position of the scapula affect shoulder joint flexion?
5. Describe the bony articulations and movements specific to shoulder joint rotation during the acceleration phase of the throwing motion and how an athlete can work toward increasing the velocity of the throw. What factors affect the velocity of the throw?
6. Using the information from this chapter and other resources, how would you strengthen the four rotator cuff muscles? Give several examples of how these muscles are used in everyday activities.

7. **Muscle analysis chart** • Shoulder girdle and shoulder joint

Complete the chart by listing the muscles primarily involved in each movement.

Shoulder girdle	Shoulder joint
Upward rotation	Abduction
Downward rotation	Adduction
Depression	Extension
Elevation	Flexion
Abduction	Horizontal adduction
	Internal rotation
Adduction	Horizontal abduction
	External rotation

8. Antagonistic muscle action chart • Shoulder joint

Complete the chart by listing the muscle(s) or parts of muscles that are antagonist in their actions to the muscles in the left column.

Agonist	Antagonist
Deltoid (anterior fibers)	
Deltoid (middle fibers)	
Deltoid (posterior fibers)	
Supraspinatus	
Subscapularis	
Teres major	
Infraspinatus/Teres minor	
Latissimus dorsi	
Pectoralis major (upper fibers)	
Pectoralis major (lower fibers)	
Coracobrachialis	

Chapter

5

LABORATORY EXERCISES

1. Locate the following parts of the humerus and scapula on a human skeleton and on a subject:
   a. Greater tubercle
   b. Lesser tubercle
   c. Neck
   d. Shaft
   e. Intertubercular groove
   f. Medial epicondyle
   g. Lateral epicondyle
   h. Trochlea
   i. Capitulum
   j. Supraspinous fossa
   k. Infraspinous fossa
   l. Spine of the scapula

2. How and where can the following muscles be palpated on a human subject?
   a. Deltoid
   b. Teres major
   c. Infraspinatus
   d. Teres minor
   e. Latissimus dorsi
   f. Pectoralis major (upper and lower)

   Note: Using the pectoralis major muscle, indicate how various actions allow muscle palpation.

3. Demonstrate and locate on a human subject the muscles that are primarily used in the following shoulder joint movements:
   a. Abduction
   b. Adduction
   c. Flexion
   d. Extension
   e. Horizontal adduction
   f. Horizontal abduction
   g. External rotation
   h. Internal rotation

4. Using an articulated skeleton, compare the relationship of the greater tubercle to the undersurface of the acromion in each of the following situations:
   a. Flexion with the humerus internally versus externally rotated
   b. Abduction with the humerus internally versus externally rotated
   c. Horizontal adduction with the humerus internally versus externally rotated

5. Pair up with a partner with the back exposed. Use your hand to grasp your partner's right scapula along the lateral border to prevent scapula movement. Have your partner slowly abduct the glenohumeral joint as much as possible. Note the difference in total abduction possible normally versus when you restrict movement of the scapula. Repeat the same exercise, except hold the inferior angle of the scapula tightly against the chest wall while you have your partner internally rotate the humerus. Note the difference in total internal rotation possible normally versus when you restrict movement of the scapula.

6. **Shoulder joint movement analysis chart**

After analyzing each of the exercises in the chart, break each into two primary movement phases, such as a lifting phase and a lowering phase. For each phase, determine the shoulder joint movements occurring, and then list the shoulder joint muscles primarily responsible for causing/controlling those movements. Beside each muscle in each movement, indicate the type of contraction as follows: I—isometric; C—concentric; E—eccentric.

Exercise	Initial movement (lifting) phase		Secondary movement (lowering) phase	
	Movement(s)	Agonist(s)—(contraction type)	Movement(s)	Agonist(s)—(contraction type)
Push-up				
Chin-up				
Bench press				
Dip				
Lat pull				
Overhead press				
Prone row				
Barbell shrugs				

7. Shoulder joint sport skill analysis chart

Analyze each skill in the chart and list the movements of the right and left shoulder joint in each phase of the skill. You may prefer to list the initial position the shoulder joint is in for the stance phase. After each movement, list the shoulder joint muscle(s) primarily responsible for causing/controlling that movement. Beside each muscle in each movement, indicate the type of contraction as follows: I—isometric; C—concentric; E—eccentric. It may be desirable to review the concepts for analysis in Chapter 8 for the various phases.

Exercise		Stance phase	Preparatory phase	Movement phase	Follow-through phase
Baseball pitch	(R)				
	(L)				
Volleyball serve	(R)				
	(L)				
Tennis serve	(R)				
	(L)				
Softball pitch	(R)				
	(L)				
Tennis backhand	(R)				
	(L)				
Batting	(R)				
	(L)				
Bowling	(R)				
	(L)				
Basketball free throw	(R)				
	(L)				

Chapter

5

References

Andrews JR, Zarins B, Wilk KE: *Injuries in baseball,* Philadelphia, 1988, Lippincott-Raven.

Bach HG, Goldberg BA: Posterior capsular contracture of the shoulder, *Journal of the American Academy of Orthopaedic Surgery* 14(5):265–277, 2006.

Field D: *Anatomy: palpation and surface markings,* ed 3, Oxford, 2001, Butterworth-Heinemann.

Fongemie AE, Buss DD, Rolnick SJ: Management of shoulder impingement syndrome and rotator cuff tears, *American Family Physician,* 57(4):667–674, 680–682, Feburary 1998.

Garth WP, et al: Occult anterior subluxations of the shoulder in noncontact sports, *American Journal of Sports Medicine* 15:579, November–December 1987.

Hislop HJ, Montgomery J: *Daniels and Worthingham's muscle testing: techniques of manual examination,* ed 8, Philadelphia, 2007, Saunders.

Loftice JW, Fleisig GS, Wilk KE, Reinold MM, Chmielewski T, Escamilla RF, Andrews JR, eds: *Conditioning program for baseball pitchers,* Birmingham, 2004, American Sports Medicine Institute.

Muscolino JE: *The muscular system manual: the skeletal muscles of the human body,* ed 3, St. Louis, 2010, Elsevier Mosby.

Myers JB, et al: Glenohumeral range of motion deficits and posterior shoulder tightness in throwers with pathologic internal impingement, *American Journal of Sports Medicine* 34:385–391, 2006.

Neumann DA: *Kinesiology of the musculoskeletal system: foundations for physical rehabilitation,* ed 2, St. Louis, 2010, Mosby.

Oatis CA: *Kinesiology: the mechanics and pathomechanics of human movement,* ed 2, Philadelphia, 2008, Lippincott Williams & Wilkins.

Perry JF, Rohe DA, Garcia AO: *The kinesiology workbook,* Philadelphia, 1992, Davis.

Rasch PJ: *Kinesiology and applied anatomy,* ed 7, Philadelphia, 1989, Lea & Febiger.

Reinold MM, Macrina LC, Wilk KE, Fleisig GS, Dun S, Barrentine SW, Ellerbusch MT, Andrews JR: Electromyographic analysis of the supraspinatus and deltoid muscles during 3 common rehabilitation exercises, *Journal of Athletic Training* 42(4): 464–469, 2007.

Seeley RR, Stephens TD, Tate P: *Anatomy & physiology,* ed 8, New York, 2008, McGraw-Hill.

Sieg KW, Adams SP: *Illustrated essentials of musculoskeletal anatomy,* ed 4, Gainesville, FL, 2002, Megabooks.

Smith LK, Weiss EL, Lehmkuhl LD: *Brunnstrom's clinical kinesiology,* ed 5, Philadelphia, 1996, Davis.

Spigelman T: Identifying and assessing glenohumeral internal-rotation deficit, *Athletic Therapy Today* 6:29–31, 2006.

Stacey E: Pitching injuries to the shoulder, *Athletic Journal* 65:44, January 1984.

Wilk KE, Reinold MM, Andrews JR, eds: *The athlete's shoulder,* ed 2, Philadelphia, 2009, Churchill Livingstone Elsevier.

For additional resources and a list of related websites, visit **www.mhhe.com/floyd19e**.

Worksheet Exercises

For in- or out-of-class assignments, or for testing, utilize this tear-out worksheet.

Worksheet 1

Using crayons or colored markers, draw and label on the worksheet the following muscles. Indicate the origin and insertion of each muscle with an "O" and an "I," respectively, and draw in the origin and insertion on the contralateral side of the skeleton.

a. Deltoid

b. Supraspinatus

c. Subscapularis

d. Teres major

e. Infraspinatus

f. Teres minor

g. Latissimus dorsi

h. Pectoralis major

i. Coracobrachialis

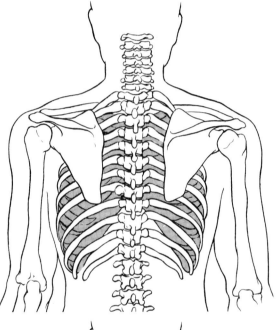

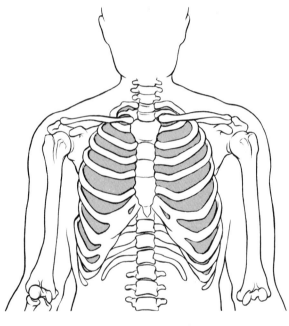

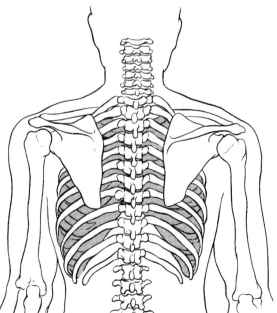

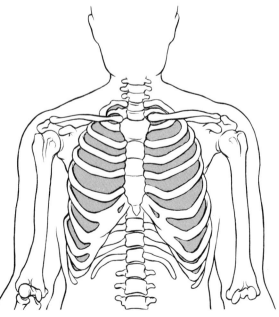

Worksheet Exercises

For in- or out-of-class assignments, or for testing, utilize this tear-out worksheet.

Worksheet 2

Label and indicate with arrows the following movements of the shoulder joint. For each motion, complete the sentence by supplying the plane in which it occurs and the axis of rotation.

a. Abduction occurs in the _____ plane about the _____ axis.

b. Adduction occurs in the _____ plane about the _____ axis.

c. Flexion occurs in the _____ plane about the _____ axis.

d. Extension occurs in the _____ plane about the _____ axis.

e. Horizontal adduction occurs in the _____ plane about the _____ axis.

f. Horizontal abduction occurs in the _____ plane about the _____ axis.

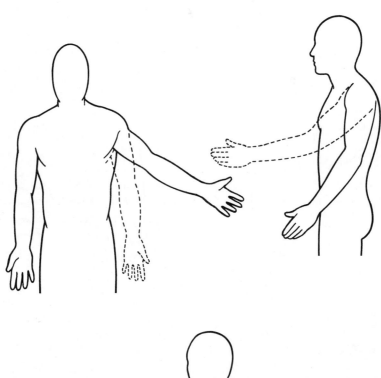

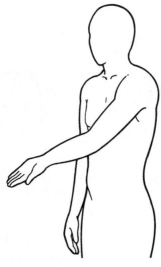

CHAPTER 6

THE ELBOW AND RADIOULNAR JOINTS

Objectives

- To identify on a human skeleton selected bony features of the elbow and radioulnar joints

- To label selected bony features on a skeletal chart

- To draw and label the muscles on a skeletal chart

- To palpate the muscles on a human subject and list their antagonists

- To list the planes of motion and their respective axes of rotation

- To organize and list the muscles that produce the primary movements of the elbow joint and the radioulnar joint

- To determine, through analysis, the elbow and radioulnar joint movements and muscles involved in selected skills and exercises

Online Learning Center Resources

Visit *Manual of Structural Kinesiology*'s **Online Learning Center** at **www.mhhe.com/floyd19e** for additional information and study material for this chapter, including:

- *Self-grading quizzes*
- *Anatomy flashcards*
- *Animations*
- *Related websites*

Chapter
6

Almost any movement of the upper extremity will involve the elbow and radioulnar joints. Quite often, these joints are grouped together because of their close anatomical relationship. The elbow joint is intimately associated with the radioulnar joint in that both bones of the radioulnar joint, the radius and ulna, share an articulation with the humerus to form the elbow joint. For this reason, some may confuse motions of the elbow with those of the radioulnar joint. In addition, radioulnar joint motion may be incorrectly attributed to the wrist joint because it appears to occur there. However, with close inspection, movements of the elbow joint can be clearly distinguished from those of the radioulnar joints, just as the radioulnar movements can be distinguished from those of the wrist. Even though the radius and ulna are both part of the articulation with the wrist, the relationship between them is not nearly as intimate as that of the elbow and radioulnar joints.

Bones

The ulna is much larger proximally than the radius (Fig. 6.1), but distally the radius is much larger than the ulna (see Fig. 7.1 on p. 170). The scapula and humerus serve as the proximal attachments for the muscles that flex and extend the elbow. The ulna and radius serve as the distal attachments for the same muscles. The scapula, humerus, and ulna serve as proximal attachments for the muscles that pronate and supinate the radioulnar joints. The distal attachments of the radioulnar joint muscles are located on the radius.

The medial condyloid ridge, olecranon process, coronoid process, and radial tuberosity are important bony landmarks for these muscles. Additionally, the medial epicondyle, lateral epicondyle, and lateral supracondylar ridge are key bony landmarks for the muscles of the wrist and hand, discussed in Chapter 7.

Joints

The elbow joint is classified as a ginglymus or hinge-type joint that allows only flexion and extension (Fig. 6.1). The elbow may actually be thought of as two interrelated joints: the humeroulnar and the radiohumeral joints (Fig. 6.2). Elbow motions

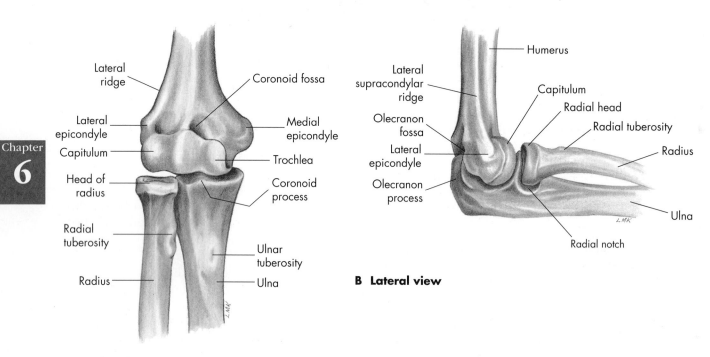

A Anterior view

B Lateral view

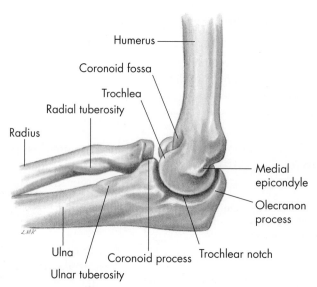

C Medial view

FIG. 6.1 • Right elbow joint. **A,** Anterior view; **B,** Lateral view; **C,** Medial view.

primarily involve movement between the articular surfaces of the humerus and ulna—specifically, the humeral trochlear fitting into the trochlear notch of the ulna. The head of the radius has a relatively small amount of contact with the capitulum of the humerus at the radiocapitellar joint. As the elbow reaches full extension, the olecranon process of the ulna is received by the olecranon fossa of the humerus. This arrangement provides increased joint stability when the elbow is fully extended.

As the elbow flexes approximately 20 degrees or more, its bony stability is somewhat unlocked, allowing for more side-to-side laxity. The stability of the elbow in flexion is more dependent on the collateral ligaments, such as the lateral or radial collateral ligament and especially the medial or ulnar collateral ligament (Fig. 6.3). The ulnar collateral ligament is critical to providing medial support to prevent the elbow from abducting (not a normal movement of the elbow) when stressed in physical activity. Many contact sports, particularly sports with throwing activities, place stress on the medial aspect of the joint, resulting in injury. Often this injury involves either acute or chronic stress to the ulnar collateral ligament, or UCL,

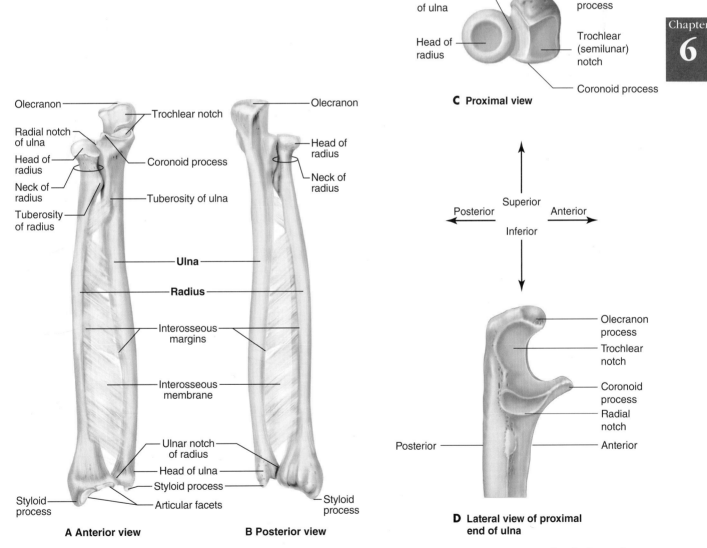

FIG. 6.2 ● Right radioulnar joint in supination. **A,** Anterior view; **B,** Posterior view; **C,** Proximal view of radioulnar joint; **D,** Lateral view of proximal end of ulna.

resulting in partial to complete tears to it. The UCL is particularly crucial to those high-velocity sporting activities, such as baseball pitching, that require optimal stability of the medial elbow. Even moderate injury to this structure can seriously impact an athlete's ability to throw at the highest levels. Compromise of this structure often requires surgery using a tendon graft such as the palmaris longus tendon to reconstruct this ligament. This surgery, often referred to as the "Tommy John procedure," is particularly common among high school, collegiate, and professional pitchers. The radial collateral and lateral ulnar collateral ligaments on the opposite side provide lateral stability and are rarely injured. Additionally, the annular ligament is located laterally, providing a sling effect around the radial head to secure its stability.

In the anatomical position, it is common for the forearm to deviate laterally from the arm from 5 to 15 degrees. This is referred to as the **carrying angle** and permits the forearms to clear the hips in the swinging movements during walking and also is important when carrying objects. Typically, the angle is slightly greater in the dominant limb than in the nondominant limb. It is also common for females to have a slightly greater carrying angle than men (Fig. 6.4).

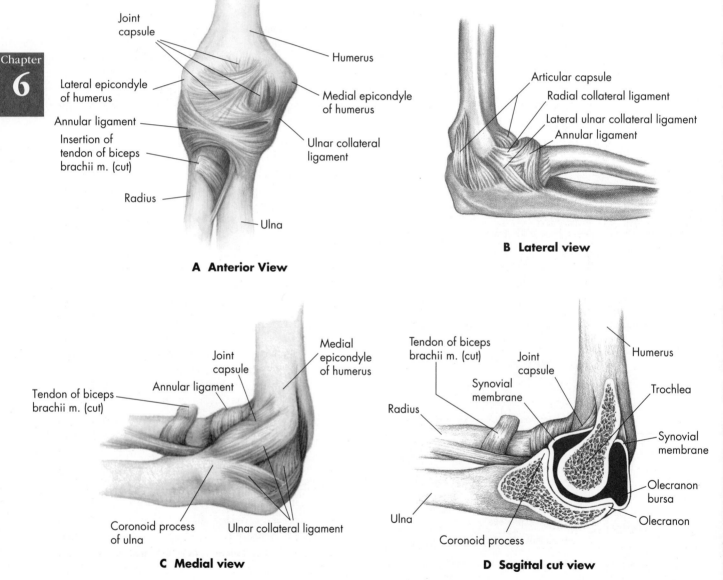

FIG. 6.3 • Right elbow with ligaments detailed. **A,** Anterior view; **B,** Lateral view; **C,** Medial view; **D,** Sagittal cut view.

The elbow is capable of moving from 0 degrees of extension to approximately 145 to 150 degrees of flexion, as detailed in Fig. 6.5. Some people, more commonly females, may hyperextend the elbow up to approximately 15 degrees.

The radioulnar joint is classified as a trochoid or pivot-type joint. The radial head rotates around in its location at the proximal ulna. This rotary movement is accompanied by rotation of the distal radius around the distal ulna. The radial head is maintained in its joint by the annular ligament. The radioulnar joint can supinate approximately 80 to 90 degrees from the neutral position. Pronation varies from 70 to 90 degrees (Fig. 6.6).

Due to the radius and ulna being held tightly together between the proximal and distal articulations by an interosseus membrane, the joint between the shafts of these bones is often referred

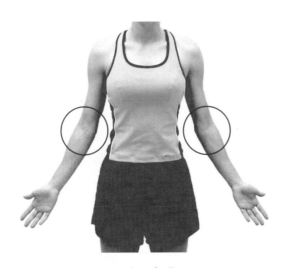

FIG. 6.4 • Carrying angle of elbow.

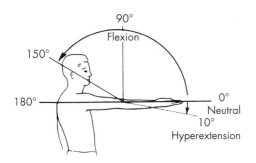

FIG. 6.5 • ROM of the elbow: flexion, extension, and hyperextension. *Flexion:* zero to 150 degrees. *Extension:* 150 degrees to zero. *Hyperextension:* measured in degrees beyond the zero starting point. This motion is not present in all persons. When it is present, it may vary from 5 to 15 degrees.

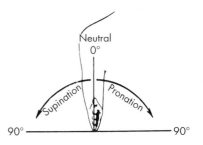

FIG. 6.6 • ROM of the forearm: pronation and supination. *Pronation:* zero to 80 or 90 degrees. *Supination:* zero to 80 or 90 degrees. *Total forearm motion:* 160 to 180 degrees. Persons may vary in the range of supination and pronation. Some may reach the 90-degree arc, and others may have only 70 degrees plus.

to as a syndesmosis type of joint. This interosseus membrane is helpful in absorbing and transmitting forces received by the hand, particularly during upper-extremity weight bearing. There is substantial rotary motion between the bones despite this classification.

Even though the elbow and radioulnar joints can and do function independently of each other, the muscles controlling each work together in synergy to perform actions at both to benefit the total function of the upper extremity. For this reason, dysfunction at one joint may affect normal function at the other.

Synergy among the glenohumeral, elbow, and radioulnar joint muscles

Just as there is synergy between the shoulder girdle and the shoulder joint in accomplishing upper-extremity activities, there is also synergy between the glenohumeral joint and the elbow joint as well as the radioulnar joints.

As the radioulnar joint goes through its ranges of motion, the glenohumeral and elbow muscles contract to stabilize or assist in the effectiveness of movement at the radioulnar joints. For example, when attempting to fully tighten (with the right hand) a screw with a screwdriver that involves radioulnar supination, we tend to externally rotate and flex the glenohumeral and elbow joints, respectively. Conversely, when attempting to loosen a tight screw with pronation, we tend to internally rotate and extend the elbow and glenohumeral joints, respectively. In either case, we depend on both the agonists and the antagonists in the surrounding joints to provide an appropriate amount of stabilization and assistance with the required task.

Movements FIGS. 6.5, 6.6, 6.7, 6.8, 6.9

Elbow movements

Flexion: Movement of the forearm to the shoulder by bending the elbow to decrease its angle

Extension: Movement of the forearm away from the shoulder by straightening the elbow to increase its angle

Radioulnar joint movements

Pronation: Internal rotary movement of the radius on the ulna that results in the hand moving from the palm-up to the palm-down position

Supination: External rotary movement of the radius on the ulna that results in the hand moving from the palm-down to the palm-up position

Elbow
flexion

Elbow
extension

Radioulnar
pronation

Radioulnar
supination

Flexion

A

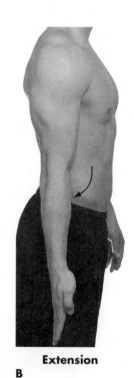

Extension

B

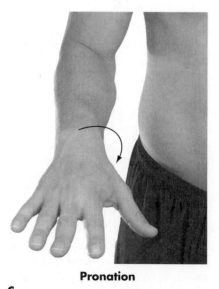

Pronation

C

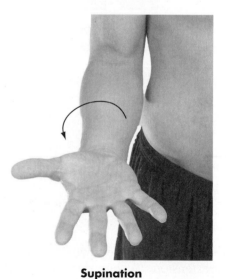

Supination

D

FIG. 6.7 • Movements of the elbow and radioulnar joint. **A,** Elbow flexion; **B,** Elbow extension; **C,** Radioulnar pronation; **D,** Radioulnar supination.

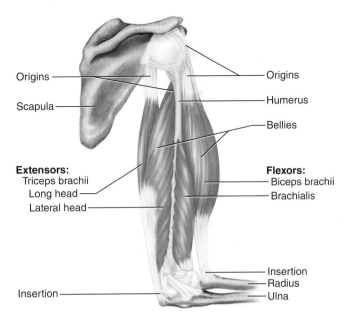

FIG. 6.8 ● Right elbow, lateral view of flexors and extensors.

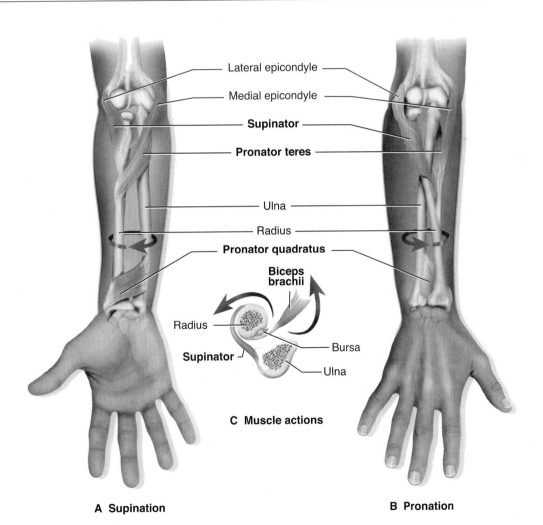

FIG. 6.9 ● Actions of the rotator muscles of the forearm. **A,** Supination; **B,** Pronation, **C,** Cross section just distal to the elbow, showing how the biceps brachii aids the supinator.

Muscles

The muscles of the elbow and radioulnar joints may be more clearly understood when separated by function. The elbow flexors, located anteriorly, are the biceps brachii, the brachialis, and the brachioradialis, with some weak assistance from the pronator teres (Figs. 6.10, 6.11, and 6.14).

The triceps brachii, located posteriorly, is the primary elbow extensor, with assistance provided by the anconeus (Figs. 6.12, 6.13, and 6.14). The pronator group, located anteriorly, consists of the pronator teres, the pronator quadratus, and the brachioradialis. The brachioradialis also assists with supination, which is controlled mainly by the supinator muscle and the biceps brachii. The supinator muscle is located posteriorly. See Table 6.1.

A common problem associated with the muscles of the elbow is "tennis elbow," which usually involves the extensor digitorum muscle near its origin on the lateral epicondyle. This condition, known technically as **lateral epicondylitis**, is quite frequently associated with gripping and lifting activities. More recently, depending upon the specific pathology this condition may be termed lateral epicondylagia or lateral epicondylosis. **Medial epicondylitis**, a somewhat less common problem frequently referred to as "golfer's elbow," is associated with the wrist flexor and pronator group near their origin on the medial epicondyle. Both of these conditions involve muscles that cross the elbow but act primarily on the wrist and hand. These muscles will be addressed in Chapter 7.

Elbow and radioulnar joint muscles—location

Anterior
    Primarily flexion and pronation
        Biceps brachii
        Brachialis
        Brachioradialis
        Pronator teres
        Pronator quadratus

Posterior
    Primarily extension and supination
        Triceps brachii
        Anconeus
        Supinator

FIG. 6.10 • Anterior upper-extremity muscles.

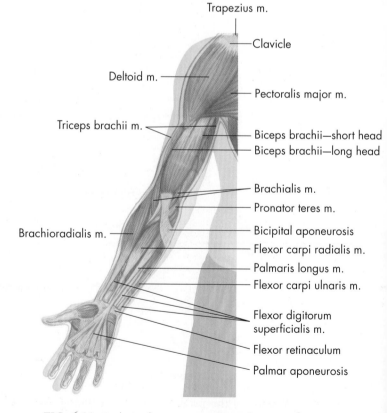

FIG. 6.11 • Anterior upper-extremity muscles.

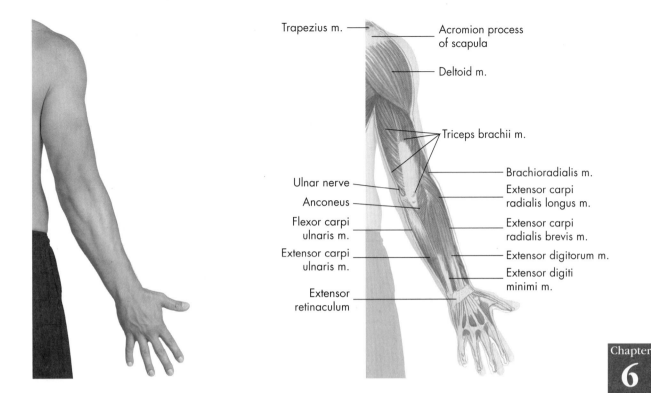

FIG. 6.12 • Posterior upper-extremity muscles.

FIG. 6.13 • Posterior upper-extremity muscles.

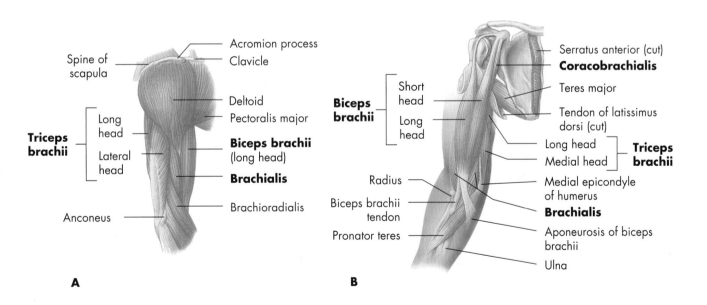

A

B

FIG. 6.14 • Muscles of the arm. **A,** Lateral view of the right shoulder and arm; **B,** Anterior view of the right shoulder and arm (deep). Deltoid, pectoralis major, and pectoralis minor muscles removed to reveal deeper structures.

TABLE 6.1 • Agonist muscles of the elbow and radioulnar joints

	Muscle	Origin	Insertion	Action	Plane of motion	Palpation	Innervation
Anterior muscles (primarily flexors and pronators)	Biceps brachii long head	Supraglenoid tubercle above the superior lip of the glenoid fossa	Tuberosity of the radius and bicipital aponeurosis (lacertus fibrosis)	Supination of the forearm	Transverse	Easily palpated on the anterior humerus; the long head and short head tendons may be palpated in the intertubercular groove and just inferomedial to the coracoid process, respectively; distally, the biceps tendon is palpated just anteromedial to the elbow joint during supination and flexion	Musculocutaneous nerve (C5, C6)
				Flexion of the elbow	Sagittal		
				Weak flexion of the shoulder joint			
				Weak abduction of the shoulder joint	Frontal		
	Biceps brachii short head	Coracoid process of the scapula and upper lip of the glenoid fossa in conjunction with the proximal attachment of the coracobrachialis		Supination of the forearm	Transverse		
				Flexion of the elbow	Sagittal		
				Weak flexion of the shoulder joint			
				Weak abduction of the shoulder joint	Frontal		
	Brachialis	Distal half of the anterior shaft of the humerus	Coronoid process of the ulna	Flexion of the elbow	Sagittal	Deep on either side of the biceps tendon during flexion/extension with forearm in partial pronation; lateral margin may be palpated between biceps brachii and triceps brachii; belly may be palpated through biceps brachii when forearm is in pronation during light flexion	Musculocutaneous nerve (C5, C6)
	Brachioradialis	Distal 2/3 of the lateral condyloid (supracondylar) ridge of the humerus	Lateral surface of the distal end of the radius at the styloid process	Flexion of the elbow	Sagittal	Anterolaterally on the proximal forearm during resisted elbow flexion with the radioulnar joint positioned in neutral	Radial nerve (C5, C6)
				Pronation from supination to neutral	Transverse		
				Supination from pronation to neutral			
	Pronator teres	Distal part of the medial condyloid ridge of the humerus and medial side of the proximal ulna	Middle 1/3 of the lateral surface of the radius	Pronation of the forearm	Transverse	Anteromedial surface of the proximal forearm during resisted mid- to full pronation	Median nerve (C6, C7)
				Weak flexion of the elbow	Sagittal		
	Pronator quadratus	Distal 1/4 of the anterior side of the ulna	Distal 1/4 of the anterior side of the radius	Pronation of the forearm	Transverse	Very deep and difficult to palpate, but with the forearm in supination, palpate immediately on either side of the radial pulse with resisted pronation	Median nerve (C6, C7)

TABLE 6.1 (continued) • Agonist muscles of the elbow and radioulnar joints

Muscle		Origin	Insertion	Action	Plane of motion	Palpation	Innervation
Posterior muscles (primarily extensors and supinators)	Triceps brachii long head	Infraglenoid tubercle below inferior lip of glenoid fossa of the scapula	Olecranon process of the ulna	Extension of the elbow joint	Sagittal	Proximally as a tendon on the posteromedial arm to underneath the posterior deltoid during resisted shoulder extension/ adduction	Radial nerve (C7, C8)
				Extension of the shoulder joint			
				Adduction of the shoulder joint	Frontal		
				Horizontal abduction of the shoulder joint	Transverse		
	Triceps brachii lateral head	Upper half of the posterior surface of the humerus		Extension of the elbow joint	Sagittal	Easily palpated on the proximal 2/3 of the posterior humerus during resisted extension	
	Triceps brachii medial head	Distal 2/3 of the posterior surface of the humerus		Extension of the elbow joint		Deep head: medially and laterally just proximal to the medial and lateral epicondyles	
	Supinator	Lateral epicondyle of the humerus and neighboring posterior part of the ulna	Lateral surface of the proximal radius just below the head	Supination of the forearm	Transverse	Position elbow and forearm in relaxed flexion and pronation, respectively; palpate deep to the brachioradialis, extensor carpi radialis longus, extensor carpi radialis brevis on the lateral aspect of the proximal radius with slight resistance to supination	Radial nerve (C6)
	Anconeus	Posterior surface of the lateral condyle of the humerus	Posterior surface of the lateral olecranon process and proximal 1/4 of the ulna	Extension of the elbow	Sagittal	Posterolateral aspect of the proximal ulna to the olecranon process during resisted extension of the elbow with the wrist in flexion	Radial nerve (C7, C8)

Note: The flexor carpi radialis, palmaris longus, flexor carpi ulnaris, and flexor digitorum superficialis assist in weak flexion of the elbow, while the extensor carpi ulnaris, extensor carpi radialis brevis, extensor carpi radialis longus, and extensor digitorum assist in weak extension of the elbow. Because they are covered in Chapter 7, they are not listed above.

Chapter

6

Nerves   FIGS. 5.13, 6.15, 6.16

The muscles of the elbow and radioulnar joints are all innervated from the median, musculocutaneous, and radial nerves of the brachial plexus. The radial nerve, originating from C5, C6, C7, and C8, provides innervation for the triceps brachii, brachioradialis, supinator, and anconeus (Fig. 6.15). More specifically, the posterior interosseous nerve, derived from the radial nerve, supplies the supinator. The radial nerve also provides sensation to the posterolateral arm, forearm, and hand. The median nerve, illustrated in Fig. 6.16, innervates the pronator teres and further branches to become the anterior interosseus nerve, which supplies the pronator quadratus. The median nerve's most important related derivations are from C6 and C7. It provides sensation to the palmar aspect of the hand and first three phalanges. The palmar aspect of the radial side of the fourth finger is also provided sensation, along with the dorsal aspect of the index and long fingers. The musculocutaneous nerve, shown in Fig. 5.13, branches from C5 and C6 and supplies the biceps brachii and brachialis. Because there is no innervation to the muscles discussed in this chapter from the ulnar nerve, it is not addressed here, but it is often injured at the elbow in various ways. See Chapter 7 for some discussion related to this nerve and injuries to it.

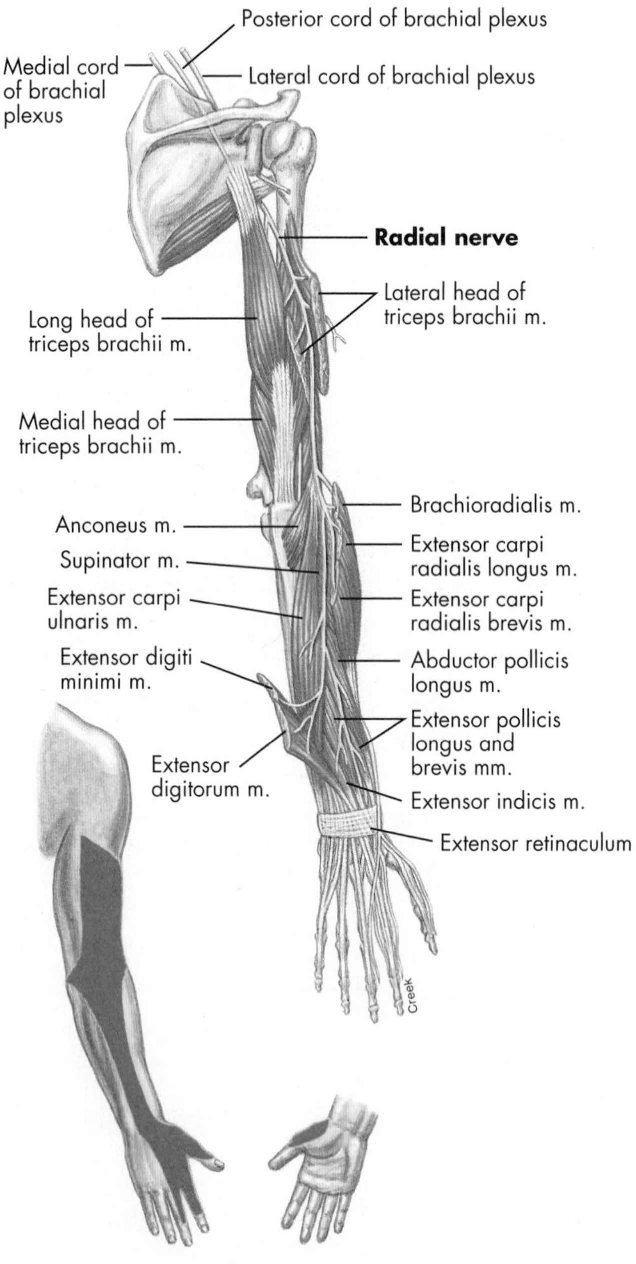

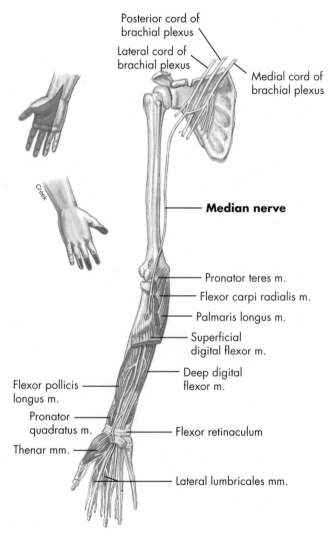

FIG. 6.15 • Muscular and cutaneous distribution of the radial nerve.

FIG. 6.16 • Muscular and cutaneous distribution of the median nerve.

Biceps brachii muscle   FIG. 6.17

(bi´seps bra´ki-i)

Origin

Long head: supraglenoid tubercle above the
  superior lip of the glenoid fossa
Short head: coracoid process of the scapula and
  upper lip of the glenoid fossa in conjunction with
  the proximal attachment of the coracobrachialis

Insertion

Tuberosity of the radius and bicipital aponeurosis
  (lacertus fibrosis)

Action

Flexion of the elbow
Supination of the forearm
Weak flexion of the shoulder joint
Weak abduction of the shoulder joint when the
  shoulder joint is in external rotation

Palpation

Easily palpated on the anterior humerus. The long
  head and short head tendons may be palpated in
  the intertubercular groove and just inferior to the
  coracoid process, respectively. Distally, the biceps
  tendon is palpated just anteromedial to the elbow
  joint during supination and flexion.

Innervation

Musculocutaneous nerve (C5, C6)

Application, strengthening, and flexibility

The biceps is commonly known as a two-joint
(shoulder and elbow), or biarticular, muscle.
However, technically it should be considered
a three-joint (multiarticular) muscle—shoulder,
elbow, and radioulnar. It is weak in actions of the
shoulder joint, although it does assist in providing
dynamic anterior stability to maintain the humeral
head in the glenoid fossa. It is more powerful in
flexing the elbow when the radioulnar joint is
supinated. It is also a strong supinator, particularly if the elbow is flexed. Palms away from the
face (pronation) decrease the effectiveness of the
biceps, partly as a result of the disadvantageous
pull of the muscle as the radius rotates. The same
muscles are used in elbow joint flexion, regardless of forearm pronation or supination.

Flexion of the forearm with a barbell in the
hands, known as "curling," is an excellent exercise to develop the biceps brachii. This movement
can be performed one arm at a time with dumb-
bells or both arms simultaneously with a barbell.
Other activities in which there is powerful flexion
of the forearm are chinning and rope climbing. See
Appendix 3 for more commonly used exercises for
the biceps brachii and other muscles in this chapter.

Due to the multiarticular orientation of the
biceps, all three joints must be positioned appropriately to achieve optimal stretching. The elbow
must be extended maximally with the shoulder in
full extension. The biceps may also be stretched
by beginning with full elbow extension and pro-
gressing into full horizontal abduction at approximately 70 to 110 degrees of shoulder abduction. In
all cases, the forearm should be fully pronated to
achieve maximal lengthening of the biceps brachii.

Elbow
flexion

Radioulnar
supination

Shoulder
flexion

Shoulder
abduction

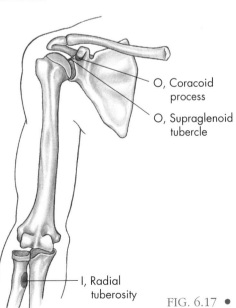

Long head
Short head } Biceps brachii

Flexion

I, Bicipital
aponeurosis

Supination

O, Coracoid
process

O, Supraglenoid
tubercle

I, Radial
tuberosity

FIG. 6.17 • Biceps brachii muscle,
anterior view. O, Origin; I, Insertion.

Elbow
flexion

Brachialis muscle FIG. 6.18

(braˊki-aˊlis)

Origin
Distal half of the anterior shaft of the humerus

Insertion
Coronoid process of the ulna

Action
True flexion of the elbow

Palpation
Deep on either side of the biceps tendon during flexion/extension with forearm in partial pronation. The lateral margin may be palpated between the biceps brachii and the triceps brachii, and the belly may be palpated through the biceps brachii when the forearm is in pronation during light flexion.

Innervation
Musculocutaneous nerve and sometimes branches from radial and median nerves (C5, C6)

Application, strengthening, and flexibility

The brachialis muscle is used along with other flexor muscles, regardless of pronation or supination. It pulls on the ulna, which does not rotate, thus making this muscle the only pure flexor of this joint.

The brachialis muscle is called into action whenever the elbow flexes. It is exercised along with elbow curling exercises, as described for the biceps brachii, pronator teres, and brachioradialis muscles. Elbow flexion activities with the forearm pronated isolate the brachialis to some extent by reducing the effectiveness of the biceps brachii. Since the brachialis is a pure flexor of the elbow, it can be stretched maximally only by extending the elbow with the shoulder relaxed and flexed. Forearm positioning should not affect the stretch on the brachialis unless the forearm musculature itself limits elbow extension, in which case the forearm is probably best positioned in neutral.

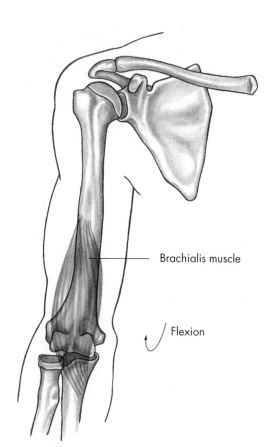

Brachialis muscle

Flexion

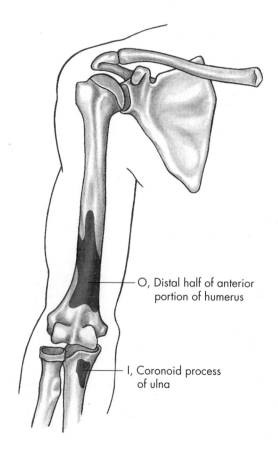

O, Distal half of anterior portion of humerus

I, Coronoid process of ulna

FIG. 6.18 ● Brachialis muscle, anterior view. O, Origin; I, Insertion.

Brachioradialis muscle   FIG. 6.19

(bra´ki-o-ra´di-a´lis)

Origin

Distal two-thirds of the lateral condyloid (supracondylar) ridge of the humerus

Insertion

Lateral surface of the distal end of the radius at the styloid process

Action

Flexion of the elbow

Pronation of the forearm from supinated position to neutral

Supination of the forearm from pronated position to neutral

Palpation

Anterolaterally on the proximal forearm during resisted elbow flexion with the radioulnar joint positioned in neutral

Innervation

Radial nerve (C5, C6)

Application, strengthening, and flexibility

The brachioradialis is one of three muscles, sometimes known as the mobile wad of three, on the lateral forearm. The other two muscles are the extensor carpi radialis brevis and extensor carpi radialis longus, to which it lies directly anterior. The brachioradialis muscle acts as a flexor best in a midposition or neutral position between pronation and supination. In a supinated position of the forearm, it tends to pronate as it flexes. In a pronated position, it tends to supinate as it flexes. This muscle is favored in its action of flexion when the neutral position between pronation and supination is assumed, as previously suggested. Its insertion at the end of the radius makes it a strong elbow flexor. Its ability as a supinator decreases as the radioulnar joint moves toward neutral. Similarly, its ability to pronate decreases as the forearm reaches neutral. Because of its action of rotating the forearm to a neutral thumb-up position, it is referred to as the hitchhiker muscle, although it has no action at the thumb. As you will see in Chapter 7, nearly all the muscles originating off the lateral epicondyle have some action as weak elbow extensors. This is not the case with the brachioradialis, due to its line of pull being anterior to the elbow's axis of rotation.

The brachioradialis may be strengthened by performing elbow curls against resistance, particularly with the radioulnar joint in the neutral position. In addition, the brachioradialis may be developed by performing pronation and supination movements through the full range of motion against resistance.

The brachioradialis is stretched by maximally extending the elbow with the shoulder in flexion and the forearm in either maximal pronation or maximal supination.

Elbow flexion

Radioulnar pronation

Chapter **6**

Radioulnar supination

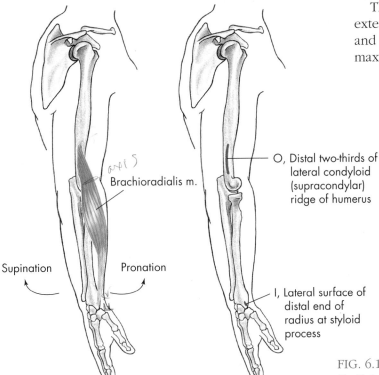

Brachioradialis m.

Supination   Pronation

O, Distal two-thirds of lateral condyloid (supracondylar) ridge of humerus

I, Lateral surface of distal end of radius at styloid process

FIG. 6.19 • Brachioradialis muscle, lateral view. O, Origin; I, Insertion.

Triceps brachii muscle FIG. 6.20

(tri´seps bra´ki-i)

Elbow
extension

Shoulder
extension

Shoulder
adduction

Shoulder
horizontal
abduction

Origin

Long head: infraglenoid tubercle below inferior lip of glenoid fossa of the scapula

Lateral head: upper half of the posterior surface of the humerus

Medial head: distal two-thirds of the posterior surface of the humerus

Insertion

Olecranon process of the ulna

Action

All heads: extension of the elbow

Long head: extension, adduction, and horizonal abduction of the shoulder joint

Palpation

Posterior arm during resisted extension from a flexed position and distally just proximal to its insertion on the olecranon process

Long head: proximally as a tendon on the postero-medial arm to underneath the posterior deltoid during resisted shoulder extension/adduction

Lateral head: easily palpated on the proximal two-thirds of the posterior humerus

Medial head (deep head): medially and laterally just proximal to the medial and lateral epicondyles

Innervation

Radial nerve (C7, C8)

Application, strengthening, and flexibility

Typical action of the triceps brachii is shown in push-ups when there is powerful extension of the elbow. It is used in hand balancing and in any pushing movement involving the upper extremity. The long head is an important extensor of the shoulder joint.

Two muscles extend the elbow—the triceps brachii and the anconeus. Push-ups demand strenuous contraction of these muscles. Dips on the parallel bars are more difficult to perform. Bench pressing a barbell or a dumbbell is an excellent exercise. Overhead presses and triceps curls (elbow extensions from an overhead position) emphasize the triceps.

The triceps brachii should be stretched with both the shoulder and the elbow in maximal flexion.

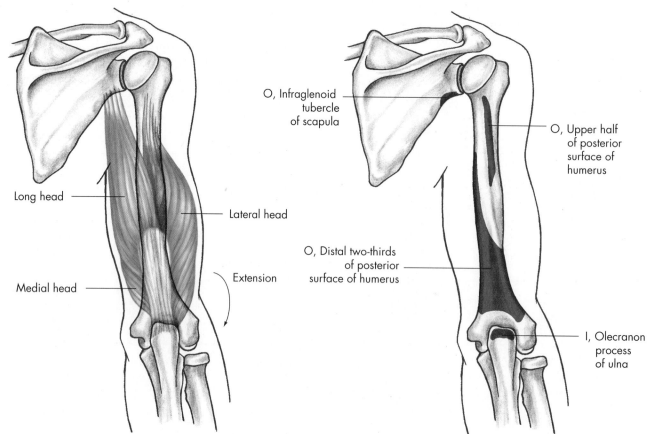

FIG. 6.20 • Triceps brachii muscle, posterior view. O, Origin; I, Insertion.

Anconeus muscle   FIG. 6.21
(an-ko´ne-us)

Origin
Posterior surface of the lateral condyle of the
humerus

Insertion
Posterior surface of the lateral olecranon process
and proximal one-fourth of the ulna

Action
Extension of the elbow

Palpation
Posterolateral aspect of the proximal ulna to the
olecranon process during resisted extension of the
elbow with the wrist in flexion

Innervation
Radial nerve (C7, C8)

Application, strengthening, and flexibility
The chief function of the anconeus muscle is to
pull the synovial membrane of the elbow joint out
of the way of the advancing olecranon process
during extension of the elbow. It contracts along
with the triceps brachii. It is strengthened with
any elbow extension exercise against resistance.
Maximal elbow flexion stretches the anconeus.

Elbow
extension

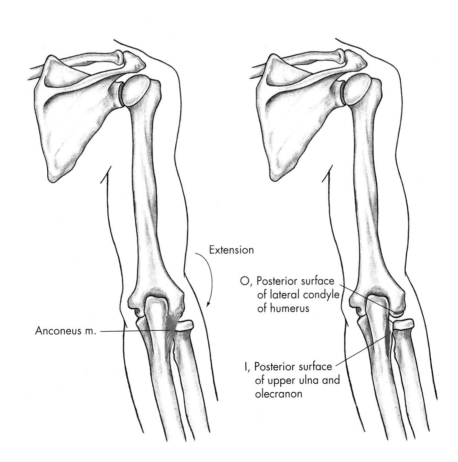

Extension

O, Posterior surface
of lateral condyle
of humerus

Anconeus m.

I, Posterior surface
of upper ulna and
olecranon

FIG. 6.21 ● Anconeus muscle, posterior view. O, Origin; I, Insertion.

Pronator teres muscle   FIG. 6.22

(pro-na´tor te´rez)

Origin

Distal part of the medial condyloid ridge of the
humerus and medial side of the proximal ulna

Insertion

Middle third of the lateral surface of the radius

Action

Pronation of the forearm
Weak flexion of the elbow

Palpation

Anteromedial surface of the proximal forearm
during resisted mid to full pronation

Innervation

Median nerve (C6, C7)

Application, strengthening, and flexibility

Typical movement of the pronator teres muscle is
with the forearm pronating as the elbow flexes.
Movement is weaker in flexion with supination.
The use of the pronator teres alone in movement
tends to bring the back of the hand to the face
as it contracts. Pronation of the forearm with a
dumbbell in the hand localizes action and develops the pronator teres muscle. Strengthening this
muscle begins with holding a hammer in the hand
with the hammer head suspended from the ulnar
side of the hand while the forearm is supported
on a desk or table. The hammer should be hanging toward the floor, with the forearm pronated to
the palm-down position.

The elbow must be fully extended while taking the forearm into full supination to stretch the
pronator teres.

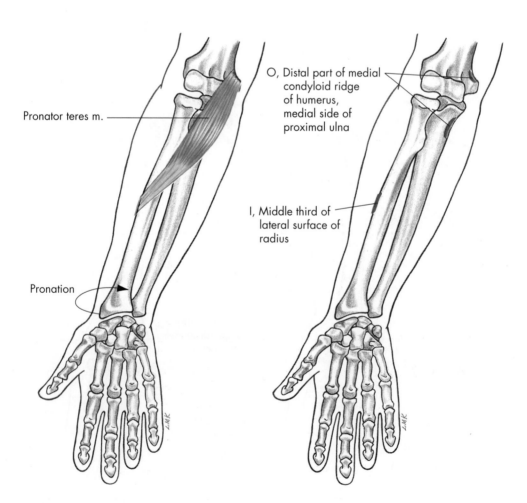

Pronator teres m.

O, Distal part of medial
condyloid ridge
of humerus,
medial side of
proximal ulna

I, Middle third of
lateral surface of
radius

Pronation

FIG. 6.22 • Pronator teres muscle, anterior view. O, Origin; I, Insertion.

Pronator quadratus muscle FIG. 6.23

(pro-na'tor kwad-ra'tus)

Origin

Distal fourth of the anterior side of the ulna

Insertion

Distal fourth of the anterior side of the radius

Action

Pronation of the forearm

Palpation

The pronator quadratus, because of its proximity and appearance in some anatomical drawings, is sometimes confused with the flexor retinaculum. See Fig. 6.16. It is very deep and difficult to palpate, but with the forearm in supination it may be palpated immediately on either side of the radial pulse with resisted pronation.

Innervation

Median nerve (palmar interosseous branch) (C6, C7)

Application, strengthening, and flexibility

The pronator quadratus muscle works in pronating the forearm in combination with the triceps in extending the elbow. It is commonly used in turning a screwdriver, as in taking out a screw (with the right hand), when extension and pronation are needed. It is used also in throwing a screwball, when extension and pronation are needed. It may be developed with similar pronation exercises against resistance, as described for the pronator teres. The pronator quadratus is best stretched by using a partner to grasp the wrist and passively take the forearm into extreme supination.

Radioulnar pronation

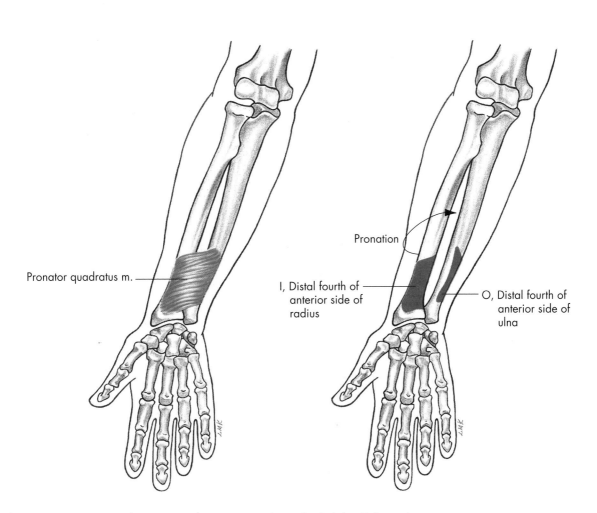

Pronator quadratus m.

Pronation

I, Distal fourth of anterior side of radius

O, Distal fourth of anterior side of ulna

FIG. 6.23 ● Pronator quadratus muscle, anterior view. O, Origin; I, Insertion.

Supinator muscle FIG. 6.24

(su´pi-na´tor)

Radioulnar
supination

Origin

Lateral epicondyle of the humerus and neighboring
posterior part of the ulna

Insertion

Lateral surface of the proximal radius just below the
head

Action

Supination of the forearm

Palpation

Position the elbow and forearm in relaxed flexion
and pronation, respectively, and palpate deep to
the brachioradialis, extensor carpi radialis longus,
extensor carpi radialis brevis on the lateral aspect
of the proximal radius with slight resistance to
supination

Innervation

Radial nerve (C6)

Chapter 6

Application, strengthening, and flexibility

The supinator muscle is called into play when
movements of extension and supination are
required, as when turning a screwdriver. The
curve in throwing a baseball calls this muscle
into play as the elbow is extended just before ball
release. It is most isolated in activities that require
supination with elbow extension, because the
biceps brachii assist with supination most when
the elbow is flexed.

The hands should be grasped and the forearms
extended, in an attempt to supinate the forearms
against the grip of the hands. This localizes, to a
degree, the action of the supinator.

The hammer exercise used for the pronator
teres muscle may be modified to develop the
supinator. In the beginning, the forearm is supported and the hand is free off the table edge.
The hammer is again held suspended out of the
ulnar side of the hand hanging toward the floor.
The forearm is then supinated to the palm-up
position to strengthen this muscle.

The supinator is stretched when the forearm is
maximally pronated.

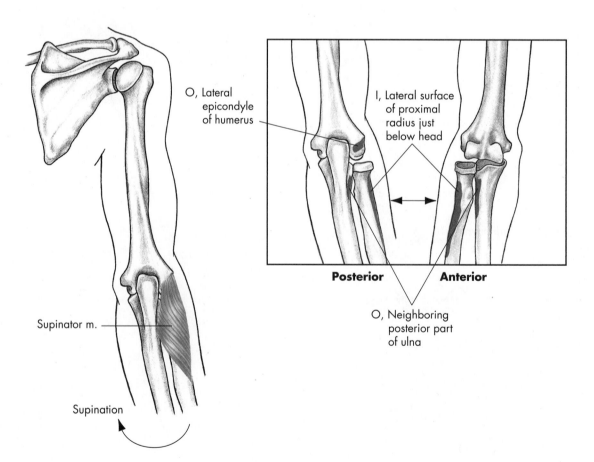

FIG. 6.24 • Supinator muscle, posterior view. O, Origin; I, Insertion.

REVIEW EXERCISES

1. List the planes in which each of the following elbow and radioulnar joint movements occurs. List the respective axis of rotation for each movement in each plane.
   a. Flexion
   b. Extension
   c. Pronation
   d. Supination
2. Discuss the difference between chinning with the palms toward the face and chinning with the palms away from the face. Consider this muscularly and anatomically.
3. Analyze and list the differences in elbow and radioulnar joint muscle activity between turning a doorknob clockwise and pushing the door open, and turning the knob counterclockwise, and pulling the door open.
4. The elbow joint is the distal insertion of which biarticular muscles? Describe what motions each is involved in at the elbow and the superior joint of muscular origin.
5. Lifting a television set as you help your roommate move in requires appropriate lifting techniques and an effective angle of pull. Describe the angle of pull chosen at the elbow joint and why it is chosen instead of other angles.
6. List the muscles involved with "tennis elbow" and describe specifically how you would encourage someone to work on both the strength and the flexibility of these muscles.

7. **Muscle analysis chart** • Elbow and radioulnar joints

Fill in the chart by listing the muscles primarily involved in each movement.	
Flexion	Extension
Pronation	Supination

8. **Antagonistic muscle action chart** • Elbow and radioulnar joints

Fill in the chart by listing the muscle(s) or parts of muscles that are antagonist in their actions to the muscles in the left column.

Agonist	Antagonist
Biceps brachii	
Brachioradialis	
Brachialis	
Pronator teres	
Supinator	
Triceps brachii	
Anconeus	

LABORATORY EXERCISES

1. Locate the following parts of the humerus, radius, and ulna on a human skeleton and on a subject:
   a. Skeleton
      1. Medial epicondyle
      2. Lateral epicondyle
      3. Lateral supracondylar ridge
      4. Trochlea
      5. Capitulum
      6. Olecranon fossa
      7. Olecranon process
      8. Coronoid process
      9. Coronoid fossa
      10. Tuberosity of the radius
      11. Ulnar tuberosity
   b. Subject
      1. Medial epicondyle
      2. Lateral epicondyle
      3. Lateral supracondylar ridge
      4. Proximal radioulnar joint
      5. Radiocapitellar joint
      6. Olecranon process
      7. Olecranon fossa

2. How and where can the following muscles be palpated on a human subject?
   a. Biceps brachii
   b. Brachioradialis
   c. Brachialis
   d. Pronator teres
   e. Supinator
   f. Triceps brachii
   g. Anconeus

3. Palpate and list the muscles primarily responsible for the following movements as you demonstrate each:
   a. Flexion
   b. Extension
   c. Pronation
   d. Supination

Chapter 6

4. Elbow and radioulnar joint movement analysis chart

After analyzing each exercise in the chart, break each into two primary movement phases, such as a lifting phase and a lowering phase. For each phase, determine the elbow and radioulnar joint movements occurring, and then list the elbow and radioulnar joint muscles primarily responsible for causing/controlling those movements. Beside each muscle in each movement, indicate the type of contraction as follows: I—isometric; C—concentric; E—eccentric.

Exercise	Initial movement (lifting) phase		Secondary movement (lowering) phase	
	Movement(s)	Agonist(s)—(contraction type)	Movement(s)	Agonist(s)—(contraction type)
Push-up				
Chin-up				
Bench press				
Dip				
Lat pull				
Overhead press				
Prone row				

5. Elbow and radioulnar joint sport skill analysis chart

Analyze each skill in the chart and list the movements of the right and left elbow and radioulnar joints in each phase of the skill. You may prefer to list the initial positions that the elbow and radioulnar joints are in for the stance phase. After each movement, list the elbow and radioulnar joint muscle(s) primarily responsible for causing/controlling that movement. Beside each muscle in each movement, indicate the type of contraction as follows: I—isometric; C—concentric; E—eccentric. It may be desirable to review the concepts for analysis in Chapter 8 for the various phases.

Exercise		Stance phase	Preparatory phase	Movement phase	Follow-through phase
Baseball pitch	(R)				
	(L)				
Volleyball serve	(R)				
	(L)				
Tennis serve	(R)				
	(L)				
Softball pitch	(R)				
	(L)				
Tennis backhand	(R)				
	(L)				
Batting	(R)				
	(L)				
Bowling	(R)				
	(L)				
Basketball free throw	(R)				
	(L)				

Chapter
6

References

Andrews JR, Zarins B, Wilk KE: *Injuries in baseball,* Philadelphia, 1998, Lippincott-Raven.

Back BR Jr, et al: Triceps rupture: a case report and literature review, *American Journal of Sports Medicine* 15:285, May–June 1987.

Gabbard CP, et al: Effects of grip and forearm position on flex arm hang performance, *Research Quarterly for Exercise and Sport,* July 1983.

Guarantors of Brain: *Aids to the examination of the peripheral nervous system,* ed 4, London, 2000, Saunders.

Herrick RT, Herrick S: Ruptured triceps in powerlifter presenting as cubital tunnel syndrome—a case report, *American Journal of Sports Medicine* 15:514, September–October 1987.

Hislop HJ, Montgomery J: *Daniels and Worthingham's muscle testing: techniques of manual examination,* ed 8, Philadelphia, 2007, Saunders.

Loftice JW, Fleisig GS, Wilk KE, Reinold MM, Chmielewski T, Escamilla RF, Andrews JR (eds): *Conditioning program for baseball pitchers,* Birmingham, 2004, American Sports Medicine Institute.

Magee DJ: *Orthopedic physical assessment,* ed 5, Philadelphia, 2008, Saunders.

Muscolino JE: *The muscular system manual: the skeletal muscles of the human body,* ed 3, St. Louis, 2010, Elsevier Mosby.

Oatis CA: *Kinesiology: the mechanics and pathomechanics of human movement,* ed 2, Philadelphia, 2008, Lippincott Williams & Wilkins.

Rasch PJ: *Kinesiology and applied anatomy,* ed 7, Philadelphia, 1989, Lea & Febiger.

Shier D, Butler J, Lewis R: *Hole's human anatomy & physiology,* ed 12, New York, 2010, McGraw-Hill.

Sieg KW, Adams SP: *Illustrated essentials of musculoskeletal anatomy,* ed 4, Gainesville, FL, 2002, Megabooks.

Sisto DJ, et al: An electromyographic analysis of the elbow in pitching, *American Journal of Sports Medicine* 15:260, May–June 1987.

Smith LK, Weiss EL, Lehmkuhl LD: *Brunnstrom's clinical kinesiology,* ed 5, Philadelphia, 1996, Davis.

Springer SI: Racquetball and elbow injuries, *National Racquetball* 16:7, March 1987.

Van De Graaff KM: *Human anatomy,* ed 6, Dubuque, IA, 2002, McGraw-Hill.

Van Roy P, Baeyens JP, Fauvart D, Lanssiers R, Clarijs JP: Arthrokinematics of the elbow: study of the carrying angle, *Ergonomics* 48(11–14):1645–1656, 2005.

Wilk KE, Reinold MM, Andrews JR (eds): *The athlete's shoulder,* ed 2, Philadelphia, 2009, Churchill Livingstone Elsevier.

Yilmaz E, Karakurt L, Belhan O, Bulut M, Serin E, Avci M: Variation of carrying angle with age, sex, and special reference to side, *Orthopedics* 28(11):1360–1363, 2005.

For additional resources and a list of related websites, visit **www.mhhe.com/floyd19e**.

Worksheet Exercises

For in- or out-of-class assignments, or for testing, utilize this tear-out worksheet.

Worksheet 1

Using crayons or colored markers, draw and label on the worksheet the following muscles. Indicate the origin and insertion of each muscle with an "O" and an "I," respectively.

a. Biceps brachii
b. Brachioradialis
c. Brachialis
d. Pronator teres

e. Supinator
f. Triceps brachii
g. Pronator quadratus
h. Anconeus

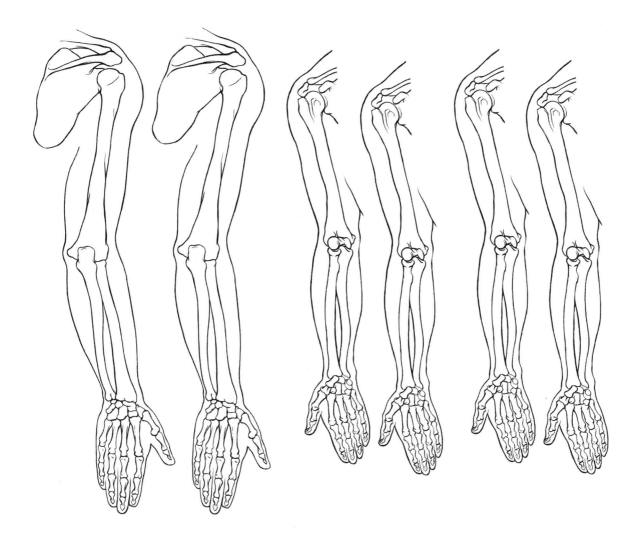

Worksheet Exercises

For in- or out-of-class assignments, or for testing, utilize this tear-out worksheet.

Worksheet 2

Label and indicate by arrows the following movements of the elbow and radioulnar joints. Then below, for each motion, list the agonist muscle(s), the plane in which the motion occurs, and its axis of rotation.

1. Elbow joints
   a. Flexion
   b. Extension

2. Radioulnar joints
   a. Pronation
   b. Supination

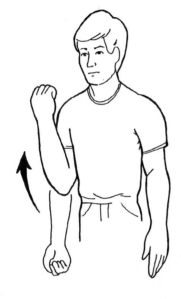
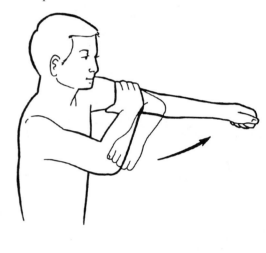

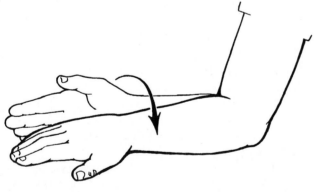
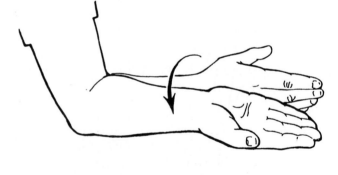

1. a. The _____ muscle(s) cause(s) _____ to occur in the

   _____ plane about the _____ axis.

   b. The _____ muscle(s) cause(s) _____ to occur in the

   _____ plane about the _____ axis.

2. a. The _____ muscle(s) cause(s) _____ to occur in the

   _____ plane about the _____ axis.

   b. The _____ muscle(s) cause(s) _____ to occur in the

   _____ plane about the _____ axis.

THE WRIST AND HAND JOINTS

Objectives

- To identify on a human skeleton selected bony features of the wrist, hand, and fingers

- To label selected bony features on a skeletal chart

- To draw and label the muscles on a skeletal chart

- To palpate the muscles on a human subject while demonstrating their actions

- To list the planes of motion and their respective axes of rotation

- To organize and list the muscles that produce the primary movements of the wrist, hand, and fingers

- To determine, through analysis, the wrist and hand movements and muscles involved in selected skills and exercises

Online Learning Center Resources

Visit *Manual of Structural Kinesiology*'s **Online Learning Center** at **www.mhhe.com/floyd19e** for additional information and study material for this chapter, including:

- *Self-grading quizzes*
- *Anatomy flashcards*
- *Animations*
- *Related websites*

The importance to us of the joints of the wrist, hand, and fingers is often overlooked in comparison with the larger joints needed for ambulation. This should not be the case, because even though the fine motor skills characteristic of this area are not essential in some sports, many sports with skilled activities require precise functioning of the wrist and hand. Several sports, such as archery, bowling, golf, baseball, and tennis, require the combined use of all these joints. Beyond this, appropriate function in the joints and muscles of our hands is critical for daily activities throughout our life.

Because of the numerous muscles, bones, and ligaments, along with relatively small joint size, the functional anatomy of the wrist and hand is complex and overwhelming to some. This complexity may be simplified by relating the functional anatomy to the major actions of the joints: flexion, extension, abduction, and adduction of the wrist and hand.

A large number of muscles are used in these movements. Anatomically and structurally, the human wrist and hand have highly developed, complex mechanisms capable of a variety of movements—a result of the arrangement of 29 bones, more than 25 joints, and more than 30 muscles, of which 18 are intrinsic muscles (both origin and insertion found in the hand).

For most who use this text, an extensive knowledge of these intrinsic muscles is not necessary. However, athletic trainers, physical therapists, occupational therapists, chiropractors, anatomists, physicians, and nurses require a more extensive knowledge. The intrinsic muscles are listed, illustrated, and discussed to a limited degree at the end of this chapter. References at the end of this chapter provide additional sources from which to gain further information.

Our discussion is limited to a review of the muscles, joints, and movements involved in gross motor activities. The muscles discussed are those of the forearm and the extrinsic muscles of the wrist, hand, and fingers. The larger, more important extrinsic muscles of each joint are included, providing a basic knowledge of this area. The prescription of exercises for strengthening these muscles will be somewhat redundant, as there are primarily only four movements accomplished by their combined actions. One exercise that will strengthen many of these muscles is fingertip push-ups.

Bones

The wrist and hand contain 29 bones, including the radius and ulna (Fig. 7.1). Eight carpal bones in two rows of four bones form the wrist. The proximal row consists, from the radial (thumb) side to the ulnar (little finger) side, of the scaphoid (boat-shaped) or navicular as it is commonly called, the lunate (moon-shaped), the triquetrum (three-cornered), and the pisiform (pea-shaped) bones. The distal row, from the radial to the ulnar side, consists of the trapezium (greater multangular), trapezoid (lesser multangular), capitate (head-shaped), and hamate (hooked) bones. These bones form a three-sided arch that is concave on the palmar side. This bony arch is spanned by the transverse carpal and volar carpal ligaments creating the **carpal tunnel**, which is frequently a source of problems known as carpal tunnel syndrome (see Fig. 7.8). Of these carpal bones, the

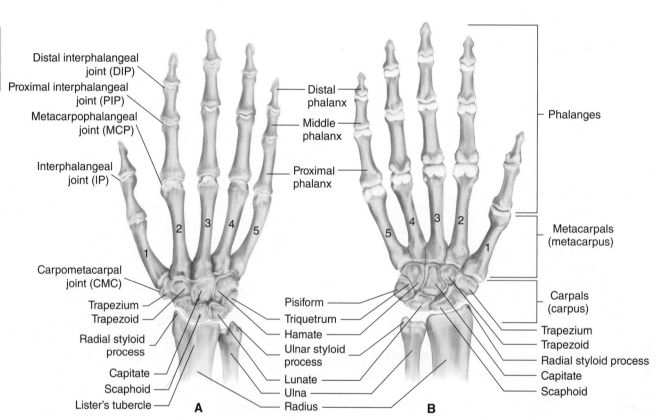

FIG. 7.1 • The right hand. **A,** Posterior (dorsal) view; **B,** Anterior (palmar) view.

scaphoid is by far the most commonly fractured, usually by severe wrist hyperextension from falling on the outstretched hand. Unfortunately, this particular fracture is often dismissed as a sprain after initial injury, only to cause significant problems in the long term if not properly treated. Treatment often requires precise immobilization for periods longer than that of many fractures and/or surgery. Five metacarpal bones, numbered one to five from the thumb to the little finger, join the wrist bones. There are 14 phalanxes (digits), three for each phalange except the thumb, which has only two. They are indicated as proximal, middle, and distal from the metacarpals. Additionally, the thumb has a sesamoid bone within its flexor tendon, and other sesamoids may occur in the fingers.

The medial epicondyle, medial condyloid ridge, and coronoid process serve as a point of origin for many of the wrist and finger flexors, whereas the lateral epicondyle and lateral supracondylar ridge serve as the point of origin for many extensors of the wrist and fingers (Figs. 6.1 and 6.3). Distally, the key bony landmarks for the muscles involved in wrist motion are the base of the second, third, and fifth metacarpals and the pisiform and hamate. The muscles of the fingers, which are also involved in wrist motion, insert on the base of the proximal, middle, and distal phalanxes (Figs. 7.1 and 7.2). The base of the first metacarpal and the proximal and distal phalanxes of the thumb serve as key insertion points for the muscles involved in thumb motion (Fig. 7.1). The hand consists of three distinct regions: the wrist, the palm, and the finger digits. The palm of the hand may be further separated into the thenar, hypothenar, and mid-palmar or intermediate regions.

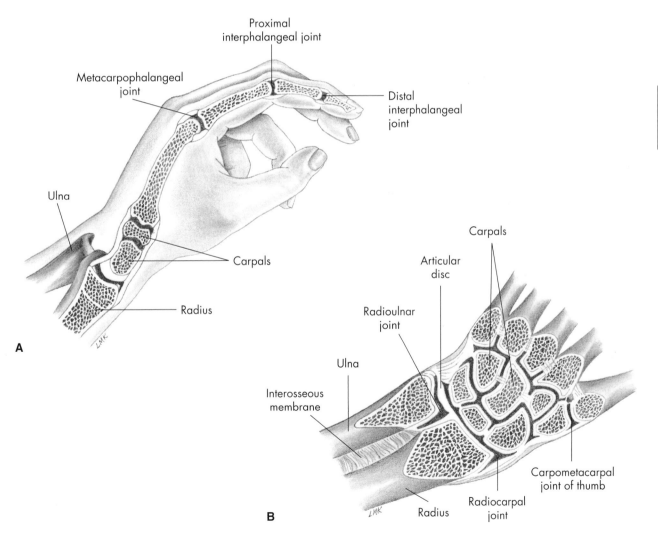

FIG. 7.2 ● Left wrist and hand joint structures. **A,** Medial view; **B,** Posterior view with frontal section through wrist.

Joints

The wrist joint is classified as a condyloid-type joint, allowing flexion, extension, abduction (radial deviation), and adduction (ulnar deviation) (Fig. 7.2). Wrist motion occurs primarily between the distal radius and the proximal carpal row, consisting of the scaphoid, lunate, and triquetrum. As a result, the wrist is often referred to as the radiocarpal joint. The joint allows 70 to 90 degrees of flexion and 65 to 85 degrees of extension. The wrist can abduct 15 to 25 degrees and adduct 25 to 40 degrees (Fig. 7.3).

Each finger has three joints. The metacarpophalangeal (MCP) joints are classified as condyloid. In these joints, 0 to 40 degrees of extension and 85 to 100 degrees of flexion are possible. The proximal interphalangeal (PIP) joints, classified as ginglymus, can move from full extension to approximately 90 to 120 degrees of flexion. The distal interphalangeal (DIP) joints, also classified as ginglymus, can flex 80 to 90 degrees from full extension (Fig. 7.4).

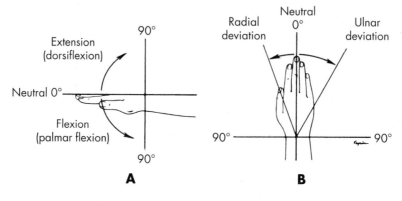

FIG. 7.3 • ROM of the wrist. **A,** Flexion and extension. *Flexion* (palmar flexion): zero to ± 80 degrees. *Extension* (dorsiflexion): zero to ± 70 degrees; **B,** Radial and ulnar deviation. *Radial deviation* zero to 20 degrees. *Ulnar deviation:* zero to 30 degrees.

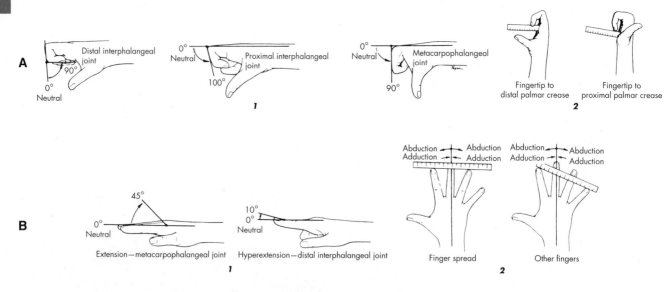

FIG. 7.4 • ROM of the fingers. **A,** Flexion. *1,* Motion can be estimated or measured in degrees. *2,* Motion can be estimated by a ruler as the distance from the tip of the finger to the distal palmar crease *(left)* (measures flexion of the middle and distal joints) and to the proximal palmar crease *(right)* (measures the distal, middle, and proximal joints of the fingers); **B,** Extension, abduction, and adduction. *1,* Extension and hyperextension. *2,* Abduction and adduction. These motions take place in the plane of the palm away from and to the long, or middle, finger of the hand. Abduction of the middle finger occurs when it moves laterally toward the thumb, and adduction occurs when it moves medially toward the little finger. The spread of fingers can be measured from the tip of the index finger to the tip of the little finger *(right)*. Individual fingers spread from tip to tip of indicated fingers *(left)*.

The thumb has only two joints, both of which are classified as ginglymus. The MCP joint moves from full extension into 40 to 90 degrees of flexion. The interphalangeal (IP) joint can flex 80 to 90 degrees. The carpometacarpal (CMC) joint of the thumb is a unique saddle-type joint having 50 to 70 degrees of abduction. It can flex approximately 15 to 45 degrees and extend 0 to 20 degrees (Fig. 7.5).

Although there are far too many ligaments in the wrist and hand to allow a detailed discussion, injuries to the collateral ligaments of the metacarpophalangeal and proximal interphalangeal joints are very common because of the medial and lateral stresses they often encounter. The wrist, hand, and fingers depend heavily on the ligaments to provide support and static stability. Some of the finger ligaments are detailed in Fig. 7.6.

Movements

The common actions of the wrist are flexion, extension, abduction, and adduction (Fig. 7.7, A–D). The fingers can only flex and extend (Fig. 7.7, E and F), except at the metacarpophalangeal joints, where abduction and adduction (Fig. 7.7, G and H) are controlled by the intrinsic hand muscles. In the hand, the middle phalange is regarded as the reference point by which to differentiate abduction and adduction. Abduction of the index and middle fingers occurs when they move laterally toward the radial side of the forearm. Abduction of the ring and little fingers occurs when they move medially toward the ulnar side of the hand. Movement medially of the index and middle fingers toward the ulnar side of the forearm is adduction. Ring and little finger adduction occurs when these fingers move laterally toward the radial side of the hand. The thumb is abducted when it moves away from the palm and is adducted when it moves toward the palmar aspect of the second metacarpal. These

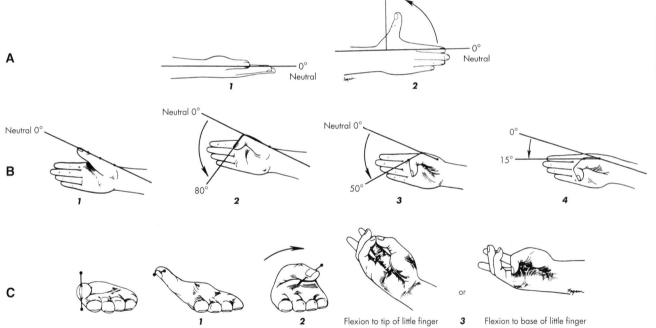

FIG. 7.5 • ROM of the thumb. **A,** Abduction. *1,* Zero starting position: the extended thumb alongside the index finger, which is in line with the radius. *Abduction* is the angle created between the metacarpal bones of the thumb and the index finger. This motion may take place in two planes. *2,* Radial abduction or *extension* takes place parallel to the plane of the palm; **B,** Flexion. *1,* Zero starting position: the extended thumb. *2,* Flexion of the interphalangeal joint: zero to ±80 degrees. *3,* Flexion of the metacarpophalangeal joint: zero to ±50 degrees. *4,* Flexion of the carpometacarpal joint: zero to ±15 degrees; **C,** Opposition. Zero starting position (*far left*): the thumb in line with the index fingers. *Opposition* is a composite motion consisting of three elements: (*1*) abduction, (*2*) rotation, and (*3*) flexion. Motion is usually considered complete when the tip of the thumb touches the tip of the fifth finger. Some consider the arc of opposition complete when the tip of the thumb touches the base of the fifth finger. Both methods are illustrated.

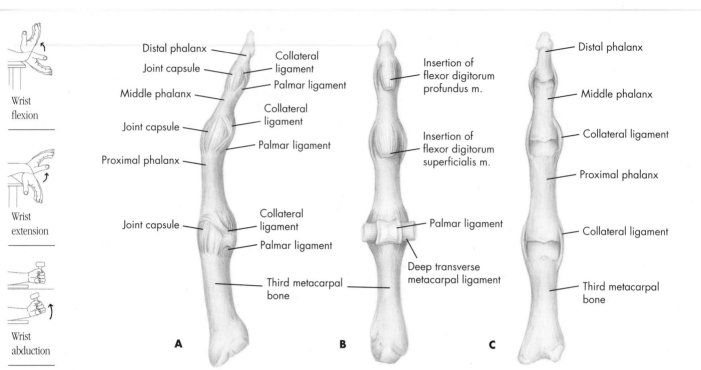

FIG. 7.6 ● Metacarpophalangeal and interphalangeal joints of left long finger. **A,** Lateral view; **B,** Anterior (palmar) view; **C,** Posterior view.

movements, together with pronation and supination of the forearm, make possible the many fine, coordinated movements of the forearm, wrist, and hand.

Flexion (palmar flexion): Movement of the palm of the hand and/or the phalanges toward the anterior or volar aspect of the forearm

Extension (dorsiflexion): Movement of the back of the hand and/or the phalanges toward the posterior or dorsal aspect of the forearm; sometimes referred to as hyperextension

Abduction (radial deviation, radial flexion): Movement of the thumb side of the hand toward the lateral

aspect or radial side of the forearm; also, movement of the fingers away from the middle finger

Adduction (ulnar deviation, ulnar flexion): Movement of the little finger side of the hand toward the medial aspect or ulnar side of the forearm; also, movement of the fingers back together toward the middle finger

Opposition: Movement of the thumb across the palmar aspect to oppose any or all of the phalanges

Reposition: Movement of the thumb as it returns to the anatomical position from opposition with the hand and/or fingers

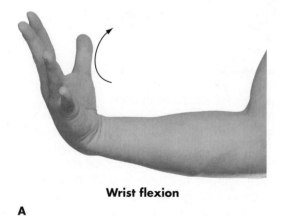

Wrist flexion

A

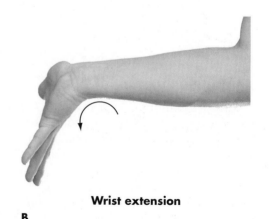

Wrist extension

B

FIG. 7.7 ● Right wrist and hand movements. **A,** Wrist flexion; **B,** Wrist extension.

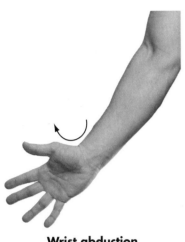

**Wrist abduction
(radial deviation)**

C

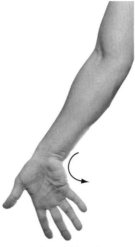

**Wrist adduction
(ulnar deviation)**

D

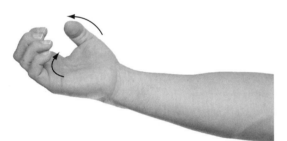

Flexion of fingers and thumb, opposition

E

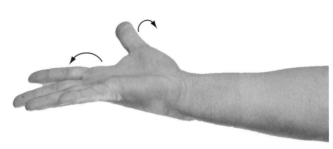

Extension of fingers and thumb, reposition

F

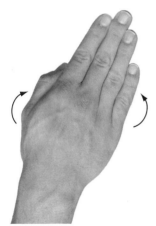

**Adduction of metacarpophalangeal
joints and the thumb**

G

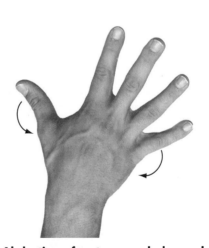

**Abduction of metacarpophalangeal
joints and the thumb**

H

2nd–5th
MCP, PIP,
and DIP
flexion

Thumb
CMC flexion

2nd–5th
MCP, PIP,
and DIP
extension

**Chapter
7**

Thumb
CMC
extension

FIG. 7.7 (continued) • Right wrist and hand movements. **C,** Wrist abduction; **D,** Right wrist adduction;
E, Flexion of the fingers and thumb, opposition; **F,** Extension of the fingers and thumb, reposition;
G, Adduction of metacarpophalangeal joints and the thumb; **H,** Abduction of metacarpophalangeal joints
and the thumb.

Muscles TABLE 7.1

The extrinsic muscles of the wrist and hand can be grouped according to function and location (Table 7.1). There are six muscles that move the wrist but do not cross the hand to move the fingers and thumb. The three wrist flexors in this group are the flexor carpi radialis, flexor carpi ulnaris, and palmaris longus—all of which have their origin on the medial epicondyle of the humerus. The extensors of the wrist have their origins on the lateral epicondyle and include the extensor carpi radialis longus, extensor carpi radialis brevis, and extensor carpi ulnaris (Figs. 6.11 and 6.13).

Another nine muscles function primarily to move the phalanges but are also involved in wrist joint actions because they originate on the forearm and cross the wrist. These muscles generally are weaker in their actions on the wrist. The flexor digitorum superficialis and the flexor digitorum profundus are finger flexors; however, they also assist in wrist flexion along with the flexor pollicis longus, which is a thumb flexor. The extensor digiorum, the extensor indicis, and the extensor digiti minimi are finger extensors but also assist in wrist extension, along with the extensor pollicis longus and extensor pollicis brevis, which extend the thumb. The abductor pollicis longus abducts the thumb and assists in wrist abduction.

TABLE 7.1 • Agonist muscles of the wrist and hand joints

	Muscle	Origin	Insertion	Action	Plane of motion	Palpation	Innervation
Anterior muscles (wrist flexors)	Flexor carpi radialis	Medial epicondyle of humerus	Base of 2nd and 3rd metacarpals on palmar surface	Flexion of the wrist	Sagittal	Anterior distal forearm and wrist surface, slightly lateral, in line with the 2nd and 3rd metacarpals with resisted flexion and abduction	Median nerve (C6, C7)
				Abduction of the wrist	Frontal		
				Weak flexion of the elbow	Sagittal		
				Weak pronation of the forearm	Transverse		
	Palmaris longus	Medial epicondyle of humerus	Palmar aponeurosis of the 2nd, 3rd, 4th, and 5th metacarpals	Flexion of the wrist	Sagittal	Anteromedial and central aspect of the anterior forearm just proximal to the wrist, particularly with slight wrist flexion and opposition of thumb to the 5th finger	Median nerve (C6, C7)
				Weak flexion of the elbow			
	Flexor carpi ulnaris	Medial epicondyle of humerus and posterior aspect of proximal ulna	Base of 5th metacarpal (palmar surface), pisiform, and hamate	Flexion of the wrist	Sagittal	Anteromedial surface of the forearm, a few inches below the medial epicondyle of the humerus to just proximal to the wrist, with resisted flexion/adduction	Ulnar nerve (C8, T1)
				Adduction of the wrist	Frontal		
				Weak flexion of the elbow	Sagittal		

TABLE 7.1 (continued) • Agonist muscles of the wrist and hand joints

	Muscle	Origin	Insertion	Action	Plane of motion	Palpation	Innervation
Anterior muscles (wrist and phalangeal flexors)	Flexor digitorum superficialis	Medial epicondyle of humerus Ulnar head: medial coronoid process Radial head: upper 2/3 of anterior border of the radius just distal to the radial tuberosity	Each tendon splits and attaches to the sides of the middle phalanx of the four fingers on the palmar surface	Flexion of the fingers at the metacarpo-phalangeal and proximal inter-phalangeal joints	Sagittal	In depressed area between palmaris longus and flexor carpi ulnaris tendons, particularly when making a fist but keeping the distal interphalangeals extended and with slightly resisted wrist flexion; also on anterior mid-forearm during same activity	Median nerve (C7, C8, T1)
				Flexion of the wrist			
				Weak flexion of the elbow			
	Flexor digitorum profundus	Proximal 3/4 of anterior and medial ulna	Base of distal phalanges of four fingers	Flexion of the four fingers at the metacar-pophalangeal, proximal inter-phalangeal, and distal inter-phalangeal joints	Sagittal	Deep to the flexor digitorum superficialis, but on anterior mid-forearm while flexing the distal interphalangeal joints and keeping the proximal interphalangeal joints extended; over the palmar surface of the 2nd, 3rd, 4th, and 5th metacarpo-phalangeal joints during finger flexion against resistance	Median nerve (C8, T1) to 2nd and 3rd fingers; ulnar nerve (C8, T1) to 4th and 5th fingers
				Flexion of the wrist			
	Flexor pollicis longus	Middle anterior surface of the radius and ante-rior medial bor-der of the ulna just distal to the coronoid pro-cess; occasionally a small head is present attaching on the medial epicondyle of the humerus	Base of distal phalanx of thumb on pal-mar surface	Flexion of the thumb carpo-metacarpal, metacarpo-phalangeal, and interphalangeal joints	Sagittal	Anterior surface of the thumb on the proximal phalanx, and just lateral to the palmaris longus and medial to the flexor carpi radi-alis on the anterior distal forearm, especially during active flexion of the thumb interphalan-geal joint	Median nerve palmar interos-seous branch (C8, T1)
				Flexion of the wrist			
				Abduction of the wrist	Frontal		

Chapter
7

TABLE 7.1 (continued) • Agonist muscles of the wrist and hand joints

	Muscle	Origin	Insertion	Action	Plane of motion	Palpation	Innervation
Posterior muscles (wrist extensors)	Extensor carpi ulnaris	Lateral epicondyle of humerus and middle 1/2 of the posterior border of the ulna	Base of 5th metacarpal on dorsal surface	Extension of the wrist	Sagittal	Just lateral to the ulnar styloid process and crossing the posteromedial wrist, particularly with wrist extension/ adduction	Radial nerve (C6, C7, C8)
				Adduction of the wrist	Frontal		
				Weak extension of the elbow	Sagittal		
	Extensor carpi radialis brevis	Lateral epicondyle of humerus	Base of 3rd metacarpal on dorsal surface	Extension of the wrist	Sagittal	Just proximal to the dorsal aspect of the wrist and approximately 1 cm medial to the radial styloid process, the tendon may be felt during extension and traced to base of 3rd metacarpal, particularly when making a fist; proximally and posteriorly, just medial to the bulk of the brachioradialis	Radial nerve (C6, C7)
				Abduction of the wrist	Frontal		
				Weak flexion of the elbow	Sagittal		
	Extensor carpi radialis longus	Distal third of lateral supracondylar ridge of humerus and lateral epicondyle of the humerus	Base of 2nd metacarpal on dorsal surface	Extension of the wrist	Sagittal	Just proximal to the dorsal aspect of the wrist and approximately 1 cm medial to the radial styloid process, the tendon may be felt during extension and traced to base of 2nd metacarpal, particularly when making a fist; proximally and posteriorly, just medial to the bulk of the brachioradialis	Radial nerve (C6, C7)
				Abduction of the wrist	Frontal		
				Weak flex-ion of the elbow	Sagittal		
				Weak pronation	Transverse		

Chapter 7

TABLE 7.1 (continued) • **Agonist muscles of the wrist and hand joints**

	Muscle	Origin	Insertion	Action	Plane of motion	Palpation	Innervation
Posterior muscles (wrist and phalangeal extensors)	Extensor digitorum	Lateral epicondyle of humerus	Four tendons to bases of middle and distal phalanxes of four fingers on dorsal surface	Extension of the 2nd, 3rd, 4th, and 5th phalanges at the metacarpophalangeal joints	Sagittal	With all four fingers extended, on the posterior surface of the distal forearm immediately medial to extensor pollicis longus tendon and lateral to the extensor carpi ulnaris and extensor digiti minimi, then dividing into four separate tendons that are over the dorsal aspect of the hand and metacarpophalangeal joints	Radial nerve (C6, C7, C8)
				Extension of the wrist			
				Weak extension of the elbow			
	Extensor indicis	Middle to distal 1/3 of posterior ulna	Base of the middle and distal phalanxes of the 2nd phalange on dorsal surface	Extension of the index finger at the metacarpophalangeal joint	Sagittal	With forearm pronated on the posterior aspect of the distal forearm and dorsal surface of the hand just medial to the extensor digitorum tendon of the index finger with extension of the index fingers and flexion of the 3rd, 4th, and 5th fingers	Radial nerve (C6, C7, C8)
				Weak wrist extension			
				Weak supination	Transverse		
	Extensor digiti minimi	Lateral epicondyle of humerus	Base of the middle and distal phalanxes of the 5th phalange on dorsal surface	Extension of the little finger at the metacarpophalangeal joint	Sagittal	Passing over the dorsal aspect of the distal radioulnar joint, particularly with relaxed flexion of other fingers and alternating 5th finger extension and relaxation; dorsal surface of forearm immediately medial to the extensor digitorum and lateral to the extensor carpi ulnaris	Radial nerve (C6, C7, C8)
				Weak wrist extension			
				Weak elbow extension			
	Extensor pollicis longus	Posterior lateral surface of the lower middle ulna	Base of distal phalanx of thumb on dorsal surface	Extension of the thumb at the carpometacarpal, metacarpophalangeal, and interphalangeal joints	Sagittal	Dorsal aspect of the hand to its insertion on the base of the distal phalanx; also on posterior surface of lower forearm between radius and ulnar just proximal to the extensor indicis and medial to the extensor pollicis brevis and abductor pollicis longus with forearm pronated and fingers in relaxed flexion while actively extending the thumb	Radial nerve (C6, C7, C8)
				Extension of the wrist			
				Abduction of the wrist	Frontal		
				Weak supination	Transverse		

Chapter
7

TABLE 7.1 (continued) • Agonist muscles of the wrist and hand joints

	Muscle	Origin	Insertion	Action	Plane of motion	Palpation	Innervation
Posterior muscles (wrist and phalangeal extensors)	Extensor pollicis brevis	Posterior surface of lower middle radius	Base of proximal phalanx of thumb on dorsal surface	Extension of the thumb at the carpometacarpal and metacarpophalangeal joints	Sagittal	Just lateral to the extensor pollicis longus tendon on the dorsal side of the hand to its insertion on the proximal phalanx with extension of the thumb carpometacarpal and metacarpophalangeal and flexion of the interphalangeal joints	Radial nerve (C6, C7)
				Weak wrist extension			
				Abduction of the wrist	Frontal		
Posterior muscles	Abductor pollicis longus	Posterior aspect of radius and midshaft of the ulna	Base of 1st metacarpal on dorsal lateral surface	Abduction of the thumb at the carpometacarpal joint	Frontal	Lateral aspect of the wrist joint just proximal to the 1st metacarpal	Radial nerve (C6, C7)
				Abduction of the wrist			
				Extension of the thumb at the carpometacarpal joint	Sagittal		
				Weak wrist extension			
				Weak supination	Transverse		

All of the wrist flexors generally have their origins on the anteromedial aspect of the proximal forearm and the medial epicondyle of the humerus, whereas their insertions are on the anterior aspect of the wrist and hand. All of the flexor tendons except for the flexor carpi ulnaris and palmaris longus pass through the carpal tunnel, along with the median nerve (Fig. 7.8). Conditions leading to swelling and inflammation in this area can result in increased pressure in the carpal tunnel, which interferes with normal function of the median nerve, leading to reduced motor and sensory function of its distribution. Known as **carpal tunnel syndrome**, this condition is particularly common with repetitive use of the hand and wrist in manual labor and clerical work such as typing and keyboarding. Often, slight modifications in work habits and the positions of the hand and wrist during these activities can be preventive. Additionally, flexibility exercises for the wrist and finger flexors may be helpful.

The wrist extensors generally have their origins on the posterolateral aspect of the proximal forearm and the lateral humeral epicondyle, whereas their insertions are located on the posterior aspect of the wrist and hand. The flexor and extensor tendons at the distal forearm immediately proximal to the wrist are held in place on the palmar and dorsal aspects by transverse bands of tissue. These bands, known respectively as the flexor and extensor retinaculum, prevent these tendons from bowstringing during flexion and extension.

The wrist abductors are the flexor carpi radialis, extensor carpi radialis longus, extensor carpi radialis brevis, abductor pollicis longus, extensor pollicis longus, and extensor pollicis brevis. These muscles generally cross the wrist joint anterolaterally and posterolaterally to insert on the radial side of the hand. The flexor carpi ulnaris and extensor carpi ulnaris adduct the wrist and cross the wrist joint anteromedially and posteromedially to insert on the ulnar side of the hand.

Chapter
7

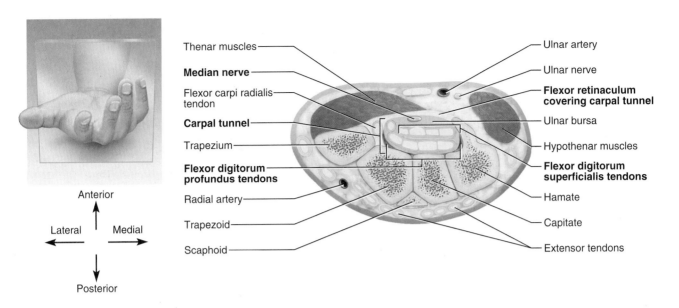

Thenar muscles

Median nerve

Flexor carpi radialis tendon

Carpal tunnel

Trapezium

Flexor digitorum profundus tendons

Radial artery

Trapezoid

Scaphoid

Ulnar artery

Ulnar nerve

Flexor retinaculum covering carpal tunnel

Ulnar bursa

Hypothenar muscles

Flexor digitorum superficialis tendons

Hamate

Capitate

Extensor tendons

Anterior

Lateral — Medial

Posterior

FIG. 7.8 ● Cross section of the right wrist, viewed as if from the distal end of a person's right forearm extended toward you with the palm up. Note how the flexor tendons and median nerve are confined in a tight space between the carpal bones and the flexor retinaculum.

The intrinsic muscles of the hand (see Table 7.2 and Fig. 7.26) have their origins and insertions on the bones of the hand. Grouping the intrinsic muscles into three groups according to location is helpful in understanding and learning these muscles. On the radial side are four muscles of the thumb—the opponens pollicis, the abductor pollicis brevis, the flexor pollicis brevis, and the adductor pollicis. On the ulnar side are three muscles of the little finger—the opponens digiti minimi, the abductor digiti minimi, and the flexor digiti minimi brevis. In the remainder of the hand are 11 muscles, which can be further grouped as the 4 lumbricals, the 3 palmar interossei, and the 4 dorsal interossei.

Wrist and hand muscles—location

Anteromedial at the elbow and forearm and anterior at the hand (Fig. 7.9, A–C)

Primarily wrist flexion

    Flexor carpi radialis

    Flexor carpi ulnaris

    Palmaris longus

Primarily wrist and phalangeal flexion

    Flexor digitorum superficialis

    Flexor digitorum profundus

    Flexor pollicis longus

Posterolateral at the elbow and forearm and posterior at the hand (Fig. 7.9, D)

Primarily wrist extension

    Extensor carpi radialis longus

    Extensor carpi radialis brevis

    Extensor carpi ulnaris

Primarily wrist and phalangeal extension

    Extensor digitorum

    Extensor indicis

    Extensor digiti minimi

    Extensor pollicis longus

    Extensor pollicis brevis

    Abductor pollicis longus

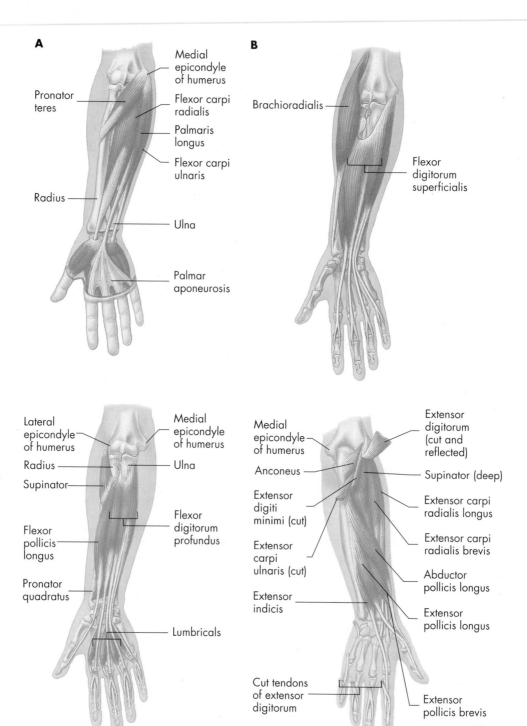

FIG. 7.9 • Muscles of the forearm. **A,** Anterior view of the right forearm (superficial). Brachioradialis muscle is removed; **B,** Anterior view of the right forearm (deeper than *A*). Pronator teres, flexor carpi radialis and ulnaris, and palmaris longus muscles are removed; **C,** Anterior view of the right forearm (deeper than *A* or *B*). Brachioradialis, pronator teres, flexor carpi radialis and ulnaris, palmaris longus, and flexor digitorum superficialis muscles are removed; **D,** Deep muscles of the right posterior forearm, with extensor digitorum, extensor digiti minimi, and extensor carpi ulnaris muscles cut to reveal deeper muscles.

Nerves

The muscles of the wrist and hand are all innervated from the radial, median, and ulnar nerves of the brachial plexus, as illustrated in Figs. 6.5, 6.6, and 7.10. The radial nerve, originating from C6, C7, and C8, provides innervation for the extensor carpi radialis brevis and extensor carpi radialis longus. It then branches to become the posterior interosseous nerve, which supplies the extensor carpi ulnaris, extensor digitorum, extensor digiti minimi, abductor pollicis longus, extensor pollicis longus, extensor pollicis brevis, and extensor indicis. The median nerve, arising from C6, C7, C8, and T1, innervates the flexor carpi radialis, palmaris longus, and flexor digitorum superficialis. It then branches to become the anterior interosseous nerve, which innervates the flexor digitorum profundus for the index and long fingers as well as the flexor pollicis longus. Regarding the intrinsic muscles of the hand, the median nerve innervates the abductor pollicis brevis, flexor pollicis brevis (superficial head), opponens pollicis, and first and second lumbrical. The ulnar nerve, branching from C8 and T1, supplies the flexor digitorum profundus for the fourth and fifth fingers and the flexor carpi ulnaris. Additionally, it innervates the remaining intrinsic muscles of the hand (the deep head of the flexor pollicis brevis, adductor pollicis, palmar interossei, dorsal interossei, third and fourth lumbrical, opponens digiti minimi, abductor digiti minimi, and flexor digiti minimi brevis). Sensation to the ulnar side of the hand, the ulnar half of the ring finger, and the entire little finger is provided by the ulnar nerve. Of all the nerves in the upper extremity, the ulnar nerve is traumatized the most. Most people have hit their "funny bone" and experienced a painful tingling sensation into the ulnar side of their forearm and fourth and fifth fingers. This is actually a contusion to the ulnar nerve at the medial elbow. Usually it subsides fairly quickly, but chronic contusions or pressure over this area may lead to a hypersensitivity of this nerve, making it more easily irritated.

Additionally, throwing athletes may experience a traction injury to this nerve with the medial stress placed on the elbow during pitching or in conjunction with an ulnar collateral ligament sprain. The ulnar nerve also may become inflamed from subluxing or slipping out of its groove, especially in chronic cases.

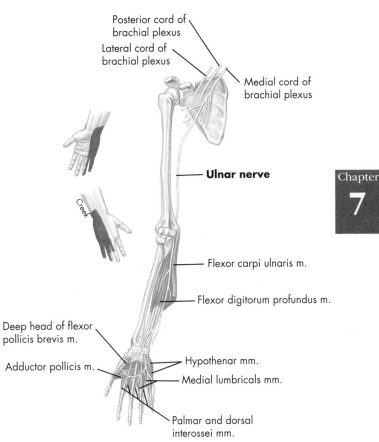

FIG. 7.10 • Muscular and cutaneous distribution of the ulnar nerve.

Flexor carpi radialis muscle FIG. 7.11

(fleks´or kar´pi ra´di-a´lis)

Wrist flexion

Origin

Medial epicondyle of the humerus

Insertion

Base of the second and third metacarpals, anterior (palmar surface)

Wrist abduction

Action

Flexion of the wrist
Abduction of the wrist
Weak flexion of the elbow
Weak pronation of the forearm

Palpation

Anterior surface of the wrist, slightly lateral, in line with the second and third metacarpals with resisted flexion and abduction

Innervation

Median nerve (C6, C7)

Application, strengthening, and flexibility

The flexor carpi radialis, flexor carpi ulnaris, and palmaris longus are the most powerful of the wrist flexors. They are brought into play during any activity that requires wrist curling or stabilization of the wrist against resistance, particularly if the forearm is supinated.

The flexor carpi radialis may be developed by performing wrist curls against a handheld resistance. This may be accomplished when the supinated forearm is supported by a table, with the hand and wrist hanging over the edge to allow full range of motion. The extended wrist is then flexed or curled up to strengthen this muscle. See Appendix 3 for more commonly used exercises for the flexor carpi radialis and other muscles in this chapter.

To stretch the flexor carpi radialis, the elbow must be fully extended with the forearm supinated while a partner passively extends and adducts the wrist.

Elbow flexion

Radioulnar pronation

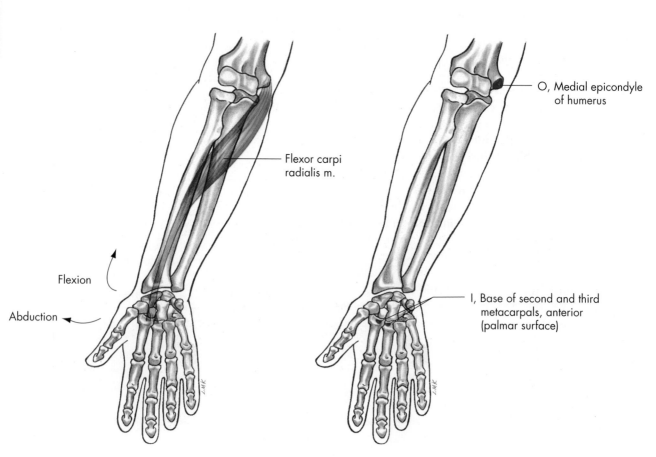

FIG. 7.11 • Flexor carpi radialis muscle, anterior view. O, Origin; I, Insertion.

Palmaris longus muscle   FIG. 7.12
(pal-ma′ris lon′gus)

Origin
Medial epicondyle of the humerus

Insertion
Palmar aponeurosis of the second, third, fourth, and fifth metacarpals

Action
Flexion of the wrist
Weak flexion of the elbow

Palpation
The palmaris longus is absent in either one or both forearms in some people. Anteromedial and central aspect of the anterior forearm just proximal to the wrist, particularly with slight wrist flexion and opposition of thumb to the fifth finger

Innervation
Median nerve (C6, C7)

Application, strengthening, and flexibility
Unlike the flexor carpi radialis and flexor carpi ulnaris, which are not only wrist flexors but also abductors and adductors, respectively, the palmaris longus is involved only in wrist flexion from the anatomical position because of its central location on the anterior forearm and wrist. It can, however, assist in abducting the wrist from an extremely adducted position back to neutral and assist in adducting the wrist from an extremely abducted position back to neutral. It may also assist slightly in forearm pronation because of its slightly lateral insertion in relation to its origin on the medial epicondyle. It may also be strengthened with any type of wrist-curling activity, such as those described for the flexor carpi radialis muscle.

Maximal elbow and wrist extension stretches the palmaris longus.

Wrist
flexion

Elbow
flexion

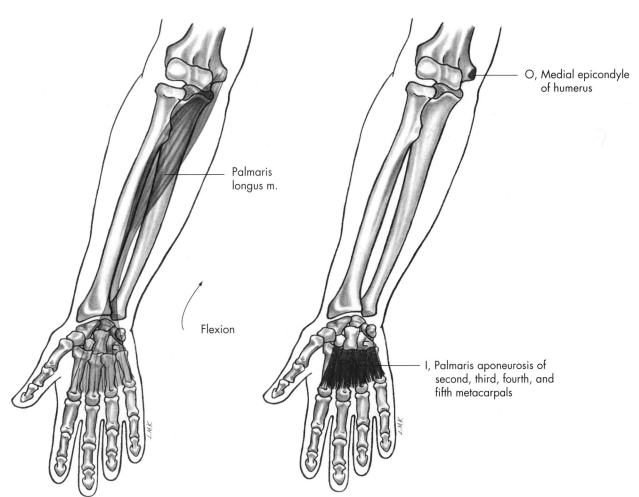

O, Medial epicondyle
of humerus

Palmaris
longus m.

Flexion

I, Palmaris aponeurosis of
second, third, fourth, and
fifth metacarpals

FIG. 7.12 ● Palmaris longus muscle, anterior view. O, Origin; I, Insertion.

Flexor carpi ulnaris muscle FIG. 7.13

(fleks´or kar´pi ul-na´ris)

Wrist
flexion

Origin

Medial epicondyle of the humerus
Posterior aspect of the proximal ulna

Insertion

Pisiform, hamate, and base of the fifth metacarpal
 (palmar surface)

Action

Flexion of the wrist
Adduction of the wrist, together with the extensor
 carpi ulnaris muscle
Weak flexion of the elbow

Wrist
adduction

Palpation

Anteromedial surface of the forearm, a few inches
 below the medial epicondyle of the humerus to
just proximal to the wrist, with resisted flexion/
adduction

Innervation

Ulnar nerve (C8, T1)

Application, strengthening, and flexibility

The flexor carpi ulnaris is very important in wrist
flexion or curling activities. In addition, it is one
of only two muscles involved in wrist adduction
or ulnar flexion. It may be strengthened with any
type of wrist-curling activity against resistance,
such as those described for the flexor carpi radialis muscle.

To stretch the flexor carpi ulnaris, the elbow
must be fully extended with the forearm supinated while a partner passively extends and
abducts the wrist.

Chapter

7

Elbow
flexion

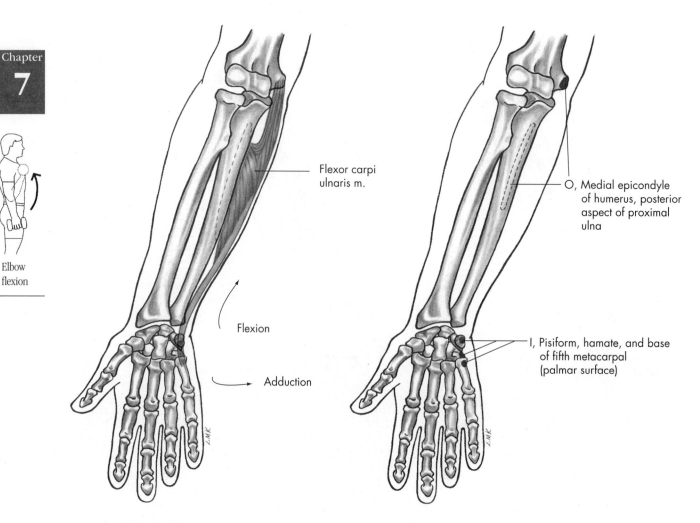

FIG. 7.13 • Flexor carpi ulnaris muscle, anterior view. O, Origin; I, Insertion.

Extensor carpi ulnaris muscle FIG. 7.14

(eks-ten′sor kar′pi ul-na′ris)

Origin

Lateral epicondyle of the humerus
Middle two-fourths of the posterior border of the
ulna

Insertion

Base of the fifth metacarpal (dorsal surface)

Action

Extension of the wrist
Adduction of the wrist together with the flexor carpi
ulnaris muscle
Weak extension of the elbow

Palpation

Just lateral to the ulnar styloid process and crossing the posteromedial wrist, particularly with wrist
extension/adduction

Innervation

Radial nerve (C6–C8)

Application, strengthening, and flexibility

Besides being a powerful wrist extensor, the
extensor carpi ulnaris muscle is the only muscle
other than the flexor carpi ulnaris involved in wrist
adduction or ulnar deviation. The extensor carpi
ulnaris, the extensor carpi radialis brevis, and the
extensor carpi radialis longus are the most powerful of the wrist extensors. These muscles act as
antagonists to wrist flexion to allow the finger
flexors to function more effectively in gripping.
Any activity requiring wrist extension or stabilization of the wrist against resistance, particularly if
the forearm is pronated, depends greatly on the
strength of these muscles. They are often brought
into play with the backhand in racquet sports.

Wrist
extension

Wrist
adduction

The extensor carpi ulnaris may be developed
by performing wrist extension against a handheld resistance. This may be accomplished with
the pronated forearm being supported by a table
with the hand hanging over the edge to allow full
range of motion. The wrist is then moved from
the fully flexed position to the fully extended
position against the resistance.

Stretching the extensor carpi ulnaris requires
the elbow to be extended with the forearm pronated while the wrist is passively flexed and
slightly abducted.

Elbow
extension

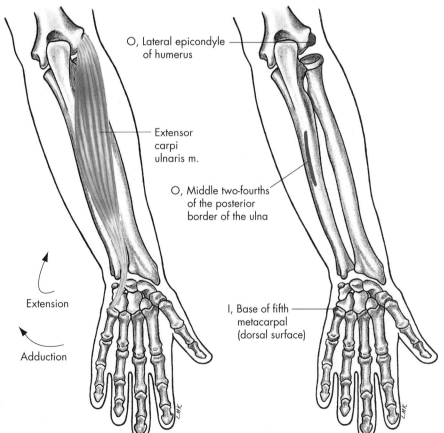

O, Lateral epicondyle
of humerus

Extensor
carpi
ulnaris m.

O, Middle two-fourths
of the posterior
border of the ulna

Extension

Adduction

I, Base of fifth
metacarpal
(dorsal surface)

FIG. 7.14 • Extensor carpi ulnaris
muscle, posterior view. O, Origin;
I, Insertion.

Wrist
extension

Wrist
abduction

Extensor carpi radialis brevis muscle   FIG. 7.15

(eks-ten´sor kar´pi ra´di-a´lis bre´vis)

Origin

Lateral epicondyle of the humerus

Insertion

Base of the third metacarpal (dorsal surface)

Action

Extension of the wrist
Abduction of the wrist
Weak flexion of the elbow

Palpation

Dorsal side of the forearm, and difficult to distinguish from the extensor carpi radialis longus and the extensor digitorum

Just proximal to the dorsal aspect of the wrist and approximately 1 cm medial to the radial styloid process, the tendon may be felt during extension and traced to the base of the third metacarpal, particularly when making a fist; proximally and posteriorly, just medial to the bulk of the brachioradialis

Innervation

Radial nerve (C6, C7)

Application, strengthening, and flexibility

The extensor carpi radialis brevis is important in any sports activity that requires powerful wrist extension, such as golf or tennis. Wrist extension exercises, such as those described for the extensor carpi ulnaris, are appropriate for development of the muscle.

Stretching the extensor carpi radialis brevis and longus requires the elbow to be extended with the forearm pronated while the wrist is passively flexed and slightly adducted.

Elbow
flexion

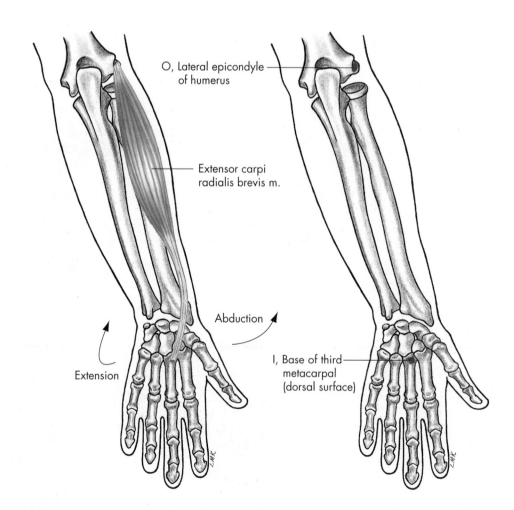

O, Lateral epicondyle of humerus

Extensor carpi radialis brevis m.

Abduction

Extension

I, Base of third metacarpal (dorsal surface)

FIG. 7.15 • Extensor carpi radialis brevis muscle, posterior view. O, Origin; I, Insertion.

Extensor carpi radialis longus muscle   FIG. 7.16

(eks-ten´sor kar´pi ra´di-a´lis lon´gus)

Origin

Distal third of lateral supracondylar ridge of the humerus and lateral epicondyle of the humerus

Insertion

Base of the second metacarpal (dorsal surface)

Action

Extension of the wrist
Abduction of the wrist
Weak flexion of the elbow
Weak pronation to neutral from a fully supinated position

Palpation

Just proximal to the dorsal aspect of the wrist and approximately 1 cm medial to the radial styloid process, the tendon may be felt during extension and traced to the base of the second metacarpal, particularly when making a fist; proximally and posteriorly, just medial to the bulk of the brachioradialis

Innervation

Radial nerve (C6, C7)

Application, strengthening, and flexibility

The extensor carpi radialis longus, like the extensor carpi radialis brevis, is important in any sports activity that requires powerful wrist extension. In addition, both muscles are involved in abduction of the wrist. The extensor carpi radialis longus may be developed with the same wrist extension exercises as described for the extensor carpi ulnaris muscle.

The extensor carpi radialis longus is stretched in the same manner as the extensor carpi radialis brevis.

Wrist extension

Wrist abduction

Chapter
7

Elbow flexion

Radioulnar pronation

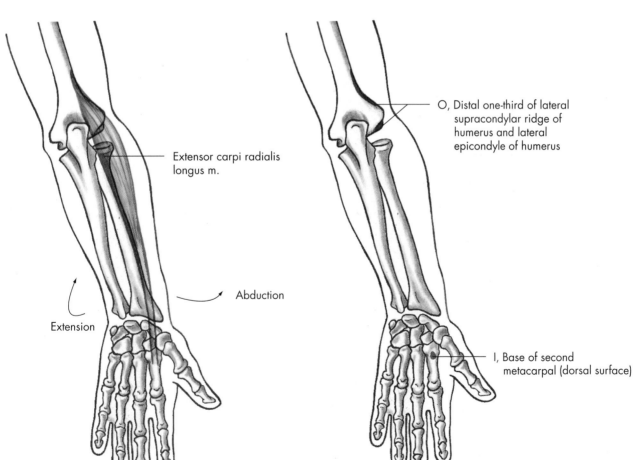

Extensor carpi radialis longus m.

Abduction

Extension

O, Distal one-third of lateral supracondylar ridge of humerus and lateral epicondyle of humerus

I, Base of second metacarpal (dorsal surface)

FIG. 7.16 • Extensor carpi radialis longus muscle, posterior view. O, Origin; I, Insertion.

Flexor digitorum superficialis muscle   FIG. 7.17

(fleks´or dij-i-to´rum su´per-fish-e-al´is)

2nd–5th
PIP
flexion

Origin

Medial epicondyle of the humerus
Ulnar head: medial coronoid process
Radial head: upper two-thirds of anterior border of
the radius just distal to the radial tuberosity

2nd–5th
MCP
flexion

Insertion

Each tendon splits and attaches to the sides of the
middle phalanx of the four fingers (palmar surface)

Action

Flexion of the fingers at the metacarpophalangeal
and proximal interphalangeal joints
Flexion of the wrist
Weak flexion of the elbow

Palpation

In depressed area between palmaris longus and
flexor carpi ulnaris tendons, particularly when
making a fist but keeping the distal interphalangeal
extended and slightly resisted wrist flexion; also on
anterior mid-forearm during the same activity

Innervation

Median nerve (C7, C8, T1)

Application, strengthening, and flexibility

The flexor digitorum superficialis muscle, also
known as the flexor digitorum sublimis, divides
into four tendons on the palmar aspect of the wrist
and hand to insert on each of the four fingers. The
flexor digitorum superficialis and the flexor digitorum profundus are the only muscles involved in
flexion of all four fingers. Both of these muscles
are vital in any type of gripping activity.

Squeezing a sponge rubber ball in the palm of
the hand, along with other gripping and squeezing
activities, can be used to develop these muscles.

The flexor digitorum superficialis is stretched
by passively extending the elbow, wrist, metacarpophalangeal, and proximal interphalangeal joints
while maintaining the forearm in full supination.

2nd–5th
MCP and
PIP flexion

Wrist
flexion

Elbow
flexion

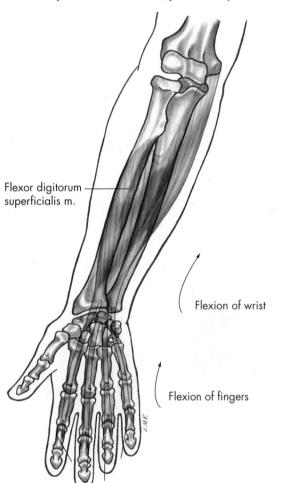

Flexor digitorum
superficialis m.

Flexion of wrist

Flexion of fingers

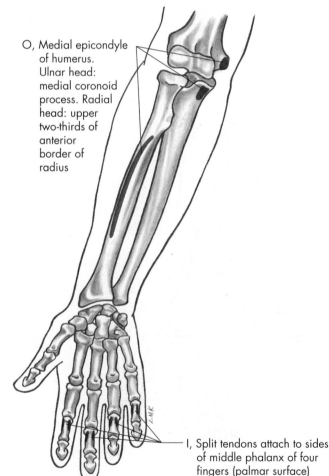

O, Medial epicondyle
of humerus.
Ulnar head:
medial coronoid
process. Radial
head: upper
two-thirds of
anterior
border of
radius

I, Split tendons attach to sides
of middle phalanx of four
fingers (palmar surface)

FIG. 7.17 ● Flexor digitorum superficialis muscle, anterior view. O, Origin; I, Insertion.

Flexor digitorum profundus muscle FIG. 7.18

(fleks´or dij-i-to´rum pro-fun´dus)

Origin
Proximal three-fourths of the anterior and medial ulna

Insertion
Base of the distal phalanxes of the four fingers

Action
Flexion of the four fingers at the metacarpophalangeal, proximal interphalangeal, and distal interphalangeal joints
Flexion of the wrist

Palpation
Difficult to distinguish, deep to the flexor digitorum superficialis, but on anterior mid-forearm while flexing the distal interphalangeal joints and keeping the proximal interphalangeal joints in extension; over the palmar surface of the second, third, fourth, and fifth metacarpophalangeal joints during finger flexion against resistance

Innervation
Median nerve (C8, T1) to the second and third fingers
Ulnar nerve (C8, T1) to the fourth and fifth fingers

Application, strengthening, and flexibility
Both the flexor digitorum profundus muscle and the flexor digitorum superficialis muscle assist in wrist flexion because of their palmar relationship to the wrist. The flexor digitorum profundus is used in any type of gripping, squeezing, or hand-clenching activity, such as gripping a racket or climbing a rope.

The flexor digitorum profundus muscle may be developed through these activities, in addition to the strengthening exercises described for the flexor digitorum superficialis muscle.

The flexor digitorum profundus is stretched similarly to the flexor digitorum superficialis, except that the distal interphalangeal joints must be passively extended in addition to the wrist, metacarpophalangeal, and proximal interphalangeal joints while maintaining the forearm in full supination.

2nd–5th DIP flexion

2nd–5th PIP flexion

2nd–5th MCP flexion

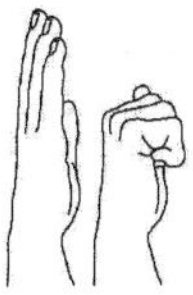

2nd–5th MCP and PIP flexion

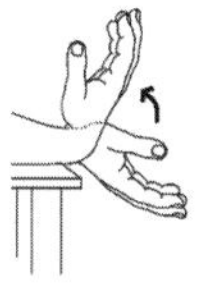

Wrist flexion

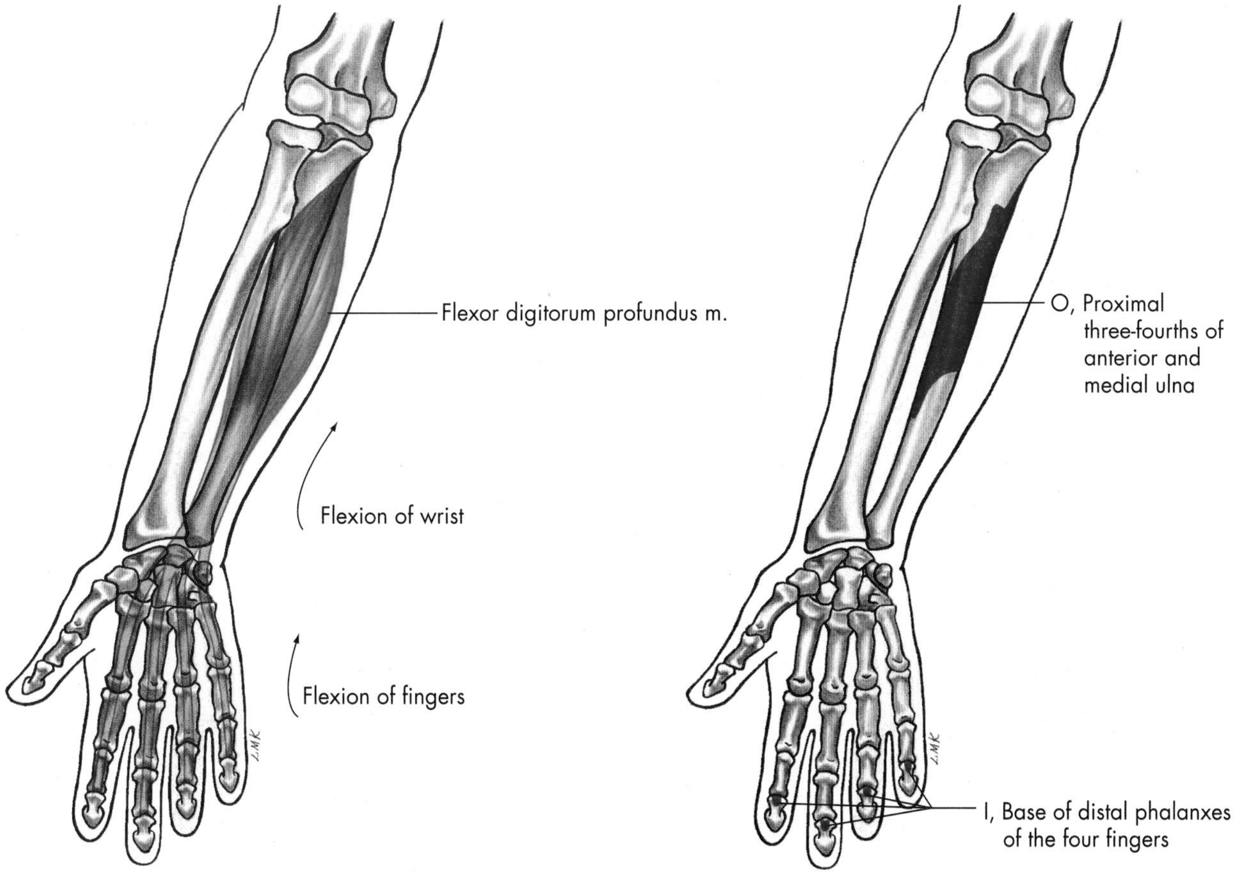

Flexor digitorum profundus m.

Flexion of wrist

Flexion of fingers

O, Proximal three-fourths of anterior and medial ulna

I, Base of distal phalanxes of the four fingers

FIG. 7.18 ● Flexor digitorum profundus muscle, anterior view. O, Origin; I, Insertion.

Flexor pollicis longus muscle   FIG. 7.19

(fleks´or pol´i-sis lon´gus)

Origin

Middle anterior surface of the radius and the anterior medial border of the ulna just distal to the coronoid process; occasionally a small head is present attaching on the medial epicondyle of the humerus

Insertion

Base of the distal phalanx of the thumb (palmar surface)

Action

Flexion of the thumb carpometacarpal, metacarpophalangeal, and interphalangeal joints
Flexion of the wrist
Abduction of the wrist

Palpation

Anterior surface of the thumb on the proximal phalanx, and just lateral to the palmaris longus and medial to the flexor carpi radialis on the anterior distal forearm, especially during active flexion of the thumb interphalangeal joint

Innervation

Median nerve, palmar interosseous branch (C8, T1)

Application, strengthening, and flexibility

The primary function of the flexor pollicis longus muscle is flexion of the thumb, which is vital in gripping and grasping activities of the hand. Because of its palmar relationship to the wrist, it provides some assistance in wrist flexion.

It may be strengthened by pressing a sponge rubber ball into the hand with the thumb and by many other gripping or squeezing activities.

The flexor pollicis longus is stretched by passively extending the entire thumb while simultaneously maintaining maximal wrist extension.

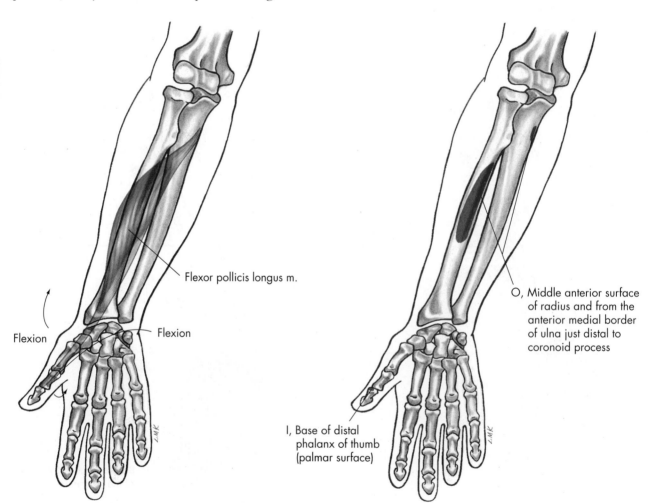

FIG. 7.19 • Flexor pollicis longus muscle, anterior view. O, Origin; I, Insertion.

Extensor digitorum muscle FIG. 7.20

(eks-ten´sor dij-i-to´rum)

Origin

Lateral epicondyle of the humerus

Insertion

Four tendons to bases of middle and distal
phalanxes of the four fingers (dorsal surface)

Action

Extension of the second, third, fourth, and fifth
phalanges at the metacarpophalangeal joints
Extension of the wrist
Weak extension of the elbow

Palpation

With all four fingers extended, on the posterior
surface of the distal forearm immediately medial
to the extensor pollicis longus tendon and lateral
to the extensor carpi ulnaris and extensor digiti
minimi, then dividing into four separate tendons
that are over the dorsal aspect of the hand and
metacarpophalangeal joints

Innervation

Radial nerve (C6–C8)

Application, strengthening, and flexibility

The extensor digitorum, also known as the extensor digitorum communis, is the only muscle
involved in extension of all four fingers. This muscle divides into four tendons on the dorsum of the
wrist to insert on each of the fingers. It also assists
with wrist extension movements. It may be developed by applying manual resistance to the dorsal
aspect of the flexed fingers and then extending
the fingers fully. When performed with the wrist
in flexion, this exercise increases the workload on
the extensor digitorum.

To stretch the extensor digitorum, the fingers
must be maximally flexed at the metacarpophalangeal, proximal interphalangeal, and distal interphalangeal joints while the wrist is fully flexed.

2nd–5th
MCP
extension

2nd–5th
MCP, PIP,
and DIP
extension

Wrist
extension

Chapter

7

Elbow
extension

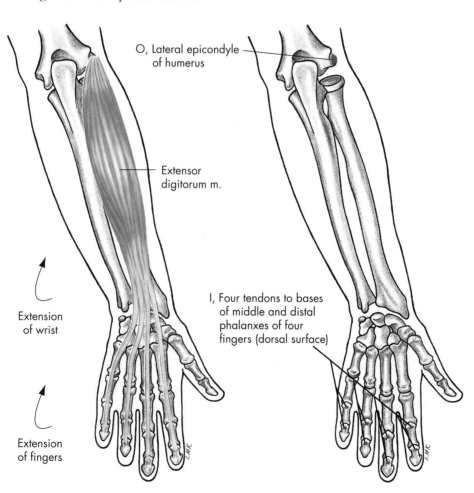

O, Lateral epicondyle
of humerus

Extensor
digitorum m.

Extension
of wrist

Extension
of fingers

I, Four tendons to bases
of middle and distal
phalanxes of four
fingers (dorsal surface)

FIG. 7.20 • Extensor digitorum muscle, posterior view. O, Origin; I, Insertion.

www.mhhe.com/floyd19e   **193**

Extensor indicis muscle   FIG. 7.21

(eks-ten´sor in´di-sis)

2nd MCP extension

Wrist extension

Radioulnar supination

Origin

Between middle and distal one-third of the posterior ulna

Insertion

Base of the middle and distal phalanxes of the second phalange (dorsal surface)

Action

Extension of the index finger at the metacarpophalangeal joint

Weak wrist extension

Weak supination of the forearm from a pronated position

Palpation

With forearm pronated on the posterior aspect of the distal forearm and dorsal surface of the hand just medial to the extensor digitorum tendon of the index finger with extension of the index fingers and flexion of the third, fourth, and fifth fingers

Innervation

Radial nerve (C6–C8)

Application, strengthening, and flexibility

The extensor indicis muscle is the pointing muscle. That is, it is responsible for extending the index finger, particularly when the other fingers are flexed. It also provides weak assistance to wrist extension and may be developed through exercises similar to those described for the extensor digitorum.

The extensor indicis is stretched by passively taking the index finger into maximal flexion at its metacarpophalangeal, proximal interphalangeal, and distal interphalangeal joints while fully flexing the wrist.

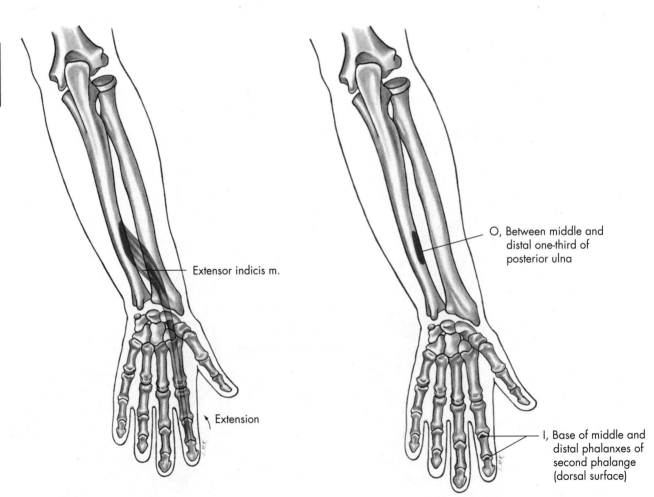

Extensor indicis m.

Extension

O, Between middle and distal one-third of posterior ulna

I, Base of middle and distal phalanxes of second phalange (dorsal surface)

FIG. 7.21 ● Extensor indicis muscle, posterior view. O, Origin; I, Insertion.

Extensor digiti minimi muscle   FIG. 7.22
(eks-ten´sor dij´i-ti min´im-i)

Origin
Lateral epicondyle of the humerus

Insertion
Base of the middle and distal phalanxes of the fifth phalange (dorsal surface)

Action
Extension of the little finger at the metacarpophalangeal joint
Weak wrist extension
Weak elbow extension

Palpation
Passing over the dorsal aspect of the distal radioulnar joint, particularly with relaxed flexion of other fingers and alternating fifth finger extension and relaxation; dorsal surface of forearm immediately medial to the extensor digitorum and lateral to the extensor carpi ulnaris

Innervation
Radial nerve (C6–C8)

Application, strengthening, and flexibility
The primary function of the extensor digiti minimi muscle is to assist the extensor digitorum in extending the little finger. Because of its dorsal relationship to the wrist, it also provides weak assistance in wrist extension. It is strengthened with the same exercises described for the extensor digitorum.

The extensor digiti minimi is stretched by passively taking the little finger into maximal flexion at its metacarpophalangeal, proximal interphalangeal, and distal interphalangeal joints while fully flexing the wrist.

5th MCP extension

Wrist extension

Elbow extension

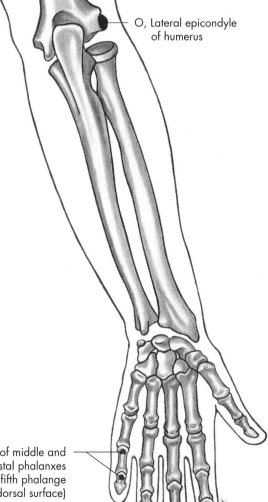

Extensor digiti minimi m.

Extension

O, Lateral epicondyle of humerus

I, Base of middle and distal phalanxes of fifth phalange (dorsal surface)

FIG. 7.22 • Extensor digiti minimi muscle, posterior view. O, Origin; I, Insertion.

Extensor pollicis longus muscle FIG. 7.23

(eks-ten´sor pol´i-sis lon´gus)

Thumb
CMC
extension

Origin

Posterior lateral surface of the lower middle ulna

Insertion

Base of the distal phalanx of the thumb (dorsal surface)

Thumb
MCP
extension

Action

Extension of the thumb at the carpometacarpal,
  metacarpophalangeal, and interphalangeal joints
Extension of the wrist
Abduction of the wrist
Weak supination of the forearm from a pronated
  position

Thumb IP
extension

Palpation

Dorsal aspect of the hand to its insertion on the base
  of the distal phalanx; also on the posterior surface
  of the lower forearm between radius and ulna just
  proximal to the extensor indicis and medial to
  the extensor pollicis brevis and abductor pollicis
  longus with the forearm pronated and fingers in
  relaxed flexion while actively extending the thumb

Wrist
extension

Wrist
abduction

Radioulnar
supination

Innervation

Radial nerve (C6–C8)

Application, strengthening, and flexibility

The primary function of the extensor pollicis longus muscle is extension of the thumb, although it does provide weak assistance in wrist extension.

It may be strengthened by extending the flexed thumb against manual resistance. It is stretched by passively taking the entire thumb into maximal flexion at its carpometacarpal, metacarpophalangeal, and interphalangeal joints while fully flexing the wrist with the forearm in pronation.

The tendons of the extensor pollicis longus and extensor pollicis brevis, along with the tendon of the abductor pollicis longus, form the "anatomical snuffbox," the small depression that develops between these two tendons when they contract. The name *anatomical snuffbox* originates from tobacco users placing their snuff in this depression. Deep in the snuffbox the scaphoid bone can be palpated and is often a site of point tenderness when it is fractured.

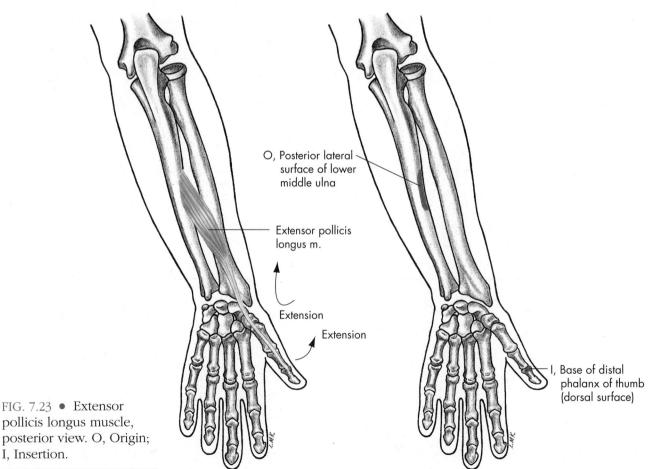

O, Posterior lateral surface of lower middle ulna

Extensor pollicis longus m.

Extension

Extension

I, Base of distal phalanx of thumb (dorsal surface)

FIG. 7.23 • Extensor pollicis longus muscle, posterior view. O, Origin; I, Insertion.

Extensor pollicis brevis muscle FIG. 7.24

(eks-ten´sor pol´i-sis bre´vis)

Origin

Posterior surface of the lower middle radius

Insertion

Base of the proximal phalanx of the thumb (dorsal surface)

Action

Extension of the thumb at the carpometacarpal and metacarpophalangeal joints
Wrist abduction
Weak wrist extension

Palpation

Just lateral to the extensor pollicis longus tendon on the dorsal side of the hand to its insertion on the proximal phalanx with extension of the thumb carpometacarpal and metacarpophalangeal joints and flexion of the interphalangeal joint

Innervation

Radial nerve (C6, C7)

Application, strengthening, and flexibility

The extensor pollicis brevis assists the extensor pollicis longus in extending the thumb. Because of its dorsal relationship to the wrist, it, too, provides weak assistance in wrist extension.

It may be strengthened through the same exercises described for the extensor pollicis longus muscle. It is stretched by passively taking the first carpometacarpal joint and the metacarpophalangeal joint of the thumb into maximal flexion while fully flexing and adducting the wrist.

Thumb
CMC
extension

Thumb
MCP
extension

Wrist
abduction

Chapter

7

Wrist
extension

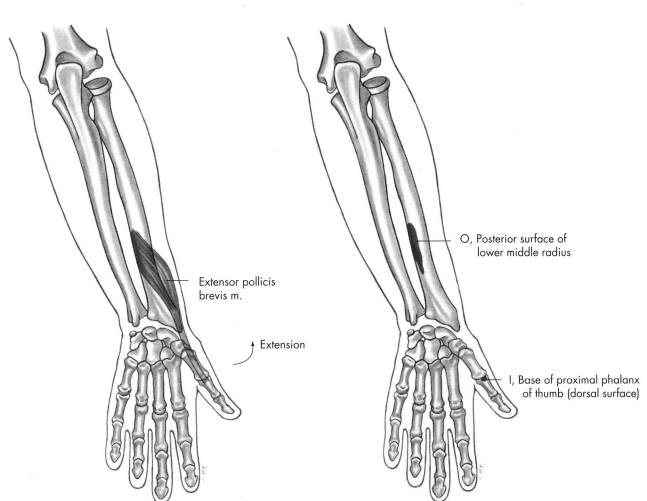

Extensor pollicis brevis m.

Extension

O, Posterior surface of lower middle radius

I, Base of proximal phalanx of thumb (dorsal surface)

FIG. 7.24 • Extensor pollicis brevis muscle, posterior view. O, Origin; I, Insertion.

Abductor pollicis longus
muscle   FIG. 7.25

(ab-duk´tor pol´i-sis lon´gus)

Origin

Posterior aspect of the radius and midshaft of the ulna

Insertion

Base of the first metacarpal (dorsal lateral surface)

Action

Abduction of the thumb at the carpometacarpal joint
Abduction of the wrist
Extension of the thumb at the carpometacarpal joint
Weak supination of the forearm from a pronated position
Weak extension of the wrist joint

Palpation

With forearm in neutral pronation/supination on the lateral aspect of the wrist joint just proximal to the first metacarpal during active thumb and wrist abduction

Innervation

Radial nerve (C6, C7)

Application, strengthening, and flexibility

The primary function of the abductor pollicis longus muscle is abduction of the thumb, although it does provide some assistance in abduction of the wrist. It may be developed by abducting the thumb from the adducted position against a manually applied resistance. Stretching of the abductor pollicis longus is accomplished by fully flexing and adducting the entire thumb across the palm with the wrist fully adducted and in slight flexion.

Wrist
abduction

Thumb
CMC
extension

Chapter
7

Radioulnar
supination

Wrist
extension

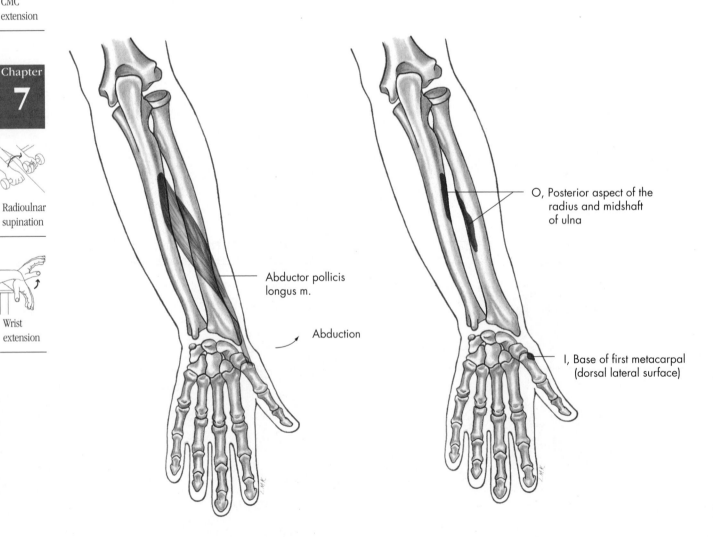

FIG. 7.25 • Abductor pollicis longus muscle, posterior view. O, Origin; I, Insertion.

Intrinsic muscles of the hand

The intrinsic hand muscles may be grouped according to location as well as according to the parts of the hand they control (Fig. 7.26). The abductor pollicis brevis, opponens pollicis, flexor pollicis brevis, and adductor pollicis make up the thenar eminence—the muscular pad on the palmar surface of the first metacarpal. The hypothenar eminence is the muscular pad that forms the ulnar border on the palmar surface of the hand and is made up of the abductor digiti minimi, flexor digiti minimi brevis, palmaris brevis, and opponens digiti minimi. The intermediate muscles of the hand consist of three palmar interossei, four dorsal interossei, and four lumbrical muscles.

Four intrinsic muscles act on the carpometacarpal joint of the thumb. The opponens pollicis is the muscle that causes opposition in the thumb metacarpal. The abductor pollicis brevis abducts the thumb metacarpal and is assisted in this action by the flexor pollicis brevis, which also flexes the thumb metacarpal. The metacarpal of the thumb is adducted by the adductor pollicis. Both the flexor pollicis brevis and the adductor pollicis flex the proximal phalanx of the thumb.

The three palmar interossei are adductors of the second, fourth, and fifth phalanges. The four dorsal interossei both flex and abduct the index, middle, and ring proximal phalanxes, in addition to assisting with extension of the middle and distal phalanxes of these fingers. The third dorsal

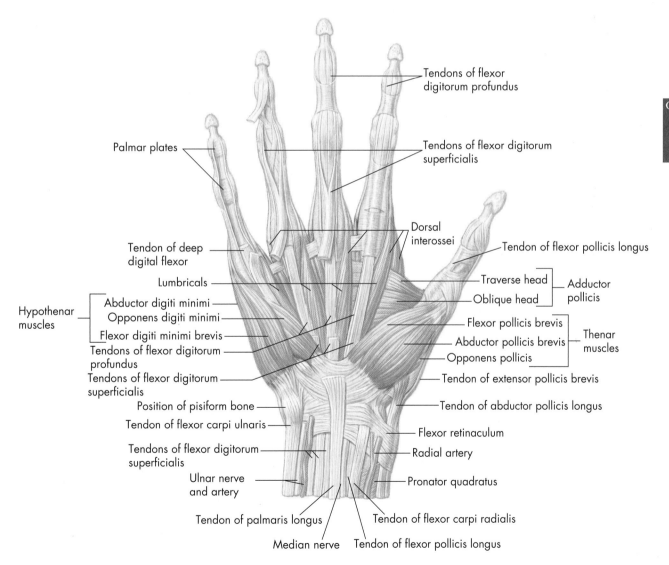

FIG. 7.26 • Intrinsic muscles of the right hand, anterior view.

interossei also adducts the middle finger. The four lumbricals flex the index, middle, ring, and little proximal phalanxes and extend the middle and distal phalanxes of these fingers.

Three muscles act on the little finger. The opponens digiti minimi causes opposition of the little finger metacarpal. The abductor digiti minimi abducts the little finger metacarpal, and the flexor digiti minimi brevis flexes this metacarpal.

Refer to Table 7.2 for further details regarding the intrinsic muscles of the hand.

Thumb CMC abduction

Thumb CMC flexion

Thumb MCP flexion

2nd–5th MCP flexion

2nd–5th MCP, PIP, and DIP extension

TABLE 7.2 • **Intrinsic muscles of the hand**

	Muscle	Origin	Insertion	Action	Palpation	Innervation
Thenar muscles	Opponens pollicis	Anterior surface of transverse carpal ligament, trapezium	Lateral border of 1st metacarpal	CMC opposition of thumb	Palmar aspect of 1st metacarpal with opposition of fingertips to thumb	Median nerve (C6, C7)
	Abductor pollicis brevis	Anterior surface of transverse carpal ligament, trapezium, scaphoid	Base of 1st proximal phalanx	CMC abduction of thumb	Radial aspect of palmar surface of 1st metacarpal with 1st CMC abduction	Median nerve (C6, C7)
	Flexor pollicis brevis	Superficial head: trapezium and transverse carpal ligament. Deep head: ulnar aspect of 1st metacarpal	Base of proximal phalanx of 1st metacarpal	CMC flexion and abduction; MCP flexion of thumb	Medial aspect of thenar eminence just proximal to 1st MCP joint with 1st MCP flexion against resistance	Superficial head: median nerve (C6, C7) Deep head: ulnar nerve (C8, T1)
	Adductor pollicis	Transverse head: anterior shaft of 3rd metacarpal Oblique head: base of 2nd and 3rd metacarpals, capitate, trapezoid	Ulnar aspect of base of proximal phalanx of 1st metacarpal	CMC adduction; MCP flexion of thumb	Palmar surface between 1st and 2nd metacarpal with 1st CMC adduction	Ulnar nerve (C8, T1)
Intermediate muscles	Palmar interossei	Shaft of 2nd, 4th, and 5th metacarpals and extensor expansions	Bases of 2nd, 4th, and 5th proximal phalanxes and extensor expansions	MCP adduction of 2nd, 4th, and 5th phalanges	Cannot be palpated	Ulnar nerve (C8, T1)
	Dorsal interossei	Two heads on shafts on adjacent metacarpals	Bases of 2nd, 3rd, and 4th proximal phalanxes and extensor expansions	MCP flexion and abduction; PIP/DIP extension of 2nd, 3rd, and 4th phalanges; MCP adduction of 3rd phalange	Dorsal surface between 1st and 2nd metacarpals, between shafts of 2nd through 5th metacarpals with active abduction/adduction of 2nd, 3rd, and 4th MCP joints	Ulnar nerve, palmar branch (C8, T1)
	Lumbricals	Flexor digitorum profundus tendon in center of palm	Extensor expansions on radial side of 2nd, 3rd, 4th, and 5th proximal phalanxes	MCP flexion and PIP/DIP extension of 2nd, 3rd, 4th, and 5th phalanges	Cannot be palpated	1st and 2nd: median nerve (C6, C7) 3rd and 4th: ulnar nerve (C8, T1)

TABLE 7.2 (continued) • Intrinsic muscles of the hand

	Muscle	Origin	Insertion	Action	Palpation	Innervation
Hypothenar muscles	Opponens digiti minimi	Hook of hamate and adjacent transverse carpal ligament	Medial border of 5th metacarpal	MCP opposition of 5th phalange	On radial aspect of hypothenar eminence with opposition of 5th phalange to thumb	Ulnar nerve (C8, T1)
	Abductor digiti minimi	Pisiform and flexor carpi ulnaris tendon	Ulnar aspect of base of 5th proximal phalanx	MCP abduction of 5th phalange	Ulnar aspect of hypothenar eminence with 5th MCP abduction	Ulnar nerve (C8, T1)
	Flexor digiti minimi brevis	Hook of hamate and adjacent transverse carpal ligament	Ulnar aspect of base of 5th proximal phalanx	MCP flexion of 5th phalange	Palmar surface of 5th metacarpal, lateral to opponens digiti minimi with 5th MCP flexion against resistance	Ulnar nerve (C8, T1)
	Palmaris brevis	Transverse carpal ligament and medial margin of palmar aponeurosis	Skin of ulnar border of palm	Tenses the skin on the ulnar side	Ulnar border of the palm of the hand	Ulnar nerve (C8, T1)

REVIEW EXERCISES

1. List the planes in which each of the following wrist, hand, and finger joint movements occurs. List the axis of rotation for each movement in each plane.
   a. Abduction
   b. Adduction
   c. Flexion
   d. Extension
2. Discuss why the thumb is the most important part of the hand.
3. How should boys and girls be taught to do push-ups? Justify your answer.
   a. Hands flat on the floor
   b. Fingertips

4. **Muscle analysis chart** • Wrist, hand, and fingers

Fill in the chart by listing the muscles primarily involved in each movement.	
Wrist and hand	
Flexion	Extension
Adduction	Abduction
Fingers, metacarpophalangeal joints	
Flexion	Extension
Abduction	Adduction

Muscle analysis chart (continued)

Fingers, proximal interphalangeal joints	
Flexion	Extension

Fingers, distal interphalangeal joints	
Flexion	Extension

Thumb	
Flexion	Extension
Abduction	Adduction

5. List the muscles involved in the little finger as you type on a computer keyboard and reach for the left tab key with the wrists properly stabilized in an ergonomic position.
6. Describe the importance of the intrinsic muscles in the hand as you reach to turn a doorknob.
7. Determine the kind of flexibility exercises that would be indicated for a patient with carpal tunnel syndrome, and explain in detail how they should be performed.

LABORATORY EXERCISES

1. Locate the following parts of the humerus, radius, ulna, carpals, and metacarpals on a human skeleton and on a subject.
   a. Skeleton
      1. Medial epicondyle
      2. Lateral epicondyle
      3. Lateral supracondylar ridge
      4. Trochlea
      5. Capitulum
      6. Coronoid process
      7. Tuberosity of the radius
      8. Styloid process—radius
      9. Styloid process—ulna
      10. First and third metacarpals
      11. Wrist bones
      12. First phalanx of third metacarpal

   b. Subject
      1. Medial epicondyle
      2. Lateral epicondyle
      3. Lateral supracondylar ridge
      4. Pisiform
      5. Scaphoid (navicular)
2. How and where can the following muscles be palpated on a human subject?
   a. Flexor pollicis longus
   b. Flexor carpi radialis
   c. Flexor carpi ulnaris
   d. Extensor digitorum communis
   e. Extensor pollicis longus
   f. Extensor carpi ulnaris
3. Demonstrate the action and list the muscles primarily responsible for the following movements at the wrist joint:
   a. Flexion
   b. Extension
   c. Abduction
   d. Adduction
4. With a laboratory partner, determine how and why maintaining full flexion of all the fingers is impossible when passively moving the wrist into maximal flexion. Is it also difficult to maintain maximal extension of all the finger joints while passively taking the wrist into full extension?

5. Wrist and hand joint exercise movement analysis chart

After analyzing each exercise in the chart, break each into two primary movement phases, such as a lifting phase and a lowering phase. For each phase, determine the wrist and hand joint movements that occur, and then list the wrist and hand joint muscles primarily responsible for causing/controlling those movements. Beside each muscle in each movement, indicate the type of contraction as follows: I—isometric; C—concentric; E—eccentric.

Exercise	Initial movement (lifting) phase		Secondary movement (lowering) phase	
	Movement(s)	Agonist(s)—(contraction type)	Movement(s)	Agonist(s)—(contraction type)
Push-up				
Chin-up				
Bench press				
Dip				
Lat pull				
Ball squeeze				
Frisbee throw				

6. Wrist and hand joint sport skill analysis chart

Analyze each skill in the chart and list the movements of the right and left wrist and hand joints in each phase of the skill. You may prefer to list the initial positions that the wrist and hand joints are in for the stance phase. After each movement, list the wrist and hand joint muscle(s) primarily responsible for causing/controlling the movement. Beside each muscle in each movement, indicate the type of contraction as follows: I—isometric; C—concentric; E—eccentric. It may be desirable to study the concepts for analysis in Chapter 8 for the various phases.

Exercise		Stance phase	Preparatory phase	Movement phase	Follow-through phase
Baseball pitch	(R)				
	(L)				
Volleyball serve	(R)				
	(L)				
Tennis serve	(R)				
	(L)				
Softball pitch	(R)				
	(L)				

Wrist and hand joint sport skill analysis chart (continued)

Exercise		Stance phase	Preparatory phase	Movement phase	Follow-through phase
Tennis backhand	(R)				
	(L)				
Batting	(R)				
	(L)				
Bowling	(R)				
	(L)				
Basketball free throw	(R)				
	(L)				

References

Gabbard CP, et al: Effects of grip and forearm position on flex arm hang performance, *Research Quarterly for Exercise and Sport,* July 1983.

Gench BE, Hinson MM, Harvey PT: *Anatomical kinesiology,* Dubuque, IA, 1995, Eddie Bowers.

Guarantors of Brain: *Aids to the examination of the peripheral nervous system,* ed 4, London, 2000, Saunders.

Hamilton N, Weimer W, Luttgens K: *Kinesiology: scientific basis of human motion,* ed 12, New York, 2012, McGraw-Hill.

Hislop HJ, Montgomery J: *Daniels and Worthingham's muscle testing: techniques of manual examination,* ed 8, Philadelphia, 2007, Saunders.

Lindsay DT: *Functional human anatomy,* St. Louis, 1996, Mosby.

Magee DJ: *Orthopedic physical assessment,* ed 5, Philadelphia, 2008, Saunders.

Muscolino JE: *The muscular system manual: the skeletal muscles of the human body,* ed 3, St. Louis, 2010, Elsevier Mosby.

Norkin CC, Levangie PK: *Joint structure and function—a comprehensive analysis,* ed 5, Philadelphia, 2011, Davis.

Norkin CC, White DJ: *Measurement of joint motion: a guide to goniometry,* ed 4, Philadelphia, 2009, Davis.

Oatis CA: *Kinesiology: the mechanics and pathomechanics of human movement,* ed 2, Philadelphia, 2008, Lippincott Williams & Wilkins.

Rasch PJ: *Kinesiology and applied anatomy,* ed 7, Philadelphia, 1989, Lea & Febiger.

Seeley RR, Stephens TD, Tate P: *Anatomy & physiology,* ed 8, New York, 2008, McGraw-Hill.

Sieg KW, Adams SP: *Illustrated essentials of musculoskeletal anatomy,* ed 4, Gainesville, FL, 2002, Megabooks.

Sisto DJ, et al: An electromyographic analysis of the elbow in pitching, *American Journal of Sports Medicine* 15:260, May–June 1987.

Smith LK, Weiss EL, Lehmkuhl LD: *Brunnstrom's clinical kinesiology,* ed 5, Philadelphia, 1996, Davis.

Springer SI: Racquetball and elbow injuries, *National Racquetball* 16:7, March 1987.

Stone RJ, Stone JA: *Atlas of the skeletal muscles,* ed 6, New York, 2009, McGraw-Hill.

Van De Graaff KM: *Human anatomy,* ed 6, Dubuque, IA, 2002, McGraw-Hill.

 For additional resources and a list of related websites, visit **www.mhhe.com/floyd19e**.

Worksheet Exercises

For in- or out-of-class assignments, or for testing, utilize this tear-out worksheet.

Worksheet 1

Using crayons or colored markers, draw and label on the worksheet the following muscles. Indicate the origin and insertion of each muscle with an "O" and an "I," respectively.

a. Flexor pollicis longus
b. Flexor carpi radialis
c. Flexor carpi ulnaris
d. Extensor digitorum
e. Extensor pollicis longus

f. Extensor carpi ulnaris
g. Extensor pollicis brevis
h. Palmaris longus
i. Extensor carpi radialis longus
j. Extensor carpi radialis brevis

k. Extensor digiti minimi
l. Extensor indicis
m. Flexor digitorum superficialis
n. Flexor digitorum profundus
o. Abductor pollicis longus

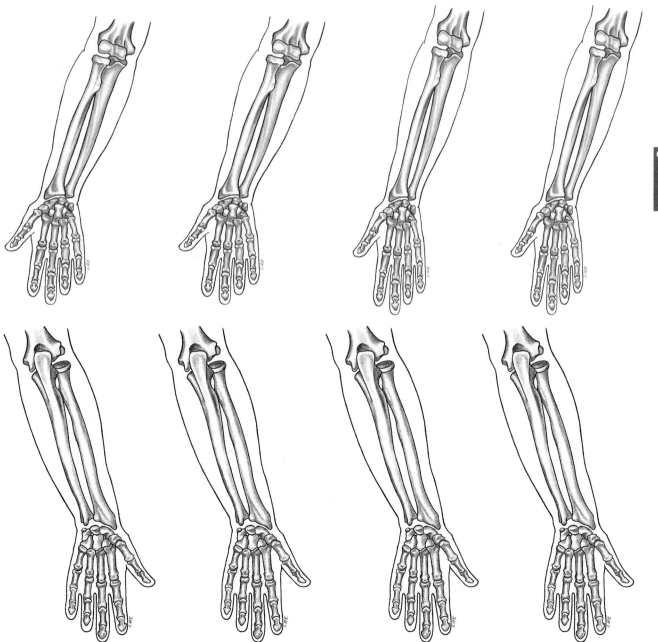

Chapter
7

Worksheet Exercises

For in- or out-of-class assignments, or for testing, utilize this tear-out worksheet.

Worksheet 2

Label and indicate with arrows the following movements of the wrist and hands. For each motion, list the agonist muscle(s), the plane in which the motion occurs, and its axis of rotation.

a. Flexion

b. Extension

c. Abduction (ulnar flexion)

d. Adduction (radial flexion)

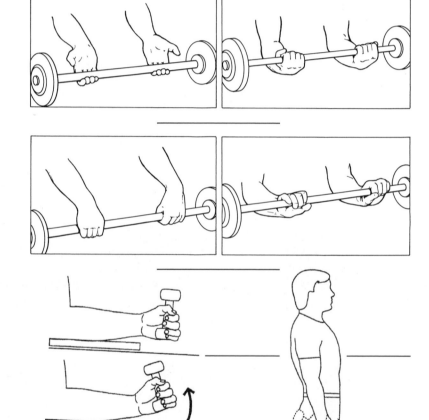

a. The _____ muscle(s) cause _____ to occur in the

_____ plane about the _____ axis.

b. The _____ muscle(s) cause _____ to occur in the

_____ plane about the _____ axis.

c. The _____ muscle(s) cause _____ to occur in the

_____ plane about the _____ axis.

d. The _____ muscle(s) cause _____ to occur in the

_____ plane about the _____ axis.

MUSCULAR ANALYSIS OF UPPER EXTREMITY EXERCISES

Objectives

- To begin analyzing sport skills in terms of phases and the various joint movements occurring in those phases

- To understand various conditioning principles and how to apply them to strengthening major muscle groups

- To analyze an exercise to determine the joint movements and the types of contractions occurring in the specific muscles involved in those movements

- To learn and understand the concept of open versus closed kinetic chain

- To learn to group individual muscles into units that produce certain joint movements

- To begin to think of exercises that increase the strength and endurance of individual muscle groups

- To learn to analyze and prescribe exercises to strengthen major muscle groups.

Online Learning Center Resources

Visit *Manual of Structural Kinesiology's* **Online Learning Center** at **www.mhhe.com/floyd19e** for additional information and study material for this chapter, including:

- *Self-grading quizzes*
- *Anatomy flashcards*
- *Animations*
- *Related websites*

Proper functioning of the upper extremities is critical for most sports activities, as well as for many activities of daily living. Strength and endurance in this part of the human body are essential for improved appearance and posture, as well as for more efficient skill performance. Unfortunately, it is often one of the body's weaker areas, considering the number of muscles involved. Specific exercises and activities to condition this area should be intelligently selected by becoming thoroughly familiar with the muscles involved.

Simple exercises may be used to begin teaching individuals how to group muscles to produce joint movement. Some of these simple introductory exercises are included in this chapter.

The early analysis of exercise makes the study of structural kinesiology more meaningful as students come to better understand the importance of individual muscles and groups of muscles in bringing about joint movements in various exercises. Chapter 13 contains analysis of exercises for the entire body, with emphasis on the trunk and lower extremities. Contrary to what most beginning students in structural kinesiology believe, muscular analysis of activities is not difficult once the basic concepts are understood.

Upper-extremity activities

Children seem to have an innate desire to climb, swing, and hang. Such movements use the muscles of the hands, wrists, elbows, and shoulder joints. But the opportunity to perform these types of activities is limited in our modern culture.

Chapter

8

Unless emphasis is placed on the development of this area of our bodies by physical education teachers in elementary schools, for both boys and girls, it will continue to be muscularly the weakest area of our bodies. Weakness in the upper extremities can impair skill development and performance in many common, enjoyable recreational activities, such as golf, tennis, softball, and racquetball. People enjoy what they can do well, and they can be taught to enjoy activities that will increase the strength and endurance of this part of the body. These and other such activities can be enjoyed throughout the entire adult life; therefore, adequate skill development built on an appropriate base of muscular strength and endurance is essential for enjoyment and prevention of injury.

Often, we perform typical strengthening exercises in the weight room, such as the bench press, overhead press, and biceps curl. These are good exercises, but they all concentrate primarily on the muscles of the anterior upper extremity. This can lead to an overdevelopment of these muscles with respect to the posterior muscles. As a result, individuals may become strong and tight anteriorly and weak and flexible posteriorly. It is for these reasons that one must be able to analyze specific strengthening exercises and determine the muscles involved so that overall muscular balance is addressed through appropriate exercise prescription.

Concepts for analysis

In analyzing activities, it is important to understand that muscles are usually grouped according to their concentric function and work in paired opposition to an antagonistic group. An example of this **aggregate** muscle grouping to perform a given joint action is seen with the elbow flexors all working together. In this example, the elbow flexors (biceps, brachii, brachialis, and brachioradialis) are concentrically contracting as an agonist group to achieve flexion. As they flex the elbow, each muscle contributes significantly to the task. They are working in opposition to their antagonists, the triceps brachii and anconeus. The triceps brachii and anconeus work together as an aggregate muscle group to cause elbow extension, but in this example they are cooperating in their lengthening to allow the flexors to perform their task. In this cooperative lengthening, the triceps and anconeus may or may not be under active tension. If there is no tension, then the lengthening is passive, caused totally by the elbow flexors. If there is active tension, then the elbow extensors are contracting eccentrically to control the amount and speed of lengthening.

An often confusing aspect is that, depending on the activity, these muscle groups can function to control the exact opposite actions by contracting eccentrically. That is, through eccentric contractions, the elbow flexors may control elbow extension, as in lowering the weight in a biceps curl, and the triceps brachii and anconeus may control elbow flexion, as in lowering the weight in a triceps extension (see Tables 8.3 and 8.4). Exercise professionals should be able to view an activity and not only determine which muscles are performing the movement but also know what type of contraction is occurring and what kinds of exercises are appropriate for developing the muscles. Chapter 2 provides a review of how muscles contract to work in groups to function in joint movement.

Analysis of movement

In analyzing various exercises and sport skills, it is essential to break down all the movements into phases. The number of phases, usually three to five, will vary depending on the skill. Practically all sport skills will have at least a preparatory phase, a movement phase, and a follow-through phase. Many will also begin with a stance phase and end with a recovery phase. The names of the phases will vary from skill to skill to fit in with the terminology used in various sports, and they may also vary depending on the body part involved. In some cases, these major phases may be divided even further, as with baseball, in which the preparatory phase for the pitching arm is broken into early cocking and late cocking.

The **stance phase** allows the athlete to assume a comfortable and appropriately balanced body position from which to initiate the sport skill. The emphasis is on setting the various joint angles in their correct positions with respect to one another and to the sport surface. Generally, with respect to the subsequent phases, the stance phase is a relatively static phase involving fairly short ranges of motion. Due to the minimal amount of movement in this phase, the majority of the joint position maintenance throughout the body will be accomplished through isometric contractions.

The **preparatory phase**, often referred to as the cocking or wind-up phase, is used to lengthen the appropriate muscles so that they will be in position to generate more force and momentum as they concentrically contract in the next phase. It is the most critical phase in achieving the desired result of the activity and becomes more dynamic as the need for explosiveness increases. Generally, to lengthen the muscles needed in the next phase, concentric contractions occur in their antagonist muscles in this phase.

The **movement phase**, sometimes known as the acceleration, action, motion, or contact phase, is the action part of the skill. It is the phase in which the summation of force is generated directly to the ball, sport object, or opponent and is usually characterized by near-maximal concentric activity in the involved muscles.

The **follow-through phase** begins immediately after the climax of the movement phase, in order to bring about negative acceleration of the involved limb or body segment. In this phase, often referred to as the deceleration phase, the velocity of the body segment progressively decreases, usually over a wide range of motion. This velocity decrease is usually attributable to high eccentric activity in the muscles that were antagonist to the muscles used in the movement phase. Generally, the greater the acceleration in the movement phase, the greater the length and importance of the follow-though phase. Occasionally, some athletes may begin the follow-through phase too soon, thereby cutting short the movement phase and achieving a less-than-desirable result in the activity.

The **recovery phase** is used after follow-through to regain balance and positioning to be ready for the next sport demand. To a degree, the muscles used eccentrically in the follow-through phase to decelerate the body or body segment will be used concentrically in recovery to bring about the initial return to a functional position.

Skill analysis can be seen with the example of a baseball pitch in Fig. 8.1. The stance phase begins when the player assumes a position with the ball in the glove before receiving the signal from the catcher. The pitcher begins the preparatory phase by extending the throwing arm posteriorly and rotating the trunk to the right in conjunction with left hip flexion. The right shoulder girdle is fully retracted in combination with abduction and maximum external rotation of the glenohumeral joint to complete this phase. Immediately following, the movement phase begins with forward movement of the arm and continues until ball release.

Stance   Preparatory

Movement

Follow-through

FIG. 8.1 • Skill analysis phases—baseball pitch. The stance phase is accomplished predominantly with isometric contractions. Movements in the preparatory phase are accomplished primarily through concentric contractions. The movement phase involves significant concentric activity. Eccentric activity is high during the follow-through phase.

At ball release, the follow-through phase begins as the arm continues moving in the same direction established by the movement phase until the velocity decreases to the point that the arm can safely change movement direction. This deceleration of the body, and especially the arm, is accomplished by high amounts of eccentric activity. At this point, the recovery phase begins, enabling the player to reposition to field the batted ball. In this example, reference has been made primarily to the throwing arm, but there are many similarities in other overhand sport skills, such as the tennis serve, javelin throw, and volleyball serve. In actual practice, the movements of each joint in the body should be analyzed with respect to the various phases.

The kinetic chain concept

As you have learned, our extremities consist of several bony segments linked by a series of joints. These bony segments and their linkage system of joints may be likened to a chain. Just as with a chain, any one link in the extremity may be moved individually without significantly affecting the other links if the chain is open or not attached at one end. However, if the chain is securely attached or closed, substantial movement of any one link cannot occur without substantial and subsequent movement of the other links.

In the body, an extremity may be seen as representing an **open kinetic chain** if the distal end of the extremity is not fixed to a relatively stable surface. This arrangement allows any one joint in the extremity to move or function separately without necessitating movement of other joints in the extremity. This does not mean that open kinetic chain activities have to involve only one joint but rather that motion at one joint does not require motion at other joints in the chain. Examples in the upper extremity of these single joint exercises include the shoulder shrug, deltoid raise (shoulder abduction), and biceps curl (Fig. 8.2, *A*). Lower-extremity examples include seated hip flexion, knee extension, and ankle dorsiflexion exercises (Fig. 8.2, *C*). In all these examples, the core of the body and the proximal segment are stabilized while the distal segment is free to move in space through a single plane. These types of exercises are known as joint-isolation exercises and are beneficial in isolating a particular joint to concentrate on specific muscle groups. However, they are not very functional in that most physical activity, particularly for the lower extremity, requires multiple-joint activity involving numerous muscle groups simultaneously. Furthermore, since the joint is stable proximally and loaded distally, shear forces are acting on the joint, with potential negative consequences.

If the distal end of the extremity is fixed, as in a pull-up, push-up, dip, squat, or dead lift, the extremity represents a **closed kinetic chain**. See Fig. 8.2, *B* and *D*. In this closed system, movement of one joint cannot occur without causing predictable movements of the other joints in the extremity. Closed-chain activities are very functional and involve the body moving in relation to the relatively fixed distal segment. The advantage of multiple-joint exercises is that several joints are involved and numerous muscle groups must participate in causing and controlling the multiple-plane movements, which strongly correlate to most physical activities. Additionally, the joint is more stable due to the joint compressive forces from weight bearing.

To state the differences another way, open-chain exercises involve the extremity being moved to or from the stabilized body, whereas closed-chain exercises involve the body being moved to or from the stabilized extremity. Table 8.1 provides a

A

B

FIG. 8.2 • Open versus closed kinetic chain activities. **A,** Open-chain activity for the upper extremity; **B,** Closed-chain activity for the upper extremity.

C

D

FIG. 8.2 (continued) • Open versus closed kinetic chain activities. **C,** Open-chain activity for the lower extremity; **D,** Closed-chain activity for the lower extremity.

TABLE 8.1 • Differences between open- and closed-chain exercises

Variable	Open-chain exercise	Closed-chain exercise
Distal end of extremity	Free in space and not fixed	Fixed to something
Movement pattern	Characterized by rotary stress in the joint (often nonfunctional)	Characterized by linear stress in the joint (functional)
Joint movements	Occur in isolation	Multiple occur simultaneously
Muscle recruitment	Isolated (minimal muscular co-contraction)	Multiple (significant muscular co-contraction)
Joint axis	Stable during movement patterns	Primarily transverse
Movement plane	Usually single	Multiple (triplanar)
Proximal segment of joint	Stable	Mobile
Distal segment of joint	Mobile	Mobile, except for most distal aspect
Motion occurs	Distal to instantaneous axis of rotation	Proximal and distal to instantaneous axis of rotation
Functionality	Often nonfunctional, especially lower extremity	Functional
Joint forces	Shear	Compressive
Joint stability	Decreased due to shear and distractive forces	Increased due to compressive forces
Stabilization	Artificial	Not artificial, rather realistic and functional
Loading	Artificial	Physiological, provides for normal proprioceptive and kinesthetic feedback

Adapted from Ellenbecker TS, Davies GJ: *Closed kinetic chain exercise: a comprehensive guide to multiple-joint exercise,* Champaign, IL, 2001, Human Kinetics.

Chapter

8

comparison of variables that differ between open- and closed-chain exercises, and Fig. 8.2 provides examples of each.

Not every exercise or activity can be classified totally as either an open- or closed-chain exercise. For example, walking and running are both open and closed due to their swing and stance phases, respectively. Another case is bicycle riding, which is mixed in that the pelvis on the seat is the stablest segment, but the feet are attached to movable pedals.

Consideration of the open versus closed kinetic chain is important in determining both the muscles and their types of contractions when analyzing sports activities. Realizing the relative differences in demands on the musculoskeletal system through detailed analysis of skilled movements is critical for determining the most appropriate conditioning exercises to improve performance. Generally, closed kinetic chain exercises are more functional and applicable to the demands of sports and physical activity. Most sports involve closed-chain activities in the lower extremities and open-chain activities in the upper extremities. However, there are many exceptions, and closed-chain conditioning exercises may be beneficial for extremities primarily involved in open-chain sporting activities. Open-chain exercises are useful in developing a specific muscle group at a single joint.

Conditioning considerations

It is not the intent of this book to thoroughly address conditioning principles, but a brief overview is provided to serve as a general reference and reminder of the importance of applying these concepts correctly when developing major muscle groups.

Overload principle

A basic physiological principle of exercise is the overload principle. It states that, within appropriate parameters, a muscle or muscle group increases in strength in direct proportion to the overload placed on it. While it is beyond the scope of this text to fully explain specific applications of the overload principle for each component of physical fitness, some general concepts follow. To improve the strength and functioning of major muscles, this principle should be applied to every large muscle group in the body, progressively throughout each year, at all age levels. In actual practice, the amount of overload applied varies significantly based on several factors. For example, an untrained person beginning a strength-training program will usually make significant gains in the amount of weight he or she is able to lift in the first few weeks of the exercise program. Most of

this increased ability is due to a refinement of neuromuscular function rather than to an actual increase in muscle tissue strength. Similarly, a well-trained person will see relatively minor improvements in the amount of weight that can be lifted over a much longer period of time. Therefore, the amount and rate of progressive overload are extremely variable and must be adjusted to match the specific needs of the individual's exercise objectives.

Overload may be modified by changing any one or a combination of three exercise variables— **frequency, intensity,** and **duration**. Frequency usually refers to the number of times per week. Intensity is usually a certain percentage of the absolute maximum, and duration usually refers to the number of minutes per exercise bout. Increasing the speed of doing the exercise, the number of repetitions, the weight, and the bouts of exercise are all ways to modify these variables and apply the overload principle. All these factors are important in determining the total exercise volume.

Overload is not always progressively increased. In certain periods of conditioning, the overload should actually be prescriptively reduced or increased to improve the total results of the entire program. This intentional variance in a training program at regular intervals is known as **periodization** and is done to bring about optimal gains in physical performance. Part of the basis for periodization is so that the athlete will be at his or her peak level during the most competitive part of the season. To achieve this, a number of variables may be manipulated, including the number of sets per exercise or repetitions per set, types of exercises, number of exercises per training session, rest periods between sets and exercises, resistance used for a set, type of muscle contraction, and number of training sessions per day and per week.

SAID principle

The **SAID** (**S**pecific **A**daptations to **I**mposed **D**emands) principle should be considered in all aspects of physiological conditioning and training. This principle, which states that the body will gradually, over time, adapt very specifically to the various stresses and overloads to which it is subjected, is applicable in every form of muscle training, as well as to the other systems of the body. For example, if an individual were to undergo several weeks of strength-training exercises for a particular joint through a limited range of motion, the specific muscles involved in performing the strengthening exercises would improve primarily in the ability to move against increased resistance through the specific range of motion used. There would be, in most cases, minimal strength gains significantly beyond the range of motion used in the training. Additionally, other components of physical fitness—such as flexibility, cardiorespiratory endurance, and muscular endurance—would be enhanced minimally, if at all. In other words, to achieve specific benefits, exercise programs must be specifically designed for the adaptation desired.

It should be recognized that this adaptation may be positive or negative, depending on whether the correct techniques are used and stressed in the design and administration of the conditioning program. Inappropriate or excessive demands placed on the body in too short a time span can result in injury. If the demands are too little or are administered too infrequently over too long a time period, less than desired improvement will occur. Conditioning programs and the exercises included in them should be analyzed to determine whether they are using the specific muscles for which they were intended in the correct manner.

Specificity

Specificity of exercise strongly relates to the discussion of the SAID principle. The components of physical fitness—such as muscular strength, muscular endurance, and flexibility—are not general body characteristics but rather are specific to each body area and muscle group. Therefore, the specific needs of the individual must be addressed when designing an exercise program. Quite often, it will be necessary to analyze an individual's exercise and skill technique to design an exercise program to meet his or her specific needs. Potential exercises to be used in the conditioning program must be analyzed to determine their appropriateness for the individual's specific needs. The goals of the exercise program should be determined regarding specific areas of the body, preferred time to physically peak, and physical fitness needs such as strength, muscular endurance, flexibility, cardiorespiratory endurance, and body composition. After establishing goals, a regimen incorporating the overload variables of frequency, intensity, and duration may be prescribed to include the entire body or specific areas in such a way as to address the improvement of the preferred physical fitness components. Regular observation and follow-up exercise analysis are necessary to ensure proper adherence to correct technique.

Muscular development

For years it was thought that a person developed adequate muscular strength, endurance, and flexibility through participation in sports activities. Now it is believed that a person needs to develop muscular strength, endurance, and flexibility in order to be able to participate safely and effectively in sports activities.

Adequate muscular strength, endurance, and flexibility of the entire body from head to toe should be developed through correct use of the appropriate exercise principles. Individuals responsible for this development need to prescribe exercises that will meet these objectives.

In schools this development should start at an early age and continue throughout the school years. Results of fitness tests such as sit-ups, the standing long jump, and the mile run reveal the need for considerable improvement in this area in children in the United States. Adequate muscular strength and endurance are important in the adult years for the activities of daily living, as well as for job-related requirements and recreational needs. Many back problems and other physical ailments could be avoided through proper maintenance of the musculoskeletal system. Refer to Chapters 4 through 7 as needed.

Analysis of upper-body exercises

Presented over the next several pages are brief analyses of several common upper-body exercises. Following and perhaps expanding on the approach used are encouraged in analyzing other upper-body activities. All muscles listed in the analysis are contracting concentrically unless specifically noted to be contracting eccentrically or isometrically.

Chapter
8

Valsalva maneuver

Many people bear down by holding their breath without thinking when attempting to lift something heavy. This bearing down, known as the Valsalva maneuver, is accomplished by exhaling against a closed epiglottis (the flap of cartilage behind the tongue that shuts the air passage when swallowing) and is thought by many to enhance lifting ability. It is mentioned here to caution against its use, because it causes a dramatic increase in blood pressure followed by an equally dramatic drop in blood pressure. Using the Valsalva maneuver can cause lightheadedness and fainting and can lead to complications in people with heart disease. Instead of using the Valsalva maneuver, people lifting should always be sure to use rhythmic and consistent breathing. It is usually advisable to exhale during the lifting or contracting phase and inhale during the lowering or recovery phase.

Shoulder pull

Description

In a standing or sitting position, the subject interlocks the fingers in front of the chest and then attempts to pull them apart (Fig. 8.3). This contraction is maintained for 5 to 20 seconds.

Analysis

In this type of exercise, there is little or no movement of the contracting muscles. In certain isometric exercises, contraction of the antagonistic muscles is as strong as contraction of the muscles attempting to produce the force for movement. The muscle groups contracting to produce a movement are designated the **agonists**. In the exercise just described, the agonists in the right upper extremity are antagonistic to the agonists in the left upper extremity, and vice versa (Table 8.2). This exercise results in isometric contractions of the wrist and hand, elbow, shoulder joint, and shoulder girdle muscles. The strength of the contraction depends on the angle of pull and the leverage of the joint involved. Thus, it is not the same at each point.

FIG. 8.3 • Shoulder pull.

Isometric exercises vary in the number of muscles contracting, depending on the type of exercise and the joints at which movement is attempted. The shoulder pull exercise produces some contraction of agonist muscles at four sets of joints. See Tables 4.1, 5.2, 6.1, and 7.1.

TABLE 8.2 • Shoulder pull

Joint	This entire exercise is designed so that all contractions presented are isometric				
	Action	Agonists	Action	Agonists
Wrist and hand	Flexion	Resisted by flexors of wrists and hand Agonists—wrist and MCP, PIP, PIP flexors	Flexion	Resisted by flexors of wrist and hand Antagonists—wrist and MCP, PIP, PIP flexors
Elbow joint	Extension	Resisted by flexors of wrist, elbow, and hand Agonists—triceps brachii, anconeus Antagonists—biceps brachii, brachialis, brachioradialis	Flexion	Resisted by extensors of wrist, elbow, and hand Agonists—biceps brachii, brachialis, brachioradialis Antagonists—triceps brachii, anconeus
Shoulder joint	Horizontal abduction	Resisted by horizontal abductors of contralateral shoulder joint Agonists—deltoid, infraspinatus, teres minor, latissimus dorsi Antagonists—contralateral deltoid, infraspinatus, teres minor, latissimus dorsi	Horizontal abduction	Resisted by horizontal abductors of contralateral shoulder joint Agonists—deltoid, infraspinatus, teres minor, latissimus dorsi Antagonists—contralateral deltoid, infraspinatus, teres minor, latissimus dorsi
Shoulder girdle	Adduction	Resisted by adductors of contralateral shoulder girdle Agonists—rhomboid and trapezius Antagonists—contralateral rhomboid and trapezius	Adduction	Resisted by adductors of contralateral shoulder girdle Agonists—rhomboid and trapezius Antagonists—contralateral rhomboid and trapezius

Arm curl

Description

With the subject in a standing position, the dumbbell is held in the hand with the palm to the front. The dumbbell is lifted until the elbow is completely flexed (Fig. 8.4). Then it is returned to the starting position.

Analysis

This open kinetic chain exercise is divided into two phases for analysis: (1) lifting phase to flexed position and (2) lowering phase to extended position (Table 8.3). *Note:* An assumption is made that no movement occurs in the shoulder joint and shoulder girdle, although many of the muscles of both the shoulder and the shoulder girdle are isometrically acting as stabilizers. Review Tables 4.1, 5.2, 6.1, and 7.1.

FIG. 8.4 • Arm curl. **A,** Beginning position in extension; **B,** Flexed position.

TABLE 8.3 • **Arm curl**

Joint	Lifting phase to flexed position		Lowering phase to extended position	
	Action	Agonists	Action	Agonists
Wrist and hand	Flexion*	Wrist and MCP, PIP, PIP flexors (isometric contraction) Flexor carpi radialis Flexor carpi ulnaris Palmaris longus Flexor digitorum profundus Flexor digitorum superficialis Flexor pollicis longus	Flexion*	Wrist and MCP, PIP, PIP flexors (isometric contraction) Flexor carpi radialis Flexor carpi ulnaris Palmaris longus Flexor digitorum profundus Flexor digitorum superficialis Flexor pollicis longus
Elbow	Flexion	Elbow flexors Biceps brachii Brachialis Brachioradialis	Extension	Elbow flexors (eccentric contraction) Biceps brachii Brachialis Brachioradialis

*The wrist is in a position of slight extension to facilitate greater active finger flexion in gripping the dumbbell. (The flexors remain isometrically contracted throughout the entire exercise, to hold the dumbbell.)

Triceps extension

Description

The subject may use the opposite hand to assist in maintaining the arm in a shoulder-flexed position. Then, grasping the dumbbell and beginning in full elbow flexion, the subject extends the elbow until the arm and forearm are straight. The shoulder joint and shoulder girdle are stabilized by the opposite hand. Consequently, no movement is assumed to occur in these areas (Fig. 8.5).

Analysis

This open kinetic chain exercise is divided into two phases for analysis: (1) lifting phase to extended position and (2) lowering phase to flexed position (Table 8.4). *Note:* An assumption is made that no movement occurs in the shoulder joint and shoulder girdle, although it is critical that many of the shoulder and shoulder girdle muscles contract isometrically to stabilize this area so that the exercise may be performed correctly. Review Tables 4.1, 5.2, 6.1, and 7.1.

FIG. 8.5 • Triceps extension. **A,** Beginning position in flexion; **B,** Extended position.

TABLE 8.4 • Triceps extension

| Joint | Lifting phase to extended position | | Lowering phase to flexed position | |
	Action	Agonists	Action	Agonists
Wrist and hand	Flexion*	Wrist and MCP, PIP, PIP flexors (isometric contraction) Flexor carpi radialis Flexor carpi ulnaris Palmaris longus Flexor digitorum profundus Flexor digitorum superficialis Flexor pollicis longus	Flexion*	Wrist and MCP, PIP, PIP flexors (isometric contraction) Flexor carpi radialis Flexor carpi ulnaris Palmaris longus Flexor digitorum profundus Flexor digitorum superficialis Flexor pollicis longus
Elbow	Extension	Elbow extensors Triceps brachii Anconeus	Flexion	Elbow extensors (eccentric contraction) Triceps brachii Anconeus

*The wrist is in a position of slight extension to facilitate greater active finger flexion in gripping the dumbbell.
(The flexors remain isometrically contracted throughout the entire exercise, to hold the dumbbell.)

Barbell press

Description

This open kinetic chain exercise is sometimes referred to as the *overhead* or *military press*. The barbell is held in a position high in front of the chest, with palms facing forward, feet comfortably spread, and back and legs straight (Fig. 8.6, *A*). From this position, the barbell is pushed upward until fully overhead (Fig. 8.6, *B*), and then it is returned to the starting position. See Tables 4.1, 5.2, 6.1, and 7.1.

Analysis

This exercise is separated into two phases for analysis: (1) lifting phase to full overhead position and (2) lowering phase to starting position (Table 8.5).

FIG. 8.6 • Barbell press. **A,** Starting position; **B,** Full overhead position.

TABLE 8.5 • Barbell press

| Joint | Lifting phase to full overhead position | | Lowering phase to starting position | |
	Action	Agonists	Action	Agonists
Wrist and hand	Flexion*	Wrist and MCP, PIP, PIP flexors (isometric contraction) Flexor carpi radialis Flexor carpi ulnaris Palmaris longus Flexor digitorum profundus Flexor digitorum superficialis Flexor pollicis longus	Flexion*	Wrist and MCP, PIP, PIP flexors (isometric contraction) Flexor carpi radialis Flexor carpi ulnaris Palmaris longus Flexor digitorum profundus Flexor digitorum superficialis Flexor pollicis longus
Elbow joint	Extension	Elbow extensors Triceps brachii Anconeus	Flexion	Elbow extensors (eccentric contraction) Triceps brachii Anconeus
Shoulder	Flexion	Shoulder joint flexors Pectoralis major (clavicular head or upper fibers) Anterior deltoid Coracobrachialis Biceps brachii	Extension	Shoulder joint flexors (eccentric contraction) Pectoralis major (clavicular head or upper fibers) Anterior deltoid Coracobrachialis Biceps brachii
Shoulder girdle	Upward rotation and elevation	Shoulder girdle upward rotators and elevators Trapezius Levator scapulae Serratus anterior	Downward rotation and depression	Shoulder girdle upward rotators and elevators (eccentric contraction) Trapezius Levator scapulae Serratus anterior

*The wrist is in a position of extension to facilitate greater active finger flexion in gripping the bar.

Chest press (bench press)

Description

The subject lies on the exercise bench in the supine position, grasps the barbell, and presses the weight upward through the full range of arm and shoulder movement (Fig. 8.7). Then the weight is lowered to the starting position. Refer to Tables 4.1, 5.2, 6.1, and 7.1.

Analysis

This open kinetic chain exercise can be divided into two phases for analysis: (1) lifting phase to up position and (2) lowering phase to starting position (Table 8.6).

A **B**

FIG. 8.7 • Chest press (bench press). **A,** Starting position; **B,** Up position.

TABLE 8.6 • Chest press (bench press)

Joint	Lifting phase to up position		Lowering phase to starting position	
	Action	Agonists	Action	Agonists
Wrist and hand	Flexion*	Wrist and MCP, PIP, PIP flexors (isometric contraction) 　Flexor carpi radialis 　Flexor carpi ulnaris 　Palmaris longus 　Flexor digitorum profundus 　Flexor digitorum superficialis 　Flexor pollicis longus	Flexion*	Wrist and MCP, PIP, PIP flexors (isometric contraction) 　Flexor carpi radialis 　Flexor carpi ulnaris 　Palmaris longus 　Flexor digitorum profundus 　Flexor digitorum superficialis 　Flexor pollicis longus
Elbow	Extension	Elbow extensors 　Triceps brachii 　Anconeus	Flexion	Elbow extensors (eccentric contraction) 　Triceps brachii 　Anconeus
Shoulder	Flexion and horizontal adduction	Shoulder flexors and horizontal adductors 　Pectoralis major 　Anterior deltoid 　Coracobrachialis 　Biceps brachii	Extension and horizontal abduction	Shoulder joint flexors and horizontal adductors (eccentric contraction) 　Pectoralis major 　Anterior deltoid 　Coracobrachialis 　Biceps brachii
Shoulder girdle	Abduction	Shoulder girdle abductors 　Serratus anterior 　Pectoralis minor	Adduction	Shoulder girdle abductors (eccentric contraction) 　Serratus anterior 　Pectoralis minor

*The wrist is in a position of slight extension to facilitate greater active finger flexion in gripping the bar.

Chin-up (pull-up)

Description

The subject grasps a horizontal bar or ladder with the palms away from the face (Fig. 8.8, *A*). From a hanging position on the bar, the subject pulls up until the chin is over the bar (Fig. 8.8, *B*) and then returns to the starting position (Fig. 8.8, *C*). The width of the grip on the chin-up bar affects the shoulder actions to a degree. A narrow grip will allow for more glenohumeral extension and flexion, whereas a wider grip, as shown in Fig. 8.8, requires more adduction and abduction, respectively. For a full review of the muscles involved in the chin-up, see Tables 4.1, 5.2, 6.1, and 7.1.

Analysis

This closed kinetic chain exercise is separated into two phases for analysis: (1) pulling-up phase to chinning position and (2) lowering phase to starting position (Table 8.7).

FIG. 8.8 • Pull-up.
A, Straight-arm hang;
B, Chin over bar;
C, Bent-arm hang on way up or down.

TABLE 8.7 • **Chin-up (pull-up)**

Chapter
8

Joint	Pulling-up phase to chinning position			Lowering phase to starting position		
	Action	Agonists		Action	Agonists	
Wrist and hand	Flexion	Wrist and MCP, PIP, PIP flexors (isometric contraction) Flexor carpi radialis Flexor carpi ulnaris Palmaris longus Flexor digitorum profundus Flexor digitorum superficialis Flexor pollicis longus		Flexion	Wrist and MCP, PIP, PIP flexors (isometric contraction) Flexor carpi radialis Flexor carpi ulnaris Palmaris longus Flexor digitorum profundus Flexor digitorum superficialis Flexor pollicis longus	
Elbow	Flexion	Elbow flexors Biceps brachii Brachialis Brachioradialis		Extension	Elbow flexors (eccentric contraction) Biceps brachii Brachialis Brachioradialis	
Shoulder	Adduction	Shoulder joint adductors Pectoralis major Posterior deltoid Latissimus dorsi Teres major Subscapularis		Abduction	Shoulder joint adductors (eccentric contraction) Pectoralis major Posterior deltoid Latissimus dorsi Teres major Subscapularis	
Shoulder girdle	Adduction, depression, and downward rotation	Shoulder girdle adductors, depressors, and downward rotators Trapezius (lower and middle) Pectoralis minor Rhomboids		Elevation, abduction, and upward rotation	Shoulder girdle adductors, depressors, and downward rotators (eccentric contraction) Trapezius (lower and middle) Pectoralis minor Rhomboids	

Latissimus pull (lat pull)

Description

From a sitting position, the subject reaches up and grasps a horizontal bar (Fig. 8.9, *A*). The bar is pulled down to a position below the chin (Fig. 8.9, *B*). Then it is returned slowly to the starting position. The width of the grip on the horizontal bar affects the shoulder actions to a degree. A narrow grip will allow for more glenohumeral extension and flexion, whereas the more common wider grip, as shown in Fig. 8.9, requires more adduction and abduction, respectively. Tables 4.1, 5.2, 6.1, and 7.1 provide a thorough listing of the muscles utilized in this exercise.

Analysis

This open kinetic chain exercise is separated into two phases for analysis: (1) pull-down phase to below the chin position and (2) return phase to starting position (Table 8.8).

A

B

FIG. 8.9 • Latissimus pull (lat pull). **A,** Starting position; **B,** Downward position.

TABLE 8.8 • Latissimus pull (lat pull)

| Joint | Pull-down phase to below the chin position | | Return phase to starting position | |
	Action	Agonists	Action	Agonists

Wrist and hand	Flexion	Wrist and MCP, PIP, PIP flexors (isometric contraction)
		  Flexor carpi radialis
		  Flexor carpi ulnaris
		  Palmaris longus
		  Flexor digitorum profundus
		  Flexor digitorum superficialis
		  Flexor pollicis longus
			Flexion	Wrist and MCP, PIP, PIP flexors (isometric contraction)
				  Flexor carpi radialis
				  Flexor carpi ulnaris
				  Palmaris longus
				  Flexor digitorum profundus
				  Flexor digitorum superficialis
				  Flexor pollicis longus

Elbow	Flexion	Elbow flexors
		  Biceps brachii
		  Brachialis
		  Brachioradialis
			Extension	Elbow flexors (eccentric contraction)
				  Biceps brachii
				  Brachialis
				  Brachioradialis

Shoulder	Adduction	Shoulder joint adductors
		  Pectoralis major
		  Latissimus dorsi
		  Teres major
		  Subscapularis
			Abduction	Shoulder joint adductors (eccentric contraction)
				  Pectoralis major
				  Latissimus dorsi
				  Teres major
				  Subscapularis

Shoulder girdle	Adduction, depression, and downward rotation	Shoulder girdle adductors, depressors, and downward rotators
		  Trapezius (lower and middle)
		  Pectoralis minor
		  Rhomboids
			Abduction, elevation, and upward rotation	Shoulder girdle adductors, depressors, and downward rotators (eccentric contraction)
				  Trapezius (lower and middle)
				  Pectoralis minor
				  Rhomboids

Push-up

Description

The subject lies on the floor in a prone position with the legs together, the palms touching the floor, and the hands pointed forward and approximately under the shoulders (Fig. 8.10, *A*). Keeping the back and legs straight, the subject pushes up to the up position and then returns to the starting position (Fig. 8.10, *B*).

The push-up is a total body exercise in that the muscles of the cervical and lumbar spine, hips, knees, ankles, and feet are active isometrically to stabilize the respective areas. Table 8.9 includes only the muscles of the upper extremity in the analysis. Tables 4.1, 5.2, 6.1, and 7.1 provide additional coverage of the muscles involved in the upper-extremity portion of the push-up.

Analysis

This closed kinetic chain exercise is separated into two phases for analysis: (1) pushing phase to up position and (2) lowering phase to starting position (Table 8.9).

Chin-ups and push-ups are excellent exercises for the shoulder area, shoulder girdle, shoulder joint, elbow joint, and wrist and hand (see Figs. 8.8 and 8.10). The use of free weights, machines, and other conditioning exercises helps develop strength and endurance for this part of the body.

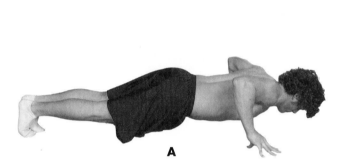

FIG. 8.10 • Push-up. **A,** Starting position; **B,** Up position.

TABLE 8.9 • **Push-up**

Joint	Pushing phase to up position		Lowering phase to starting position	
	Action	Agonists	Action	Agonists
Wrist and hand	Flexion	Wrist and MCP, PIP, PIP flexors (isometric contraction) Flexor carpi radialis Flexor carpi ulnaris Palmaris longus Flexor digitorum profundus Flexor digitorum superficialis Flexor pollicis longus	Flexion	Wrist and MCP, PIP, PIP flexors (isometric contraction) Flexor carpi radialis Flexor carpi ulnaris Palmaris longus Flexor digitorum profundus Flexor digitorum superficialis Flexor pollicis longus
Elbow	Extension	Elbow extensors Triceps brachii Anconeus	Flexion	Elbow extensors (eccentric contraction) Triceps brachii Anconeus
Shoulder	Horizontal adduction	Shoulder joint horizontal adductors Pectoralis major Anterior deltoid Biceps brachii Coracobrachialis	Horizontal abduction	Shoulder joint horizontal adductors (eccentric contraction) Pectoralis major Anterior deltoid Biceps brachii Coracobrachialis
Shoulder girdle	Abduction	Shoulder girdle abductors Serratus anterior Pectoralis minor	Adduction	Shoulder girdle abductors (eccentric contraction) Serratus anterior Pectoralis minor

Dumbbell bent-over row

Description

This open kinetic chain exercise may also be performed in the prone position and is therefore sometimes referred to as a prone row. The subject is kneeling on a bench or lying prone on a table so that the involved arm is free from contact with the floor (Fig. 8.11, *A*). When kneeling, the subject uses the contralateral arm to support the body. The dumbbell is held in the hand with the arm and shoulder hanging straight to the floor.

From this position, the subject adducts the shoulder girdle and horizontally abducts the shoulder joint (Fig. 8.11, *B*). Then the dumbbell is lowered slowly to the starting position. Tables 4.1, 5.2, 6.1, and 7.1 provide more details on the muscles used in this exercise.

Analysis

This exercise is separated into two phases for analysis: (1) pull-up phase to horizontal abducted position and (2) lowering phase to starting position (Table 8.10).

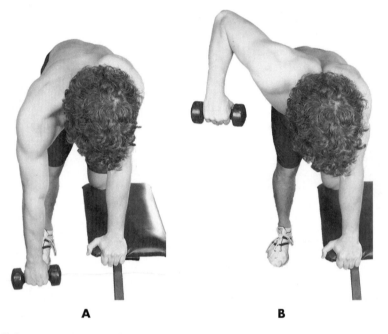

FIG. 8.11 • Dumbbell bent-over row. **A,** Starting position; **B,** Up position.

A   **B**

TABLE 8.10 • Dumbbell bent-over row

Joint	Pull-up phase to horizontal abducted position		Lowering phase to starting position	
	Action	Agonists	Action	Agonists
Hand	Flexion	MCP, PIP, PIP flexors (isometric contraction) Flexor digitorum profundus Flexor digitorum superficialis Flexor pollicis longus	Flexion	MCP, PIP, PIP flexors (isometric contraction) Flexor digitorum profundus Flexor digitorum superficialis Flexor pollicis longus
Elbow	Flexion	Passive flexion as the arm becomes parallel to the floor due to gravity	Extension	Passive extension as the arm becomes perpendicular to the floor due to gravity
Shoulder	Horizontal abduction	Shoulder joint horizontal abductors Posterior deltoid Infraspinatus Teres minor Latissimus dorsi	Horizontal adduction	Shoulder joint horizontal abductors (eccentric contraction) Posterior deltoid Infraspinatus Teres minor Latissimus dorsi
Shoulder girdle	Adduction	Shoulder girdle adductors Trapezius (lower and middle) Rhomboids	Abduction	Shoulder girdle adductors (eccentric contraction) Trapezius (lower and middle) Rhomboids

REVIEW EXERCISES

1. Analyze other conditioning exercises that involve the shoulder area, such as dips, upright rows, shrugs, dumbbell flys, and inclined presses.
2. Discuss how you would teach and train boys and girls who cannot perform one chin-up to learn to do chin-ups. Additionally, discuss how you would teach and train subjects who cannot perform one push-up to do push-ups.
3. Should boys and girls attempt to do chin-ups and push-ups to see whether they have adequate strength in the shoulder area?
4. What, if any, benefit would result from doing fingertip push-ups as opposed to push-ups with the hands flat on the floor?
5. Develop a list of exercises not found in this chapter to develop the upper-extremity muscles. Separate the list into open- and closed-chain exercises.

LABORATORY EXERCISES

1. Observe and analyze shoulder muscular activities of children on playground equipment.
2. Test yourself doing chin-ups and push-ups to determine your strength and muscular endurance in the shoulder area.
3. Stand slightly farther than arm's length from a wall with your arms by your side and hands facing forward at shoulder-level height. Perform each of the following movements fully before proceeding to the next. When finished, you should be reaching with the palm of your hand straight in front of your shoulder to attempt contact with the wall. Your elbow should be fully extended with your glenohumeral joint flexed 90 degrees.
   - Glenohumeral flexion to 90 degrees
   - Full elbow extension
   - Wrist extension to 70 degrees
   - Full shoulder girdle protraction

   Analyze the movements and muscles responsible for each movement at the shoulder girdle, glenohumeral joint, elbow, and wrist. Include the type of contraction for each muscle for each movement.
4. Face a wall and stand about 6 inches from it. Place both hands on the wall at shoulder level and put your nose and chest against the wall. Keeping your palms in place on the wall, slowly push your body from the wall as in a push-up until your chest is as far away from the wall as possible without removing your palms from the wall surface. Analyze the movements and muscles responsible for each movement at the shoulder girdle, glenohumeral joint, elbow, and wrist. Include the type of contraction for each muscle for each movement.
5. What is the difference between the two exercises in Questions 3 and 4? Can you perform the movements in Question 4 one step at a time, as you did in Question 3?

6. **Exercise analysis chart**

Analyze each exercise in the chart. Use one row for each joint involved that actively moves during the exercise. Do not include joints for which there is no active movement or joints that are maintained in one position isometrically.

Exercise	Phase	Joint, movement occurring	Force causing movement (muscle or gravity)	Force resisting movement (muscle or gravity)	Functional muscle group, type of contraction
Barbell press (overhead or military press)	Lifting phase				
	Lowering phase				

Exercise analysis chart (continued)

Exercise	Phase	Joint, movement occurring	Force causing movement (muscle or gravity)	Force resisting movement (muscle or gravity)	Functional muscle group, type of contraction
Chest press (bench press)	Lifting phase				
	Lowering phase				
Chin-up (pull-up)	Pulling-up phase				
	Lowering phase				
Latissimus pull (lat pull)	Pull-down phase				
	Return phase				

Exercise analysis chart (continued)

Exercise	Phase	Joint, movement occurring	Force causing movement (muscle or gravity)	Force resisting movement (muscle or gravity)	Functional muscle group, type of contraction
Push-up	Pushing phase				
	Lowering phase				
Dumbbell bent-over row (prone row)	Pull-up phase				
	Lowering phase				

Chapter **8**

References

Adrian M: Isokinetic exercise, *Training and Conditioning* 1:1, June 1991.

Andrews JR, Zarins B, Wilk KE: *Injuries in baseball,* Philadelphia, 1998, Lippincott-Raven.

Booher JM, Thibodeau GA: *Athletic injury assessment,* ed 4, Dubuque, IA, 2000, McGraw-Hill.

Ellenbecker TS, Davies GJ: *Closed kinetic chain exercise: a comprehensive guide to multiple-joint exercise,* Champaign, IL, 2001, Human Kinetics.

Fleck SJ: Periodized strength training: a critical review, *Journal of Strength and Conditioning Research,* 13(1):82–89, 1999.

Geisler P: Kinesiology of the full golf swing—implications for intervention and rehabilitation, *Sports Medicine Update* 11(2):9, 1996.

Hamilton N, Weimer W, Luttgens K: *Kinesiology: scientific basis of human motion,* ed 12, New York, 2012, McGraw-Hill.

Matheson O, et al: Stress fractures in athletes, *American Journal of Sports Medicine* 15:46, January–February 1987.

National Strength and Conditioning Association; Baechle TR, Earle RW: *Essentials of strength training and conditioning,* ed 2, Champaign, IL, 2000, Human Kinetics.

Northrip JW, Logan GA, McKinney WC: *Analysis of sport motion: anatomic and biomechanic perspectives,* ed 3, New York, 1983, McGraw-Hill.

Powers SK, Howley ET: *Exercise physiology: theory and application of fitness and performance,* ed 8, New York, 2012, McGraw-Hill.

Smith LK, Weiss EL, Lehmkuhl LD: *Brunnstrom's clinical kinesiology,* ed 5, Philadelphia, 1996, Davis.

Steindler A: *Kinesiology of the human body,* Springfield, IL, 1970, Charles C Thomas.

Wilk KE, Reinold MM, Andrews JR, eds: *The athlete's shoulder,* ed 2, Philadelphia, 2009, Churchill Livingstone Elsevier.

For additional resources and a list of related websites, visit **www.mhhe.com/floyd19e**.

Worksheet Exercises

For in- or out-of-class assignments, or for testing, utilize this tear-out worksheet.

Upright row exercise worksheet

List the movements that occur in each joint as the subject lifts the weight in performing upright rows and then lowers the weight. For each joint movement, list the muscles primarily responsible, and indicate whether they are contracting concentrically or eccentrically with "C" or "E."

Lifting phase		
Joint	Movement	Muscles
Wrists		
Elbows		
Shoulder joints		
Shoulder girdles		
Lowering phase		
Wrists		
Elbows		
Shoulder joints		
Shoulder girdles		

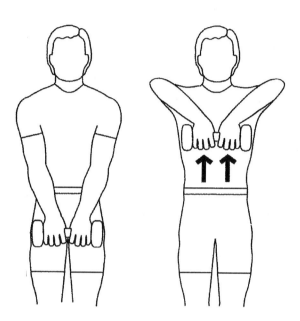

Dip exercise worksheet

List the movements that occur in each joint as the subject moves the body up and down in performing dips. For each joint movement, list the muscles primarily responsible, and indicate whether they are contracting concentrically or eccentrically with "C" or "E."

Lifting body up phase		
Joint	Movement	Muscles
Wrists		
Elbows		
Shoulder joints		
Shoulder girdles		
Lowering body down phase		
Wrists		
Elbows		
Shoulder joints		
Shoulder girdles		

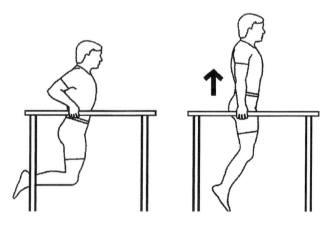

Worksheet Exercises

For in- or out-of-class assignments, or for testing, utilize this tear-out worksheet.

Upper-extremity sport skill analysis worksheet

Choose one skill in the left column to analyze and circle it. In your analysis, list the movements of each joint in each phase of the skill. You may prefer to list the initial positions that the joints are in for the stance phase. After each movement, list the muscle(s) primarily responsible for causing/controlling the movement. Beside each muscle in each movement, indicate the type of contraction as follows: I—isometric; C—concentric; E—eccentric. You might want to review Chapters 4–7.

Skill		Joint	Stance phase	Preparatory phase	Movement phase	Follow-through phase
Baseball pitch	(R)	Shoulder girdle				
		Shoulder joint				
		Elbow				
Volleyball serve		Radio-ulnar				
Tennis serve		Wrist				
Softball pitch		Fingers				
	(L)	Shoulder girdle				
Tennis backhand		Shoulder joint				
Batting		Elbow				
Bowling		Radio-ulnar				
Basketball free throw		Wrist				
		Fingers				

CHAPTER 9

THE HIP JOINT AND PELVIC GIRDLE

Objectives

- To identify on a human skeleton or subject selected bony features of the hip joint and pelvic girdle

- To label on a skeletal chart selected bony features of the hip joint and pelvic girdle

- To draw on a skeletal chart the individual muscles of the hip joint

- To demonstrate, using a human subject, all the movements of the hip joint and pelvic girdle and list their respective planes of movement and axes of motion

- To palpate on a human subject the muscles of the hip joint and pelvic girdle

- To list and organize the primary muscles that produce movement of the hip joint and pelvic girdle and list their antagonists

- To determine, through analysis, the hip movements and muscles involved in selected skills and exercises

Online Learning Center Resources

Visit *Manual of Structural Kinesiology*'s **Online Learning Center** at **www.mhhe.com/floyd19e** for additional information and study material for this chapter, including:

- *Self-grading quizzes*
- *Anatomy flashcards*
- *Animations*
- *Related websites*

The hip, or acetabular femoral joint, is a relatively stable joint due to its bony architecture, strong ligaments, and large, supportive muscles. It functions in weight bearing and locomotion, which is enhanced significantly by the hip's wide range of motion, which provides the ability to run, cross-over cut, side-step cut, jump, and make many other directional changes.

Bones   FIGS. 9.1 TO 9.3

The hip joint is a ball-and-socket joint that consists of the head of the femur connecting with the acetabulum of the pelvic girdle. The femur projects out laterally from its head toward the greater trochanter and then angles back toward the midline as it runs inferiorly to form the proximal bone of the knee. It is the longest bone in the body. The pelvic girdle consists of a right and left pelvic bone joined together posteriorly by the sacrum. The sacrum can be considered an extension of the spinal column with five fused vertebrae. Extending inferior from the sacrum is the coccyx. The pelvic bones are made up of three bones: the ilium, the ischium, and the pubis. At birth and during growth and development, they are three distinct bones. At maturity, they are fused to form one pelvic bone known as the os coxae.

The pelvic bone can be divided roughly into three areas, starting from the acetabulum:

Upper two-fifths = ilium
Posterior and lower two-fifths = ischium
Anterior and lower one-fifth = pubis

In studying the muscles of the hip and thigh, it is helpful to focus on the important bony

Chapter 9

landmarks, keeping in mind their purpose as key attachment points for the muscles. The anterior pelvis serves to provide points of origin for muscles generally involved in flexing the hip. Specifically, the tensor fasciae latae arises from the anterior iliac crest, the sartorius originates on the anterior superior iliac spine, and the rectus femoris originates on the anterior inferior iliac spine. Laterally, the gluteus medius and minimus, which abduct the hip, originate just below the iliac crest. Posteriorly, the gluteus maximus originates on the posterior iliac crest as well as the posterior

sacrum and coccyx. Posteroinferiorly, the ischial tuberosity serves as the point of origin for the hamstrings, which extend the hip. Medially, the pubis and its inferior ramus serve as the point of origin for the hip adductors, which include the adductor magnus, adductor longus, adductor brevis, pectineus, and gracilis.

The proximal thigh generally serves as a point of insertion for some of the short muscles of the hip and as the origin for three of the knee extensors. Most notably, the greater trochanter is the point of insertion for all of the gluteal

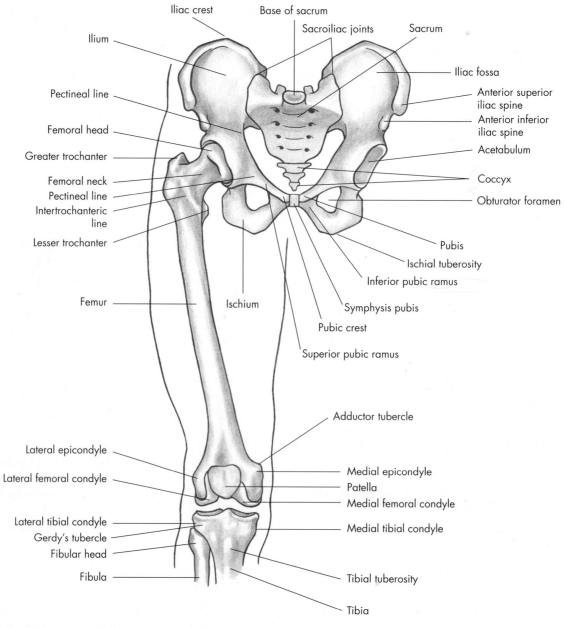

FIG. 9.1 • Right pelvis and femur, anterior view.

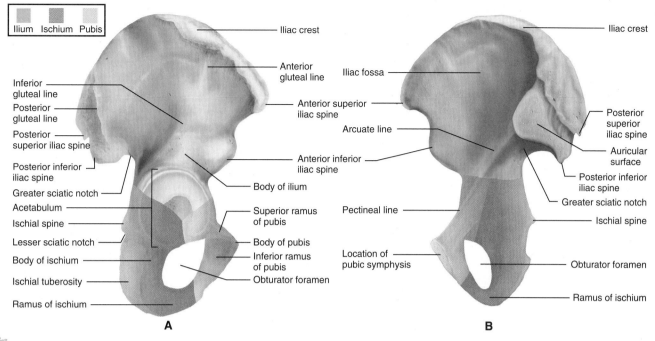

Ilium Ischium Pubis

A (Lateral view labels):

Iliac crest

Anterior gluteal line

Inferior gluteal line

Posterior gluteal line

Posterior superior iliac spine

Anterior superior iliac spine

Posterior inferior iliac spine

Anterior inferior iliac spine

Greater sciatic notch

Body of ilium

Acetabulum

Superior ramus of pubis

Ischial spine

Body of pubis

Lesser sciatic notch

Inferior ramus of pubis

Body of ischium

Obturator foramen

Ischial tuberosity

Ramus of ischium

A

B (Medial view labels):

Iliac crest

Iliac fossa

Anterior superior iliac spine

Arcuate line

Posterior superior iliac spine

Auricular surface

Posterior inferior iliac spine

Greater sciatic notch

Anterior inferior iliac spine

Ischial spine

Pectineal line

Location of pubic symphysis

Obturator foramen

Ramus of ischium

B

FIG. 9.2 • Right pelvic bone. **A,** Lateral view; **B,** Medial view.

muscles and five of the six deep external rotators. Although not palpable, the lesser trochanter serves as the bony landmark upon which the iliopsoas inserts. Anteriorly, the three vasti muscles of the quadriceps originate proximally. Posteriorly, the linea aspera serves as the insertion for the hip adductors.

Distally, the patella serves as a major bony landmark to which all four quadriceps muscles insert. The remainder of the hip muscles insert on the proximal tibia or fibula. The sartorius, gracilis, and semitendinosus insert on the upper anteromedial surface of the tibia just below the medial condyle, after crossing the knee posteromedially. The semimembranosus inserts posteromedially on the medial tibial condyle. Laterally, the biceps femoris inserts primarily on the head of the fibula, with some fibers attaching on the lateral tibial condyle. Anterolaterally, Gerdy's tubercle provides the insertion point for the iliotibial tract of the tensor fasciae latae.

Joints   FIGS. 9.1 TO 9.7

Anteriorly, the pelvic bones are joined to form the symphysis pubis, an amphiarthrodial joint. Posteriorly, the sacrum is located between the two pelvic bones and forms the sacroiliac joints. Strong ligaments unite these bones to form rigid, slightly movable joints. The bones are large and heavy and for the most part are covered by thick, heavy muscles. Very minimal oscillating-type movements can occur in these joints, as in walking or in hip flexion when lying on one's back. However, movements usually involve the entire pelvic girdle and hip joints. In walking, there is hip flexion and extension with rotation of the pelvic girdle, forward in hip flexion and backward in hip extension. Jogging and running result in faster movements and a greater range of movement.

Sport skills such as kicking a football or soccer ball are other good examples of hip and pelvic movements. Pelvic rotation helps increase the length of the stride in running; in kicking, it can result in a greater range of motion, which translates into a greater distance or more speed to the kick.

Except for the glenohumeral joint, the hip is one of the most mobile joints of the body, largely because of its multiaxial arrangement. Unlike the glenohumeral, the hip joint's bony architecture provides a great deal of stability, resulting in relatively few hip joint subluxations and dislocations.

The hip joint is classified as an enarthrodial-type joint and is formed by the femoral head inserting into the socket provided by the acetabulum of the pelvis. An extremely strong and dense ligamentous capsule, illustrated in Figs. 9.4 and 9.5, reinforces the joint, especially anteriorly.

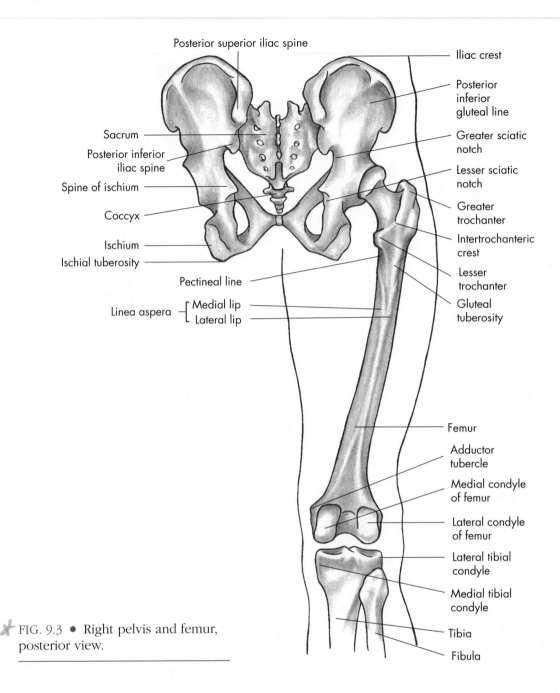

Posterior superior iliac spine

Sacrum

Posterior inferior
iliac spine

Spine of ischium

Coccyx

Ischium

Ischial tuberosity

Pectineal line

Linea aspera { Medial lip
Lateral lip

Iliac crest

Posterior
inferior
gluteal line

Greater sciatic
notch

Lesser sciatic
notch

Greater
trochanter

Intertrochanteric
crest

Lesser
trochanter

Gluteal
tuberosity

Femur

Adductor
tubercle

Medial condyle
of femur

Lateral condyle
of femur

Lateral tibial
condyle

Medial tibial
condyle

Tibia

Fibula

✱ FIG. 9.3 ● Right pelvis and femur,
posterior view.

Anteriorly, the iliofemoral, or Y, ligament prevents hip hyperextension. The teres ligament attaches from deep in the acetabulum to a depression in the head of the femur and slightly limits adduction. The pubofemoral ligament is located anteromedially and inferiorly and limits excessive extension and abduction. Posteriorly, the triangular ischiofemoral ligament extends from the ischium below to the trochanteric fossa of the femur and limits internal rotation.

Similar to the glenoid fossa of the shoulder joint, the acetabulum is lined around most of its periphery with a labrum to enhance stability and provide some shock absorption. Covering the remaining surface of the acetabulum, as well as the femoral head, is articular cartilage that may gradually degenerate with age and/or injury, leading to osteoarthritis characterized by pain, stiffness, and limited range of motion.

Because of individual differences, there is some disagreement about the exact possible range of each movement in the hip joint, but the ranges are generally 0 to 130 degrees of flexion, 0 to 30 degrees of extension, 0 to 35 degrees of abduction, 0 to 30 degrees of adduction, 0 to 45 degrees of internal rotation, and 0 to 50 degrees of external rotation (Fig. 9.8). Although rarely referred to as distinct motions, the hip, when flexed to

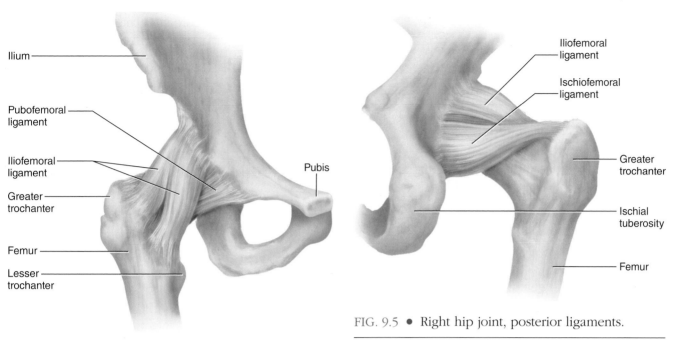

FIG. 9.4 • Right hip joint, anterior ligaments.

FIG. 9.5 • Right hip joint, posterior ligaments.

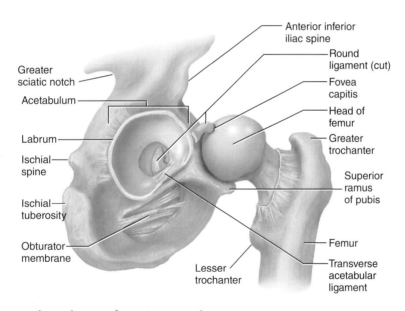

FIG. 9.6 • Right hip joint, lateral view, femur retracted.

90 degrees, can adduct and abduct in the transverse plane, similar to the glenohumeral joint. These motions include approximately 40 degrees of horizontal adduction and 60 degrees of horizontal abduction.

The pelvic girdle moves back and forth within three planes for a total of six different movements. To avoid confusion, it is important to analyze the pelvic girdle activity to determine the exact location of the movement. All pelvic girdle rotation actually results from motion at one or more of the following locations: the right hip, the left hip, the lumbar spine. Although it is not essential for movement to occur in all three of these areas, it must occur in at least one for the pelvis to rotate in any direction. Table 9.1 lists the motions at the hips and lumbar spine that can often accompany rotation of the pelvic girdle.

www.mhhe.com/floyd19e   **233**

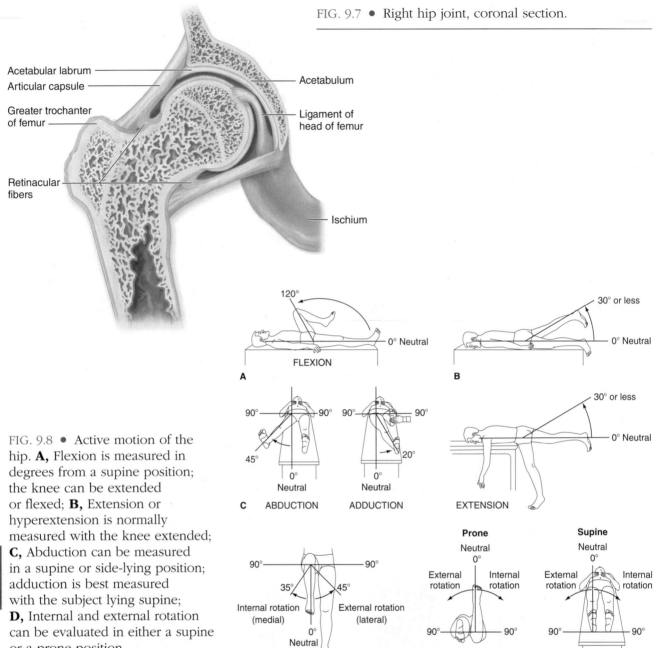

FIG. 9.7 • Right hip joint, coronal section.

Acetabular labrum
Articular capsule

Acetabulum

Greater trochanter
of femur

Ligament of
head of femur

Retinacular
fibers

Ischium

120°

30° or less

0° Neutral

FLEXION

A

0° Neutral

B

FIG. 9.8 • Active motion of the
hip. **A,** Flexion is measured in
degrees from a supine position;
the knee can be extended
or flexed; **B,** Extension or
hyperextension is normally
measured with the knee extended;
C, Abduction can be measured
in a supine or side-lying position;
adduction is best measured
with the subject lying supine;
D, Internal and external rotation
can be evaluated in either a supine
or a prone position.

90°   90°   90°   90°

45°   20°

0°   0°
Neutral   Neutral

C   ABDUCTION   ADDUCTION

30° or less

0° Neutral

EXTENSION

90°   90°

35°   45°

Internal rotation   External rotation
(medial)   (lateral)
           0°
        Neutral

D

Prone   **Supine**
Neutral   Neutral
0°   0°

External   Internal   External   Internal
rotation   rotation   rotation   rotation

90°   90°   90°   90°

ROTATION

TABLE 9.1 • Motions accompanying pelvic rotation

Pelvic rotation	Lumbar spine motion	Right hip motion	Left hip motion
Anterior rotation	Extension	Flexion	Flexion
Posterior rotation	Flexion	Extension	Extension
Right lateral rotation	Left lateral flexion	Abduction	Adduction
Left lateral rotation	Right lateral flexion	Adduction	Abduction
Right transverse rotation	Left lateral rotation	Internal rotation	External rotation
Left transverse rotation	Right lateral rotation	External rotation	Internal rotation

Chapter
9

Movements FIGS. 9.9, 9.10

Anterior and posterior pelvic rotation occur in the sagittal or anteroposterior plane, whereas right and left lateral rotation occur in the lateral or frontal plane. Right transverse (clockwise) rotation and left transverse (counterclockwise) rotation occur in the horizontal or transverse plane of motion.

Hip flexion: movement of the anterior femur from any point toward the anterior pelvis in the sagittal plane

Hip extension: movement of the posterior femur from any point toward the posterior pelvis in the sagittal plane

Hip abduction: movement of the femur in the frontal plane laterally to the side away from the midline

Hip adduction: movement of the femur in the frontal plane medially toward the midline

Hip flexion

Hip extension

Hip abduction

Hip adduction

Hip external rotation

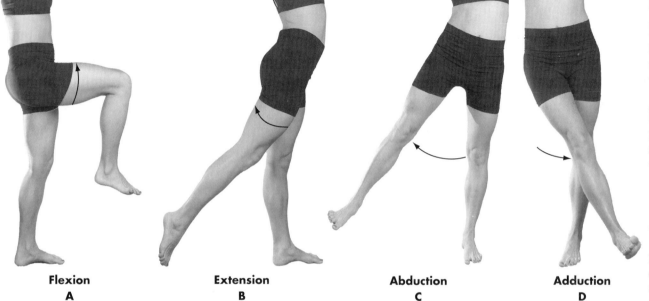

| Flexion A | Extension B | Abduction C | Adduction D |

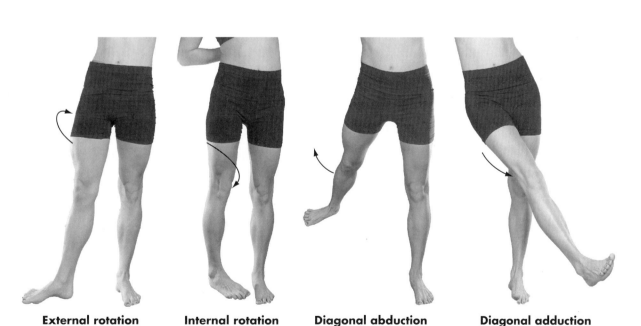

| External rotation E | Internal rotation F | Diagonal abduction G | Diagonal adduction H |

Hip internal rotation

Chapter 9

FIG. 9.9 • Movements of the hip.

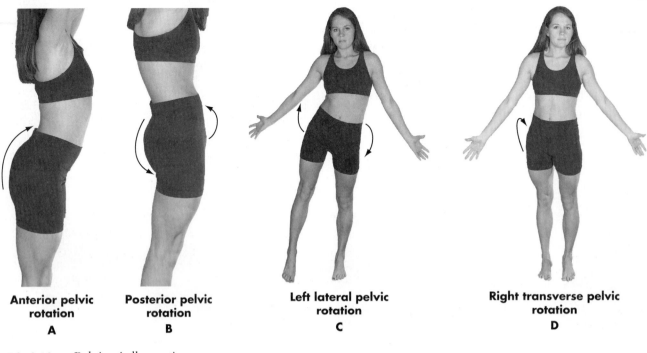

Anterior pelvic rotation	Posterior pelvic rotation	Left lateral pelvic rotation	Right transverse pelvic rotation
A	B	C	D

FIG. 9.10 ● Pelvic girdle motions.

Hip external rotation: lateral rotary movement of the femur in the transverse plane around its longitudinal axis away from the midline; lateral rotation

Hip internal rotation: medial rotary movement of the femur in the transverse plane around its longitudinal axis toward the midline; medial rotation

Hip diagonal abduction: movement of the femur in a diagonal plane away from the midline of the body

Hip diagonal adduction: movement of the femur in a diagonal plane toward the midline of the body

Hip horizontal adduction: movement of the femur in a horizontal or transverse plane toward the pelvis

Hip horizontal abduction: movement of the femur in a horizontal or transverse plane away from the pelvis

Anterior pelvic rotation: anterior movement of the upper pelvis; the iliac crest tilts forward in a sagittal plane; anterior tilt; downward rotation; accomplished by hip flexion and/or lumbar extension

Posterior pelvic rotation: posterior movement of the upper pelvis; the iliac crest tilts backward in a sagittal plane; posterior tilt; upward rotation; accomplished by hip extension and/or lumbar flexion

Left lateral pelvic rotation: in the frontal plane, the left pelvis moves inferiorly in relation to the right pelvis; either the left pelvis rotates downward or the right pelvis rotates upward; left lateral tilt; accomplished by left hip abduction, right hip adduction, and/or right lumbar lateral flexion

Right lateral pelvic rotation: in the frontal plane, the right pelvis moves inferiorly in relation to the left pelvis; either the right pelvis rotates downward or the left pelvis rotates upward; right lateral tilt; accomplished by right hip abduction, left hip adduction, and/or left lumbar lateral flexion

Left transverse pelvic rotation: in a horizontal plane of motion, rotation of the pelvis to the body's left; the right iliac crest moves anteriorly in relation to the left iliac crest, which moves posteriorly; accomplished by right hip external rotation, left hip internal rotation, and/or right lumbar rotation

Right transverse pelvic rotation: in a horizontal plane of motion, rotation of the pelvis to the body's right; the left iliac crest moves anteriorly in relation to the right iliac crest, which moves posteriorly; accomplished by left hip external rotation, right hip internal rotation, and/or left lumbar rotation

Some confusion in understanding and learning the pelvic girdle motions can be avoided by always thinking about these movements from the perspective of the person's pelvis that is actually moving. Further understanding may be gained by thinking of the person as moving his or her pelvis as if steering a vehicle, as shown in Fig. 9.11.

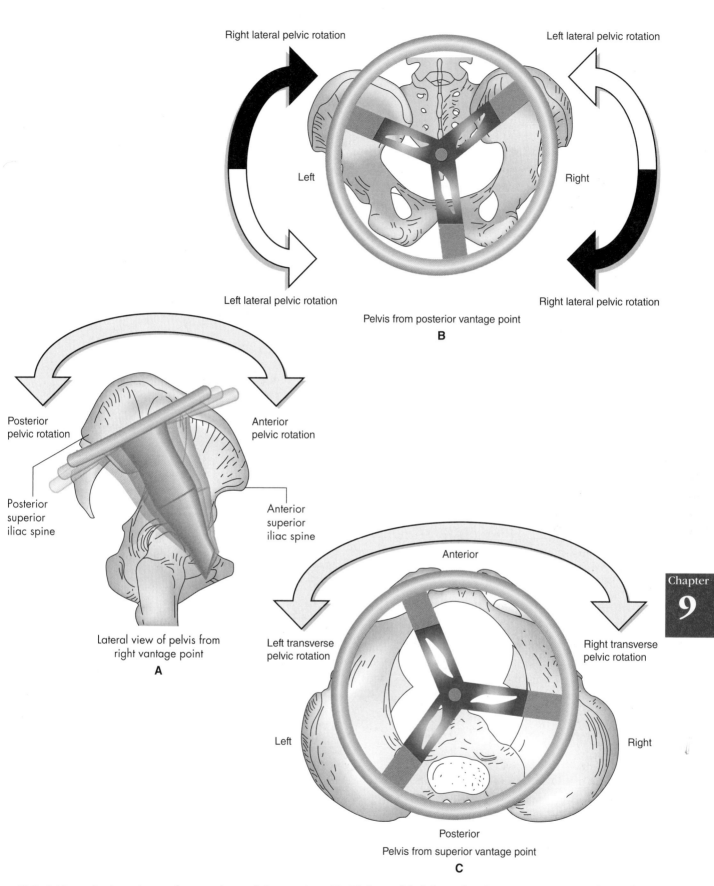

Right lateral pelvic rotation

Left lateral pelvic rotation

Left

Right

Left lateral pelvic rotation

Right lateral pelvic rotation

Pelvis from posterior vantage point

B

Posterior
pelvic rotation

Anterior
pelvic rotation

Posterior
superior
iliac spine

Anterior
superior
iliac spine

Lateral view of pelvis from
right vantage point

A

Anterior

Left transverse
pelvic rotation

Right transverse
pelvic rotation

Left

Right

Posterior

Pelvis from superior vantage point

C

FIG. 9.11 • **A,** Anterior and posterior pelvic rotation; **B,** Right and left lateral pelvic rotation; **C,** Right and left transverse pelvic rotation.

Muscles   FIGS. 9.12 TO 9.14

At the hip joint, there are seven two-joint muscles that have one action at the hip and another at the knee. The muscles actually involved in hip and pelvic girdle motions depend largely on the direction of the movement and the position of the body in relation to the earth and its gravitational forces. In addition, it should be noted that the body part that moves the most will be the part least stabilized. For example, when one is standing on both feet and contracting the hip flexors, the trunk and pelvis will rotate anteriorly, but when one is lying supine and contracting the hip flexors, the thighs will move forward into flexion on the stable pelvis.

For another example, the hip flexor muscles are used in moving the thighs toward the trunk, but the extensor muscles are used eccentrically when the pelvis and the trunk move downward slowly on the femur and concentrically when the trunk is raised on the femur—this, of course, occurs in rising to the standing position.

In the downward phase of the knee-bend exercise, the movement at the hips and knees is flexion. The muscles primarily involved are the hip and knee extensors in eccentric contraction.

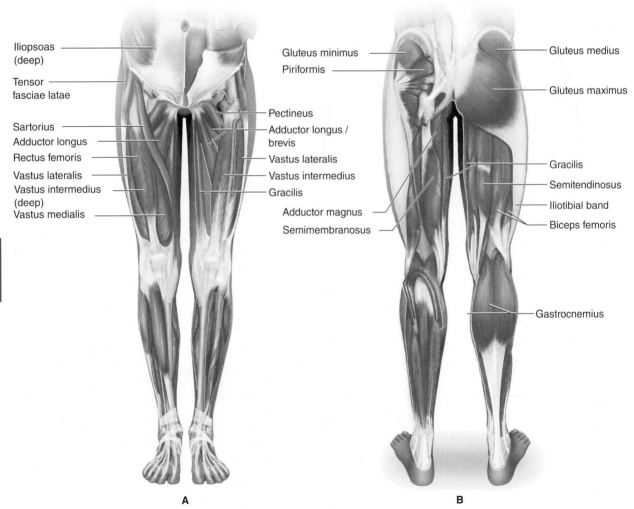

FIG. 9.12 • Superficial and deep muscles of the lower extremity. **A,** Anterior view; **B,** Posterior view.

Additionally, it is important to understand that the action a particular muscle has on the hip may vary depending on the position of the femur in relation to the pelvis at the time. As the hip moves through its rather large ranges of motion, the lines of pull of specific muscles may change significantly. This is best seen with the adductors. If the hip is in flexion, the adductors, upon concentric contraction, tend to cause extension, and if the hip is in extension they tend to cause flexion.

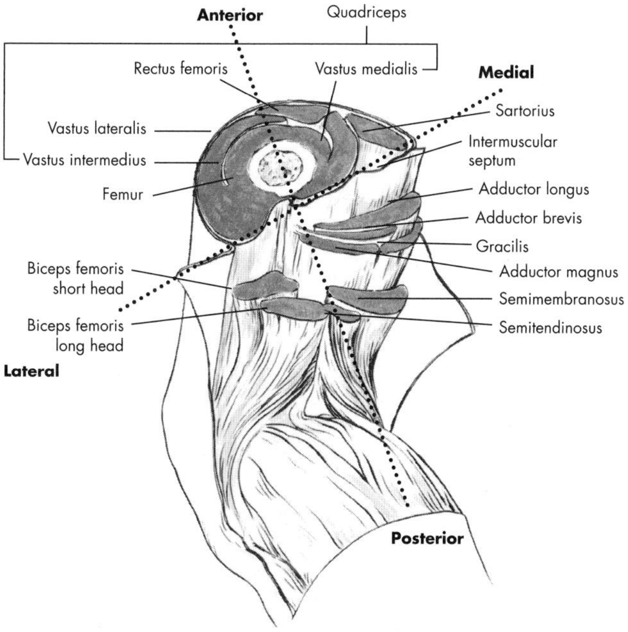

Hip joint and pelvic girdle muscles—location

Muscle location largely determines the muscle action. Seventeen or more muscles are found in the area (the six external rotators are counted as one muscle). Most hip joint and pelvic girdle muscles are large and strong.

Anterior
    Primarily hip flexion
        Iliopsoas (iliacus and psoas)
        Pectineus
        Rectus femoris[*†]
        Sartorius[†]

Lateral
    Primarily hip abduction
        Gluteus medius
        Gluteus minimus
        External rotators
        Tensor fasciae latae[†]

---

[*]Two-joint muscles; knee actions are discussed in Chapter 10.
[†]Two-joint muscles.

FIG. 9.13 • Cross section of the left thigh at the midsection.

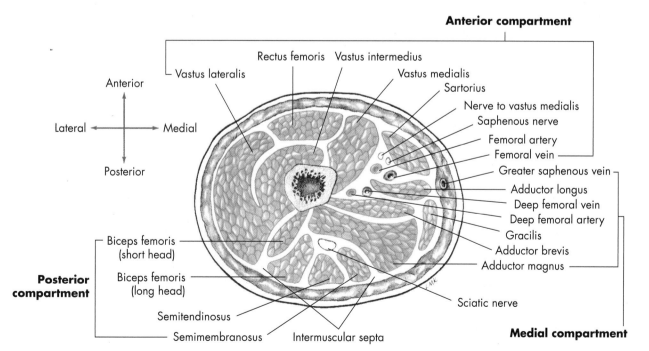

FIG. 9.14 • Transverse section of the left midthigh, detailing the anterior, posterior, and medial compartments.

Posterior
    Primarily hip extension
        Gluteus maximus
        Biceps femoris*†
        Semitendinosus*†
        Semimembranosus*†
        External rotators (six deep)

Medial
    Primarily hip adduction
        Adductor brevis
        Adductor longus
        Adductor magnus
        Gracilis†

---

*Two-joint muscles; knee actions are discussed in Chapter 10.
†Two-joint muscles.

Muscle identification

In developing a thorough and practical knowledge of the muscular system, it is essential that individual muscles be understood. Figs. 9.13, 9.15, 9.16, 9.17, and 9.18 illustrate groups of muscles that work together to produce joint movement. While viewing the muscles in these figures, correlate them with Table 9.2.

The muscles of the pelvis that act on the hip joint may be divided into two regions—the iliac and gluteal regions. The iliac region contains the iliopsoas muscle, which flexes the hip. The iliopsoas actually is two different muscles: the iliacus and the psoas major, although some include the psoas minor in discussion of the iliopsoas. The 10 muscles of the gluteal region function primarily to

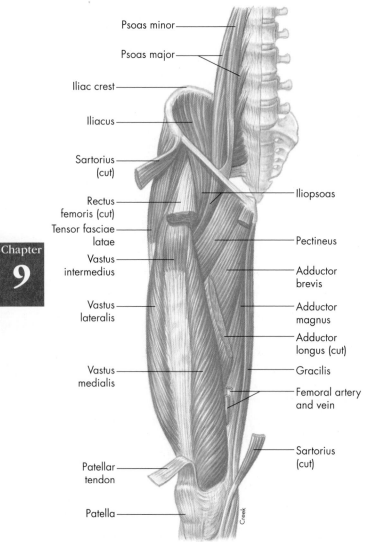

FIG. 9.15 ● Muscles of the right anterior pelvic and thigh regions.

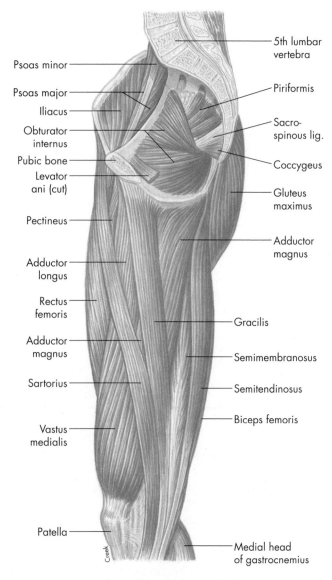

FIG. 9.16 ● Muscles of the right medial thigh.

extend and rotate the hip. Located in the gluteal region are the gluteus maximus, gluteus medius, gluteus minimi, and tensor fasciae latae and the six deep external rotators—piriformis, obturator externus, obturator internus, gemellus superior, gemellus inferior, and quadratus femoris.

The thigh is divided into three compartments by the intermuscular septa (Fig. 9.14). The anterior compartment contains the rectus femoris, vastus medialis, vastus intermedius, vastus lateralis, and sartorius. The hamstring muscle group, consisting of the biceps femoris, semitendinosus, and semimembranosus, is located in the posterior compartment. The medial compartment contains the thigh muscles primarily responsible for adduction of the hip, which are the adductor brevis, adductor longus, adductor magnus, pectineus, and gracilis.

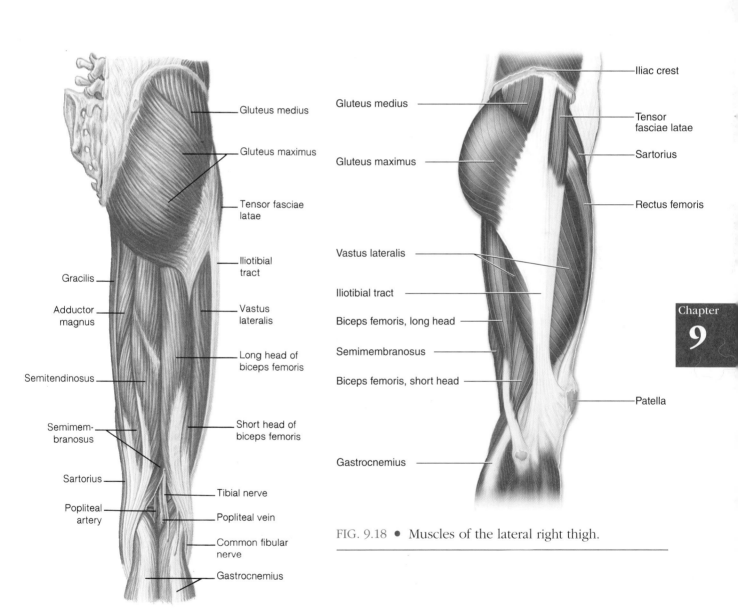

FIG. 9.17 • Muscles of the right posterior thigh.

FIG. 9.18 • Muscles of the lateral right thigh.

TABLE 9.2 • Agonist muscles of the hip joint

	Muscle	Origin	Insertion	Action	Plane of motion	Palpation	Innervation
Anterior	Iliacus	Inner surface of the ilium	Lesser trochanter of the femur and shaft just below	Flexion of the hip	Sagittal	Difficult to palpate; deep against posterior abdominal wall; with subject seated and leaning slightly forward to relax abdominal muscles, palpate psoas major deeply between iliac crest and 12th rib about halfway between ASIS and umbilicus with active hip flexion; palpate iliopsoas distal tendon on anterior aspect of hip approximately 1½ inches below center of inguinal ligament with active hip flexion/extension of supine subject, immediately lateral to pectineus and medial to sartorius	Lumbar nerve and femoral nerve (L2–L4)
				Anterior pelvic rotation			
				External rotation of the hip	Transverse		
				Transverse pelvic rotation contralaterally when ipsilateral femur is stabilized			
	Psoas major and minor	Lower borders of the transverse processes (L1–L5), sides of the bodies of the last thoracic vertebra (T12), the lumbar vertebrae (L1–L5), intervertebral fibrocartilages, and base of the sacrum	Lesser trochanter of the femur and shaft just below psoas minor; pectineal line (of pubis) and iliopectineal eminence (Psoas minor is involved only in lumbar spine movements.)	Flexion of the hip	Sagittal		
				Anterior pelvic rotation			
				Flexion of lumbar spine			
				External rotation of the hip	Transverse		
				Transverse pelvic rotation contralaterally when ipsilateral femur is stabilized			
				Lateral flexion of lumbar spine	Frontal		
				Lateral pelvic rotation to contralateral side			
	Rectus femoris	Anterior inferior iliac spine of the ilium and groove (posterior) above the acetabulum	Superior aspect of the patella and patella tendon to the tibial tuberosity	Flexion of the hip	Sagittal	Straight down anterior thigh from anterior inferior iliac spine to patella with resisted hip flexion/knee extension	Femoral nerve (L2–L4)
				Extension of the knee			
				Anterior pelvic rotation			
	Sartorius	Anterior superior iliac spine and notch just below the spine	Anterior medial surface of the tibia just below the condyle	Flexion of the hip	Sagittal	Located proximally just medial to tensor fascia latae and lateral to iliopsoas; palpate superficially from anterior superior iliac spine to medial tibial condyle during combined resistance of hip flexion/external rotation/abduction and knee flexion in a supine position	Femoral nerve (L2, L3)
				Flexion of the knee			
				Anterior pelvic rotation			
				External rotation of the thigh as it flexes the hip and knee	Transverse		
				Weak internal rotation of the knee			
				Abduction of the hip	Frontal		

TABLE 9.2 (continued) • Agonist muscles of the hip joint

	Muscle	Origin	Insertion	Action	Plane of motion	Palpation	Innervation
Medial	Pectineus	Space 1 inch wide on the front of the pubis just above the crest	Rough line leading from the lesser trochanter down to the linea aspera	Flexion of the hip	Sagittal	Difficult to distinguish from other adductors; anterior aspect of hip approximately 1½ inches below center of inguinal ligament; just lateral and slightly proximal to adductor longus and medial to iliopsoas during flexion and adduction of supine subject	Femoral nerve (L2–L4)
				Adduction of the hip	Frontal		
				External rotation of the hip	Transverse		
	Adductor brevis	Front of the inferior pubic ramus just below the origin of the adductor longus	Lower 2/3 of the pectineal line of the femur and the upper half of the medial lip of the linea aspera	Adduction of the hip	Frontal	Deep to adductor longus and superficial to adductor magnus; very difficult to palpate and differentiate from adductor longus, which is immediately inferior; proximal portion is just lateral to adductor longus	Obturator nerve (L3, L4)
				External rotation as it adducts the hip	Transverse		
				Assists in flexion of the hip	Sagittal		
	Adductor longus	Anterior pubis just below its crest	Middle 1/3 of the linea aspera	Adduction of the hip	Frontal	Most prominent muscle proximally on anteromedial thigh just inferior to the pubic bone with resisted adduction	Obturator nerve (L3, L4)
				Assists in flexion of the hip	Sagittal		
	Adductor magnus	Edge of the entire pubic ramus and the ischium and ischial tuberosity	Whole length of the linea aspera, inner condyloid ridge, and adductor tubercle	Adduction of the hip	Frontal	Medial aspect of thigh between gracilis and medial hamstrings from ischial tuberosity to adductor tubercle with resisted adduction from abducted position	Anterior: obturator nerve (L2–L4) Posterior: sciatic nerve (L4, L5, S1–S3)
				External rotation as the hip adducts	Transverse		
				Extension of the hip	Sagittal		
	Gracilis	Anteromedial edge of the descending ramus of the pubis	Anterior medial surface of the tibia below the condyle	Adduction of the hip	Frontal	A thin superficial tendon on anteromedial thigh with knee flexion and resisted adduction; just posterior to adductor longus and medial to the semitendinosus	Obturator nerve (L2–L4)
				Weak flexion of the knee	Sagittal		
				Assists with flexion of the hip			
				Internal rotation of the hip	Transverse		
				Weak internal rotation of the knee			

Chapter 9

TABLE 9.2 (continued) • Agonist muscles of the hip joint

Muscle		Origin	Insertion	Action	Plane of motion	Palpation	Innervation
Posterior	Semiten-dinosus	Ischial tuberosity	Upper anterior medial surface of the tibia just below the condyle	Flexion of the knee	Sagittal	Posteromedial aspect of distal thigh with combined knee flexion and internal rotation against resistance; just distal to ischial tuberosity in a prone position with hip internally rotated during active knee flexion	Sciatic nerve—tibial division (L5, S1, S2)
				Extension of the hip			
				Posterior pelvic rotation			
				Internal rotation of the hip	Transverse		
				Internal rotation of the flexed knee			
	Semimem-branosus	Ischial tuberosity	Posteromedial surface of the medial tibial condyle	Flexion of the knee	Sagittal	Largely covered by other muscles, tendon can be felt at posteromedial aspect of knee just deep to semitendinosus tendon with combined knee flexion and internal rotation against resistance	Sciatic nerve—tibial division (L5, S1, S2)
				Extension of the hip			
				Posterior pelvic rotation			
				Internal rotation of the hip	Transverse		
				Internal rotation of the knee			
	Biceps femoris	Long head: ischial tuberosity. Short head: lower half of the linea aspera, and lateral condyloid ridge	Head of the fibula and lateral condyle of the tibia	Flexion of the knee	Sagittal	Posterolateral aspect of distal thigh with combined knee flexion and external rotation against resistance; just distal to the ischial tuberosity in a prone position with hip internally rotated during active knee flexion	Long head: sciatic nerve—tibial division (S1–S3)

Short head: sciatic nerve—peroneal division (L5, S1, S2) |
				Extension of the hip			
				Posterior pelvic rotation			
				External rotation of the hip	Transverse		
				External rotation of the knee			
	Gluteus maximus	Posterior 1/4 of the crest of the ilium, posterior surface of the sacrum and coccyx near the ilium, and fascia of the lumbar area	Oblique ridge on the lateral surface of the greater trochanter and the iliotibial band of the fasciae latae	Extension of the hip	Sagittal	Running downward and laterally between posterior iliac crest superiorly, anal cleft medially, and gluteal fold inferiorly, emphasized with hip extension, external rotation, and abduction	Inferior gluteal nerve (L5, S1, S2)
				Posterior pelvic rotation			
				External rotation of the hip	Transverse		
				Upper fibers: assist in hip abduction	Frontal		
				Lower fibers: assist in hip adduction			

Chapter

9

TABLE 9.2 (continued) • Agonist muscles of the hip joint

Muscle		Origin	Insertion	Action	Plane of motion	Palpation	Innervation
Lateral	Gluteus medius	Lateral surface of the ilium just below the crest	Posterior and middle surfaces of the greater trochanter of the femur	Abduction of the hip	Frontal	Slightly in front of and a few inches above the greater trochanter with active elevation of opposite pelvis from a standing position or active abduction when side-lying on contralateral pelvis	Superior gluteal nerve (L4, L5, S1)
				Lateral pelvic rotation to ipsilateral side			
				Anterior fibers: internal rotation of the hip	Transverse		
				Posterior fibers: external rotation of the hip			
				Anterior fibers: flexion of the hip	Sagittal		
				Anterior fibers: anterior pelvic rotation			
				Posterior fibers: extension of the hip			
				Posterior fibers: posterior pelvic rotation			
	Gluteus minimus	Lateral surface of the ilium just below the origin of the gluteus medius	Anterior surface of the greater trochanter of the femur	Abduction of the hip	Frontal	Deep to the gluteus medius; covered by tensor fasciae latae between anterior iliac crest and greater trochanter during internal rotation and abduction	Superior gluteal nerve (L4, L5, S1)
				Lateral pelvic rotation to ipsilateral side			
				Hip internal rotation as femur is abducted	Transverse		
				Flexion of the hip	Sagittal		
				Anterior pelvic rotation			
	Tensor fasciae latae	Anterior iliac crest and surface of the ilium just below the crest	One-fourth of the way down the thigh into the iliotibial tract, which in turn inserts onto Gerdy's tubercle of the anterolateral tibial condyle	Abduction of the hip	Frontal	Anterolaterally, between anterior iliac crest and greater trochanter during flexion, internal rotation, and abduction	Superior gluteal nerve (L4, L5, S1)
				Lateral pelvic rotation to ipsilateral side			
				Flexion of the hip	Sagittal		
				Anterior pelvic rotation			
				Internal rotation of the hip as it flexes	Transverse		

Chapter 9

TABLE 9.2 (continued) • Agonist muscles of the hip joint

	Muscle	Origin	Insertion	Action	Plane of motion	Palpation	Innervation
Deep posterior	Piriformis	Anterior sacrum, posterior portions of the ischium, and obturator foramen	Superior and posterior aspect of the greater trochanter	Hip external rotation	Transverse	With subject prone and gluteus maximus relaxed, palpate deeply between posterior superior greater trochanter and sacrum while passively internally/externally rotating femur	First and second sacral nerves (S1, S2)
	Gemellus superior	Ischial spine	Posterior aspect of the greater trochanter immediately below piriformis	Hip external rotation	Transverse	With subject prone and gluteus maximus relaxed, palpate deeply between posterior superior greater trochanter and ischial spine while passively internally/externally rotating femur	Sacral nerve (L5, S1, S2)
	Gemellus inferior	Ischial tuberosity	Posterior aspect of the greater trochanter with obturator internus	Hip external rotation	Transverse	With subject prone and gluteus maximus relaxed, palpate deeply between posterior greater trochanter and ischial tuberosity while passively internally/externally rotating femur	Branches from sacral plexus (L4, L5, S1, S2)
	Obturator internus	Margin of obturator foramen	Posterior aspect of the greater trochanter with gamellus superior	Hip external rotation	Transverse	With subject prone and gluteus maximus relaxed, palpate deeply between posterior superior greater trochanter and obturator foramen while passively internally/externally rotating femur	Branches from sacral plexus (L4, L5, S1, S2)
	Obturator externus	Inferior margin of obturator foramen	Posterior aspect of the greater trochanter immediately below obturator internus	Hip external rotation	Transverse	With subject prone and gluteus maximus relaxed, palpate deeply between inferior posterior greater trochanter and obturator foramen while passively internally/externally rotating femur	Obturator nerve (L3, L4)
	Quadratus femoris	Ischial tuberosity	Intertrochanteric ridge of femur	Hip external rotation	Transverse	With subject prone and gluteus maximus relaxed, palpate deeply between inferior posterior greater trochanter and ischial tuberosity while passively internally/externally rotating femur	Branches from sacral plexus (L4, L5, S1)

Chapter

9

Nerves FIG. 9.19

The muscles of the hip and pelvic girdle are all innervated from the lumbar and sacral plexus, known collectively as the lumbosacral plexus. The lumbar plexus is formed by the anterior rami of spinal nerves L1 through L4 and some fibers from T12. The lower abdomen and the anterior and medial portions of the lower extremity are innervated by nerves arising from the lumbar plexus. The sacral plexus is formed by the anterior rami of L4, L5, and S1 through S4. The lower back, pelvis, perineum, posterior surface of the thigh and leg, and dorsal and plantar surfaces of the foot are innervated by nerves arising from the sacral plexus.

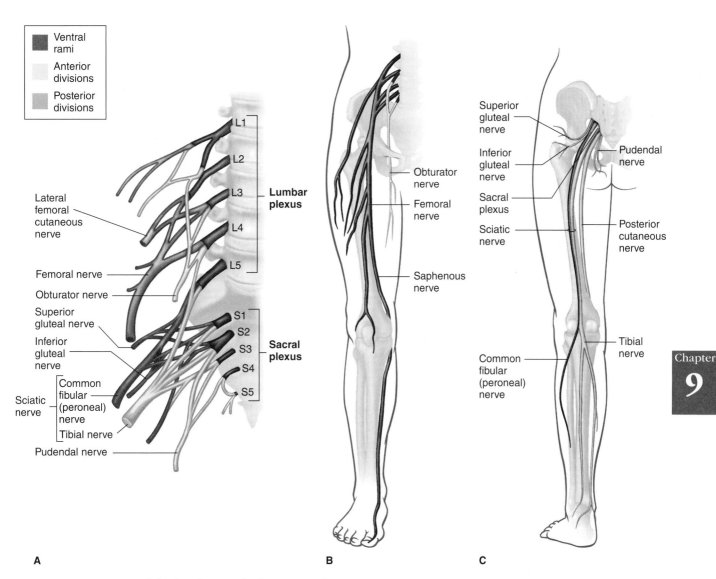

FIG. 9.19 • Nerves of the lumbosacral plexus. **A,** Close-up; **B,** Anterior view; **C,** Posterior view.

The major nerves of significance arising from the lumbar plexus to innervate the muscles of the hip are the femoral and obturator nerves. The femoral nerve (Fig. 9.20) arises from the posterior division of the lumbar plexus and innervates the anterior muscles of the thigh, including the iliopsoas, rectus femoris, vastus medialis, vastus intermedius, vastus lateralis, pectineus, and sartorius. It also provides sensation to the anterior and medial thigh and the medial leg and foot. The obturator nerve (Fig. 9.21) arises from the anterior division of the lumbar plexus and provides innervation to the hip adductors, such as the adductor brevis, adductor longus, adductor magnus, and gracilis,

as well as the obturator externus. The obturator nerve provides sensation to the medial thigh.

The nerves arising from the sacral plexus that innervate the muscles of the hip are the superior gluteal, inferior gluteal, sciatic, and branches from the sacral plexus. The superior gluteal nerve arises from L4, L5, and S1 to innervate the gluteus medius, gluteus minimus, and tensor fasciae latae. The inferior gluteal nerve arises from L5, S1, and S2 to supply the gluteus maximus. Branches from the sacral plexus innervate the piriformis (S1, S2), gemellus superior (L5, S1, S2), gemellus inferior and obturator internus (L4, L5, S1, S2), and quadratus femoris (L4, L5, S1).

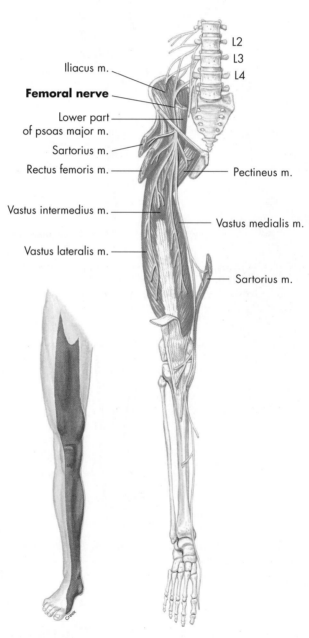

FIG. 9.20 • Muscular and cutaneous distribution of the femoral nerve.

The sciatic nerve is composed of the tibial and common peroneal (fibular) nerves, which are wrapped together in a connective tissue sheath until reaching approximately midway down the posterior thigh. The sciatic nerve tibial division (Fig. 9.22) innervates the semitendinosus, semimembranosus, biceps femoris (long head), and adductor magnus. The sciatic nerve supplies sensation to the anterolateral and posterolateral lower leg as well as to most of the dorsal and plantar aspects of the foot. The tibial division provides sensation to the posterolateral lower leg and plantar aspect of the foot, while the peroneal division provides sensation to the anterolateral lower leg and dorsum of the foot. Both of these nerves continue down the lower extremity to provide motor and sensory function to the muscles of the lower leg; this will be addressed in Chapters 10 and 11.

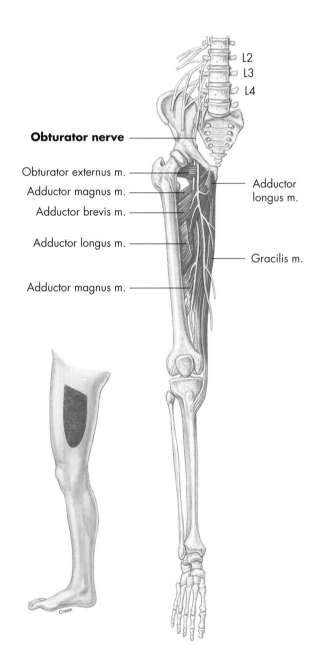

FIG. 9.21 • Muscular and cutaneous distribution of the obturator nerve.

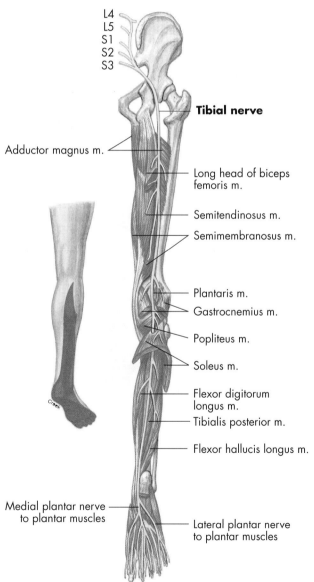

FIG. 9.22 • Muscular and cutaneous distribution of the tibial nerve.

Iliopsoas muscle   FIG. 9.23

(il´e-o-so´as)

Hip flexion

Hip external rotation

Lumbar flexion

Lumbar lateral flexion

Origin

Iliacus: inner surface of the ilium
Psoas major and minor: lower borders of the
  transverse processes (L1–L5), sides of the bodies of
the last thoracic vertebra (T12), lumbar vertebrae
(L1–L5), intervertebral fibrocartilages, and base of
the sacrum

Insertion

Iliacus and psoas major: lesser trochanter of the
femur and the shaft just below

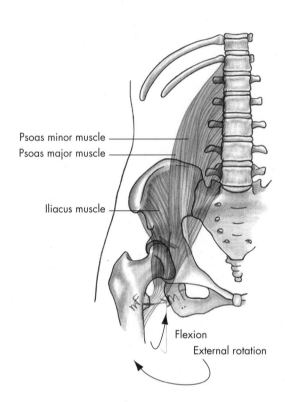

Psoas minor muscle

Psoas major muscle

Iliacus muscle

Flexion

External rotation

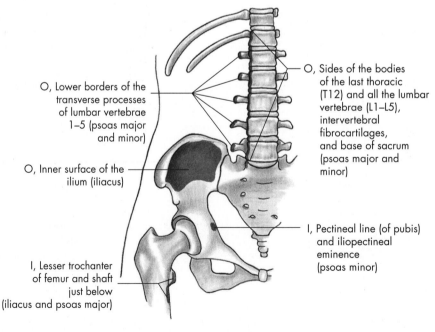

O, Lower borders of the
transverse processes
of lumbar vertebrae
1–5 (psoas major
and minor)

O, Inner surface of the
ilium (iliacus)

I, Lesser trochanter
of femur and shaft
just below
(iliacus and psoas major)

O, Sides of the bodies
of the last thoracic
(T12) and all the lumbar
vertebrae (L1–L5),
intervertebral
fibrocartilages,
and base of sacrum
(psoas major and
minor)

I, Pectineal line (of pubis)
and iliopectineal
eminence
(psoas minor)

FIG. 9.23 ● Iliopsoas muscle, anterior view. O, Origin; I, Insertion.

Psoas minor: pectineal line (of pubis) and iliopectineal eminence

Action

Flexion of the hip
Anterior pelvic rotation
External rotation of the hip
Transverse pelvis rotation contralaterally when ipsilateral femur is stabilized
Flexion of lumbar spine (psoas major and minor)
Lateral flexion of lumbar spine (psoas major and minor)
Lateral pelvic rotation to contralateral side (psoas major and minor)

Palpation

Difficult to palpate; deep against posterior abdominal wall; with subject seated and leaning slightly forward to relax abdominal muscles, palpate psoas major deeply between iliac crest and 12th rib about halfway between ASIS and umbilicus with active hip flexion; palpate iliopsoas distal tendon on anterior aspect of hip approximately 1½ inches below center of inguinal ligament with active hip flexion/extension of supine subject, immediately lateral to pectineus and medial to sartorius

Innervation

Lumbar nerve and femoral nerve (L2–L4)

Application, strengthening, and flexibility

The iliopsoas is commonly referred to as if it were one muscle, but it is actually composed of the iliacus and the psoas major. Some anatomy texts make this distinction and list each muscle individually. The psoas minor attaches on the pubis above the hip joint and therefore does not act on the hip joint. Most authorities do not include it in discussion of the iliopsoas.

The iliopsoas muscle is powerful in actions such as raising the lower extremity from the floor while in a supine position. The psoas major's origin in the lower back tends to move the lower back anteriorly or, in the supine position, pulls up the lower back as it raises the thighs. For this reason, lower back problems are often aggravated by this activity, and bilateral 6-inch leg raises are usually not recommended. The abdominals are the muscles that can be used to prevent this lower back strain by pulling up on the front of the pelvis, thus flattening the back. Leg raising is primarily hip flexion and not abdominal action. The back may be injured by strenuous and prolonged leg-raising exercises due to the iliopsoas pulling the lumbar spine into hyperextension and increasing the lordotic curve, particularly in the absence of adequate stabilization by the abdominals. The iliopsoas contracts strongly, both concentrically and eccentrically, in sit-ups, particularly if the hip is not flexed. The more flexed and/or abducted the hips are, the less the iliopsoas will be activated with abdominal strengthening exercises.

The iliopsoas may be exercised by supporting the arms on a dip bar or parallel bars and then flexing the hips to lift the legs. This may be done initially with the knees flexed in a tucked position to lessen the resistance. As the muscle becomes more developed, the knees can be straightened, which increases the resistance arm length to add more resistance. This concept of increasing or decreasing the resistance by modifying the resistance arm is explained further in Chapter 3. See Appendix 3 for more commonly used exercises for the iliopsoas and other muscles in this chapter.

To stretch the iliopsoas, which often becomes tight with excessive straight-leg sit-ups and contributes to anterior pelvic tilting, the hip must be extended so that the femur is behind the plane of the body. In order to somewhat isolate the iliopsoas, full knee flexion should be avoided. Slight additional stretch may be applied by internally rotating the hip while it is extended.

Chapter 9

Rectus femoris muscle   FIG. 9.24

Hip flexion

(rek´tus fem´or-is)

Origin

Anterior inferior iliac spine of the ilium and groove
(posterior) above the acetabulum

Knee
extension

Insertion

Superior aspect of the patella and patellar tendon to
the tibial tuberosity

Action

Flexion of the hip
Extension of the knee
Anterior pelvic rotation

Palpation

Straight down anterior thigh from anterior inferior
iliac spine to patella with resisted knee extension
and hip flexion

Innervation

Femoral nerve (L2–L4)

Application, strengthening, and flexibility

Pulling from the anterior inferior iliac spine of
the ilium, the rectus femoris muscle has the same
tendency to anteriorly rotate the pelvis (down in
front and up in back). Only the abdominal muscles can prevent this from occurring. In speaking of the hip flexor group in general, it may be
said that many people permit the pelvis to be permanently tilted forward as they get older. The
relaxed abdominal wall does not hold the pelvis
up; therefore, an increased lumbar curve results.

Generally, a muscle's ability to exert force
decreases as it shortens. This explains why the
rectus femoris muscle is a powerful extensor of
the knee when the hip is extended but is weaker
when the hip is flexed. This muscle is exercised,
along with the vastus group, in running, jumping, hopping, and skipping. In these movements,
the hips are extended powerfully by the gluteus
maximus and the hamstring muscles, which counteract the tendency of the rectus femoris muscle
to flex the hip while it extends the knee. It can
be remembered as one of the quadriceps muscle
group. The rectus femoris is developed by performing hip flexion exercises or knee extension
exercises against manual resistance.

The rectus femoris is best stretched in a sidelying position by having a partner take the knee
into full flexion and simultaneously take the hip
into extension.

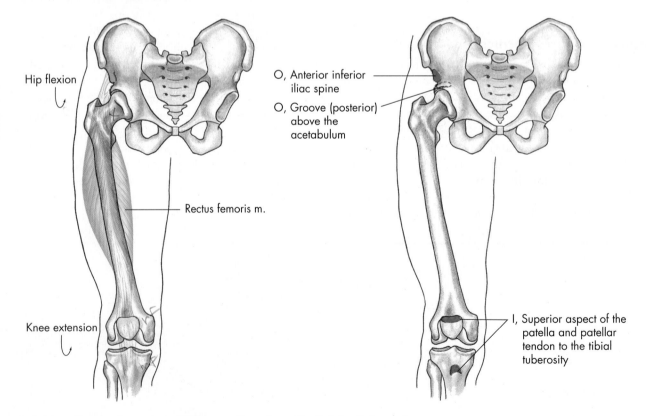

FIG. 9.24 • Rectus femoris muscle, anterior view. O, Origin; I, Insertion.

Sartorius muscle   FIG. 9.25

(sar-to´ri-us)

Origin

Anterior superior iliac spine and notch just below the spine

Insertion

Anterior medial surface of the tibia just below the condyle

Action

Flexion of the hip
Flexion of the knee
External rotation of the thigh as it flexes the hip and knee
Abduction of the hip
Anterior pelvic rotation
Weak internal rotation of the knee

Palpation

Located proximally just medial to the tensor fasciae latae and lateral to the iliopsoas; palpate superficially from the anterior superior iliac spine to the medial tibial condyle during combined resistance of hip flexion/external rotation/abduction and knee flexion in a supine position

Innervation

Femoral nerve (L2, L3)

Application, strengthening, and flexibility

Pulling from the anterior superior iliac spine and the notch just below it, the tendency again is to tilt the pelvis anteriorly (down in front) as this muscle contracts. The abdominal muscles must prevent this tendency by posteriorly rotating the pelvis (pulling up in front), thus flattening the lower back.

Hip flexion

Knee flexion

The sartorius, a two-joint muscle, is effective as a hip flexor or as a knee flexor. Sometimes referred to as the tailor's muscle, it is active in all the hip and knee movements used in assuming the sitting position of a tailor. It is weak when both flexions take place at the same time. Observe that, in attempting to cross the knees when in a sitting position, one customarily leans well back, thus raising the origin to lengthen this muscle, making it more effective in flexing and crossing the knees. With the knees held extended, the sartorius becomes a more effective hip flexor. It is the longest muscle in the body and is strengthened when hip flexion activities are performed as described for developing the iliopsoas. Stretching may be accomplished by a partner passively taking the hip into extreme extension, adduction, and internal rotation with the knee extended.

Hip external rotation

Hip abduction

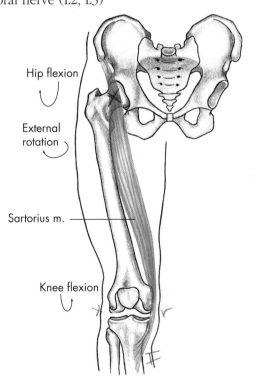

Hip flexion

External rotation

Sartorius m.

Knee flexion

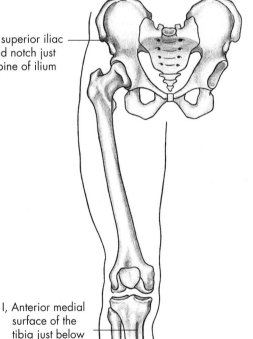

O, Anterior superior iliac spine and notch just below spine of ilium

I, Anterior medial surface of the tibia just below the condyle

Knee internal rotation

FIG. 9.25 • Sartorius muscle, anterior view. O, Origin; I, Insertion.

Pectineus muscle   FIG. 9.26

(pek-tin´e-us)

Hip flexion

Hip adduction

Hip external rotation

Origin

Space 1 inch wide on the front of the pubis just above the crest (pectineal line)

Insertion

Rough line leading from the lesser trochanter down to the linea aspera (pectineal line of femur)

Action

Flexion of the hip
Adduction of the hip
External rotation of the hip
Anterior pelvic rotation

Palpation

Difficult to distinguish from other adductors; anterior aspect of hip approximately 1½ inches below center of inguinal ligament; just lateral and slightly proximal to adductor longus and medial to iliopsoas during flexion and adduction of supine subject

Innervation

Femoral nerve (L2–L4)

Application, strengthening, and flexibility

As the pectineus contracts, it also tends to rotate the pelvis anteriorly. The abdominal muscles pulling up on the pelvis in front prevent this tilting action.

The pectineus muscle is exercised together with the iliopsoas muscle in leg raising and lowering. Hip flexion exercises and hip adduction exercises against resistance may be used for strengthening this muscle.

The pectineus is stretched by fully abducting the extended and internally rotated hip.

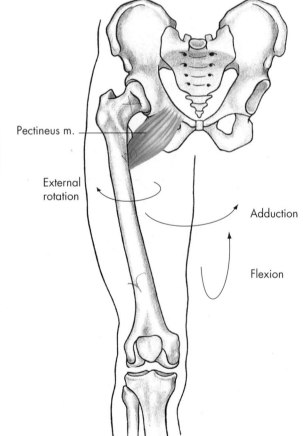

Pectineus m.

External rotation

Adduction

Flexion

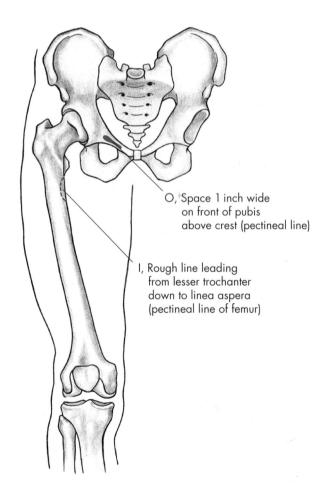

O, Space 1 inch wide on front of pubis above crest (pectineal line)

I, Rough line leading from lesser trochanter down to linea aspera (pectineal line of femur)

FIG. 9.26 ● Pectineus muscle, anterior view. O, Origin; I, Insertion.

Adductor brevis muscle FIG. 9.27
(ad-duk´tor bre´vis)

Origin
Front of the inferior pubic ramus just below the
origin of the adductor longus

Insertion
Lower two-thirds of the pectineal line of the femur
and upper half of the medial lip of the linea aspera

Action
Adduction of the hip
External rotation as it adducts the hip
Assists in flexion of the hip
Assists in anterior pelvic rotation

Palpation
Deep to the adductor longus and superficial to
the adductor magnus; very difficult to palpate
and differentiate from adductor longus, which is
immediately inferior; proximal portion is just lateral
to adductor longus

Innervation
Obturator nerve (L3, L4)

Application, strengthening, and flexibility
The adductor brevis muscle, along with the other
adductor muscles, provides powerful movement of
the thighs toward each other. Squeezing the thighs
together toward each other against resistance is
effective in strengthening the adductor brevis.
Abducting the extended and internally rotated hip
provides stretching of the adductor brevis.

Hip
adduction

Hip
external
rotation

Hip flexion

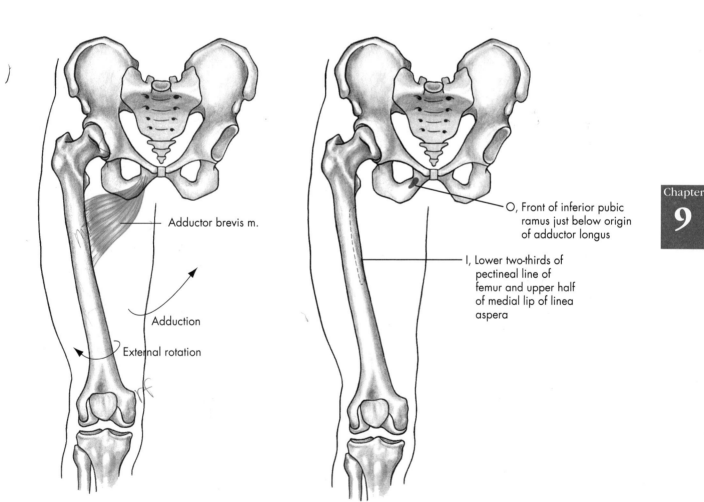

Adductor brevis m.

Adduction

External rotation

O, Front of inferior pubic
ramus just below origin
of adductor longus

I, Lower two-thirds of
pectineal line of
femur and upper half
of medial lip of linea
aspera

Chapter

9

FIG. 9.27 • Adductor brevis muscle, anterior view. O, Origin; I, Insertion.

Adductor longus muscle   FIG. 9.28

(ad-duk´tor lon´gus)

Origin

Anterior pubis just below its crest

Insertion

Middle third of the linea aspera

Action

Adduction of the hip

Assists in flexion of the hip

Assists in anterior pelvic rotation

Palpation

Most prominent muscle proximally on anteromedial thigh just inferior to the pubic bone with resisted adduction

Innervation

Obturator nerve (L3, L4)

Application, strengthening, and flexibility

The muscle may be strengthened by using the scissors exercise, which requires the subject to sit on the floor with the legs spread wide while a partner puts his or her legs or arms inside each lower leg to provide resistance. As the subject attempts to adduct his or her legs together, the partner provides manual resistance throughout the range of motion. This exercise may be used for either one or both legs. The adductor longus is stretched in the same manner as the adductor brevis.

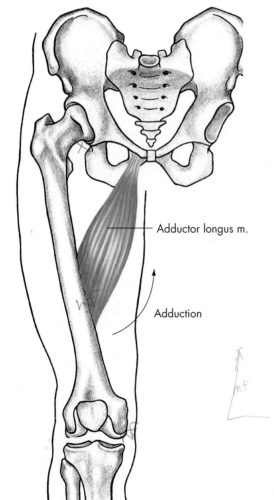

Adductor longus m.

Adduction

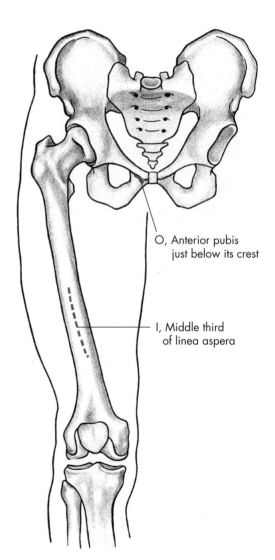

O, Anterior pubis just below its crest

I, Middle third of linea aspera

FIG. 9.28 • Adductor longus muscle, anterior view. O, Origin; I, Insertion.

Chapter 9

Adductor magnus muscle FIG. 9.29

(ad-duk´tor mag´nus)

Origin

Edge of the entire ramus of the pubis and the ischium and ischial tuberosity

Insertion

Whole length of the linea aspera, inner condyloid ridge, and adductor tubercle

Action

Adduction of the hip
External rotation as the hip adducts
Extension of the hip

Palpation

Medial aspect of thigh between gracilis and medial hamstrings from ischial tuberosity to adductor tubercle with resisted adduction from abducted position

Innervation

Anterior: obturator nerve (L2–L4)
Posterior: sciatic nerve (L4, L5, S1–S3)

Hip adduction

Hip external rotation

Application, strengthening, and flexibility

The adductor magnus muscle is used in the breaststroke kick in swimming and in horseback riding. Since the adductor muscles (adductor magnus, adductor longus, adductor brevis, and gracilis) are not heavily used in ordinary movement, some prescribed activity for them should be provided. Some modern exercise equipment is engineered to provide resistance for hip adduction movement. Hip adduction exercises such as those described for the adductor brevis and the adductor longus may be used for strengthening the adductor magnus as well. The adductor magnus is stretched in the same manner as the adductor brevis and adductor longus.

Hip extension

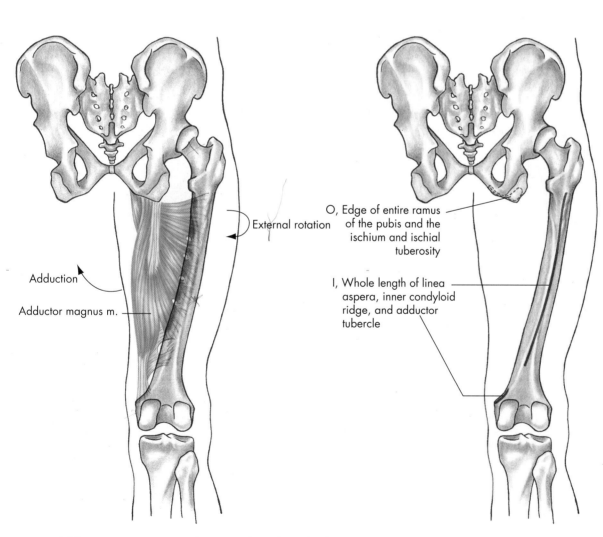

FIG. 9.29 • Adductor magnus muscle, posterior view. O, Origin; I, Insertion.

Chapter 9

Gracilis muscle   FIG. 9.30

(gras´il-is)

Hip
adduction

Knee
flexion

Hip
internal
rotation

Hip flexion

Origin

Anteromedial edge of the descending ramus of the
pubis

Insertion

Anterior medial surface of the tibia just below the
condyle

Action

Adduction of the hip
Weak flexion of the knee
Internal rotation of the hip
Assists with flexion of the hip
Weak internal rotation of the knee

Palpation

A thin tendon on anteromedial thigh with knee
flexion and resisted adduction just posterior to
adductor longus and medial to the semitendinosus

Innervation

Obturator nerve (L2–L4)

Application, strengthening, and flexibility

Also known as the adductor gracilis, this muscle
performs the same function as the other adductors
but adds some weak assistance to knee flexion.

The adductor muscles as a group (adductor
magnus, adductor longus, adductor brevis, and
gracilis) are called into action in horseback riding
and in doing the breaststroke kick in swimming.
Proper development of the adductor group prevents soreness after participation in these sports.
The gracilis is strengthened with the same exercises as described for the other hip adductors.
The gracilis may be stretched in a manner similar
to the adductors, except that the knee must be
extended.

Chapter
9

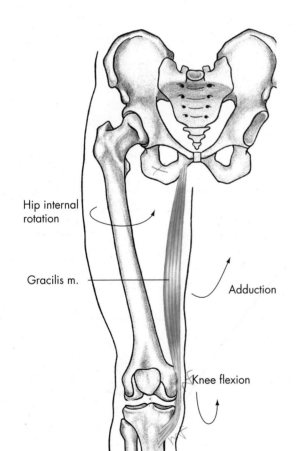

Hip internal
rotation

Gracilis m.

Adduction

Knee flexion

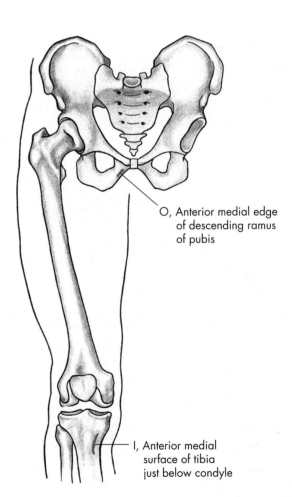

O, Anterior medial edge
of descending ramus
of pubis

I, Anterior medial
surface of tibia
just below condyle

FIG. 9.30 ● Gracilis muscle, anterior view. O, Origin; I, Insertion.

Semitendinosus muscle FIG. 9.31

(sem´i-ten-di-no´sus)

Origin

Ischial tuberosity

Insertion

Anterior medial surface of the tibia just below the condyle

Action

Flexion of the knee
Extension of the hip
Internal rotation of the hip
Internal rotation of the flexed knee
Posterior pelvic rotation

Palpation

Posteromedial aspect of distal thigh with combined knee flexion and internal rotation against resistance just distal to the ischial tuberosity in a prone position with hip internally rotated during active knee flexion

Innervation

Sciatic nerve—tibial division (L5, S1, S2)

Application, strengthening, and flexibility

This two-joint muscle is most effective when contracting to either extend the hip or flex the knee. When there is extension of the hip and flexion of the knee at the same time, both movements are weak. When the trunk is flexed forward with the knees straight, the hamstring muscles have a powerful pull on the rear pelvis and tilt it down in back by full contraction. If the knees are flexed when this movement takes place, one can observe that the work is done chiefly by the gluteus maximus muscle.

On the other hand, when the muscles are used in powerful flexion of the knees, as in hanging by the knees from a bar, the flexors of the hip come into play to raise the origin of these muscles and make them more effective as knee flexors. By full extension of the hips in this movement, the knee flexion movement is weakened. These muscles are used in ordinary walking as extensors of the hip and allow the gluteus maximus to relax in the movement.

The semitendinosus is best developed through knee flexion exercises against resistance. Commonly known as hamstring curls or leg curls, they may be performed in a prone position on a knee table or standing with ankle weights attached. This muscle is emphasized when performing hamstring curls while attempting to maintain the knee joint in internal rotation. This internally rotated position brings its insertion in alignment with its origin.

The semitendinosus is stretched by maximally extending the knee while flexing the internally rotated and slightly abducted hip.

Knee flexion

Hip extension

Hip internal rotation

Knee internal rotation

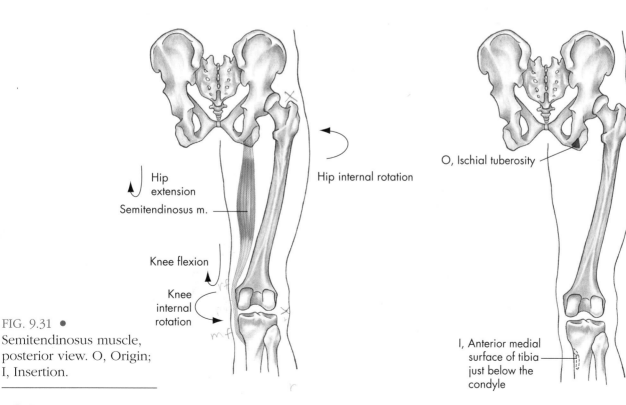

FIG. 9.31 ● Semitendinosus muscle, posterior view. O, Origin; I, Insertion.

Hip extension
Semitendinosus m.
Knee flexion
Knee internal rotation
Hip internal rotation

O, Ischial tuberosity

I, Anterior medial surface of tibia just below the condyle

Chapter 9

Knee flexion

Semimembranosus muscle   FIG. 9.32

(sem´i-mem´bra-no´sus)

Hip extension

Origin

Ischial tuberosity

Insertion

Posteromedial surface of the medial tibial condyle

Hip internal rotation

Action

Flexion of the knee
Extension of the hip
Internal rotation of the hip
Internal rotation of the flexed knee
Posterior pelvic rotation

Palpation

Largely covered by other muscles, the tendon can be felt at the posteromedial aspect of the knee just deep to the semitendinosus tendon with combined knee flexion and internal rotation against resistance

Knee internal rotation

Innervation

Sciatic nerve—tibial division (L5, S1, S2)

Application, strengthening, and flexibility

Both the semitendinosus and semimembranosus are responsible for internal rotation of the knee, along with the popliteus muscle, which is discussed in Chapter 10. Because of the manner in which they cross the joint, the muscles are very important in providing dynamic medial stability to the knee joint.

The semimembranosus is best developed by performing leg curls. Internal rotation of the knee throughout the range accentuates the activity of this muscle. The semimembranosus is stretched in the same manner as the semitendinosus.

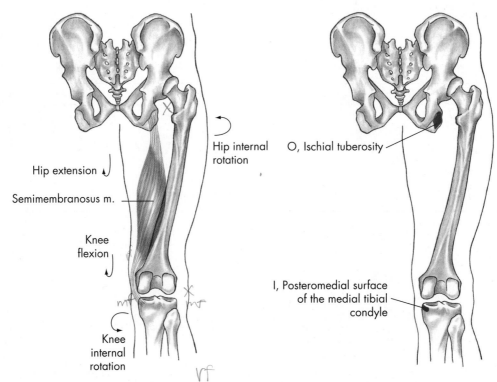

FIG. 9.32 • Semimembranosus muscle, posterior view. O, Origin; I, Insertion.

Biceps femoris muscle   FIG. 9.33

(bi´seps fem´or-is)

Origin

Long head: ischial tuberosity
Short head: lower half of the linea aspera, and lateral condyloid ridge

Insertion

Lateral condyle of the tibia and head of the fibula

Action

Flexion of the knee
Extension of the hip
External rotation of the hip
External rotation of the flexed knee
Posterior pelvic rotation

Palpation

Posterolateral aspect of distal thigh with combined knee flexion and external rotation against resistance and just distal to the ischial tuberosity in a prone position with hip externally rotated during active knee flexion

Innervation

Long head: sciatic nerve—tibial division (S1–S3) Short head: sciatic nerve—peroneal division (L5, S1, S2)

Knee flexion

Hip extension

Application, strengthening, and flexibility

The semitendinosus, semimembranosus, and biceps femoris muscles are known as the hamstrings. These muscles, together with the gluteus maximus muscle, are used in extension of the hip when the knees are straight or nearly so. Thus, in running, jumping, skipping, and hopping, these muscles are used together. The hamstrings are used without the aid of the gluteus maximus, however, when hanging from a bar by the knees. Similarly, the gluteus maximus is used without the aid of the hamstrings when the knees are flexed while the hips are being extended. This occurs when rising from a kneebend position to a standing position.

Hip external rotation

The biceps femoris is best developed through hamstring curls as described for the semitendinosus, but it is emphasized more if the knee is maintained in external rotation throughout the range of motion, which brings the origin and insertion more in line with each other. The biceps femoris is stretched by maximally extending the knee while flexing the externally rotated and slightly adducted hip.

Knee external rotation

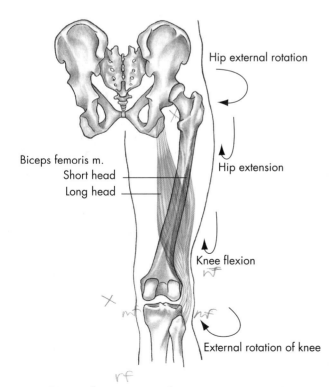

Biceps femoris m.
Short head
Long head

Hip external rotation

Hip extension

Knee flexion

External rotation of knee

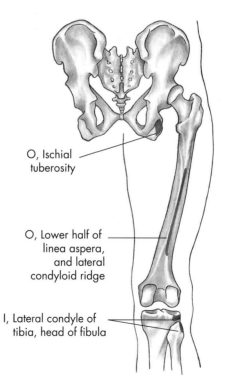

O, Ischial tuberosity

O, Lower half of linea aspera, and lateral condyloid ridge

I, Lateral condyle of tibia, head of fibula

FIG. 9.33 • Biceps femoris muscle, posterior view. O, Origin; I, Insertion.

Gluteus maximus muscle FIG. 9.34

(glu´te-us maks´i-mus)

Origin

Posterior one-fourth of the crest of the ilium, posterior surface of the sacrum and coccyx near the ilium, and fascia of the lumbar area

Insertion

Oblique ridge (gluteal tuberosity) on the lateral surface of the greater trochanter and the iliotibial band of the fasciae latae

Action

Extension of the hip
External rotation of the hip
Upper fibers: assist in hip abduction
Lower fibers: assist in hip adduction
Posterior pelvic rotation

Palpation

Running downward and laterally between posterior iliac crest superiorly, anal cleft medially, and gluteal fold inferiorly, emphasized with hip extension, external rotation, and abduction

Innervation

Inferior gluteal nerve (L5, S1, S2)

Application, strengthening, and flexibility

The gluteus maximus muscle comes into action when movement between the pelvis and the femur approaches and goes beyond 15 degrees of extension. As a result, it is not used extensively in ordinary walking. It is important in extension of the thigh with external rotation.

Strong action of the gluteus maximus muscle is seen in running, hopping, skipping, and jumping. Powerful extension of the thigh is secured in the return to standing from a squatting position, especially with weighted barbells placed on the shoulders.

Hip extension exercises from a forward-leaning or prone position may be used to develop this muscle. This muscle is most emphasized when the hip starts from a flexed position and moves to full extension and abduction, with the knee flexed 30 degrees or more to reduce the hamstrings' involvement in the action.

The gluteus maximus is stretched in the supine position with full hip flexion to the ipsilateral axilla and then to the contralateral axilla with the knee in flexion. Simultaneous internal hip rotation accentuates this stretch.

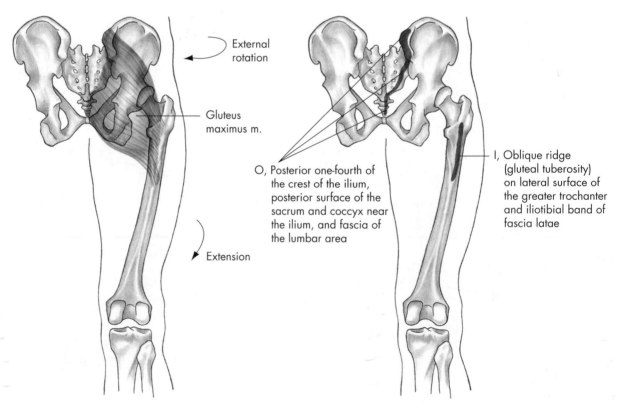

FIG. 9.34 • Gluteus maximus muscle, posterior view. O, Origin; I, Insertion.

Gluteus medius muscle   FIG. 9.35

(glu´te-us me´di-us)

Origin

Lateral surface of the ilium just below the crest

Insertion

Posterior and middle surfaces of the greater trochanter of the femur

Action

Abduction of the hip

Lateral pelvic rotation to ipsilateral side

Anterior fibers: internal rotation, flexion of the hip, and anterior pelvic rotation

Posterior fibers: external rotation, extension of the hip, and posterior pelvic rotation

Palpation

Slightly in front of and a few inches above the greater trochanter with active elevation of opposite pelvis from a standing position or active abduction when side-lying on contralateral pelvis

Innervation

Superior gluteal nerve (L4, L5, S1)

Application, strengthening, and flexibility

Typical action of the gluteus medius and gluteus minimus muscles is seen in walking. As the weight of the body is suspended on one leg, these muscles prevent the opposite pelvis from sagging. Weakness in the gluteus medius and gluteus minimus can result in a Trendelenburg gait, which is characterized by the trunk lurching to the side of the weakness when the contralateral pelvis drops. With this weakness, the individual's opposite pelvis will sag on weight bearing because the hip abductors on the weight-bearing side are not strong enough to maintain the opposite side at or near level.

Hip external rotation exercises performed against resistance can provide some strengthening of the gluteus medius, but it is best strengthened by performing side-lying leg raises or hip abduction exercises as described for the tensor fasciae latae. The gluteus medius is best stretched by moving the hip into extreme adduction in front of the opposite extremity and then behind it.

Hip abduction

Hip internal rotation

Hip flexion

Hip external rotation

Hip extension

Gluteus medius m.

Abduction

External rotation

Internal rotation

O, Lateral surface of ilium just below crest

I, Posterior and middle surfaces of greater trochanter of femur

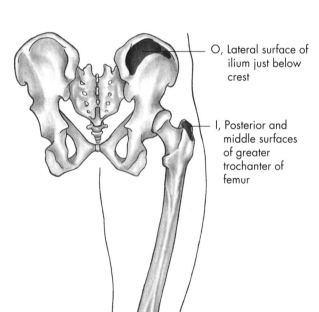

FIG. 9.35 • Gluteus medius muscle, posterior view. O, Origin; I, Insertion.

Hip
abduction

Hip
internal
rotation

Hip flexion

Gluteus minimus muscle   FIG. 9.36

(glu´te-us min´i-mus)

Origin

Lateral surface of the ilium just below the origin of
the gluteus medius

Insertion

Anterior surface of the greater trochanter of the femur

Action

Abduction of the hip
Lateral pelvic rotation to ipsilateral side
Internal rotation as the femur abducts
Flexion of the hip
Anterior pelvic rotation

Palpation

Deep to the gluteus medius; covered by tensor fasciae latae between anterior iliac crest and greater
trochanter during internal rotation and abduction

Innervation

Superior gluteal nerve (L4, L5, S1)

Application, strengthening, and flexibility

Both the gluteus minimus and the gluteus medius
are used in powerfully maintaining proper hip
abduction while running. As a result, both of
these muscles are exercised effectively in running,
hopping, and skipping, in which weight is transferred forcefully from one foot to the other. As the
body ages, the gluteus medius and gluteus minimus muscles tend to lose their effectiveness. The
spring of youth, as far as the hips are concerned,
resides in these muscles. To have great drive in
the legs, these muscles must be fully developed.

The gluteus minimus is best strengthened by
performing hip abduction exercises similar to
those described for the tensor fasciae latae and
gluteus medius muscles. It may also be developed by performing hip internal rotation exercises against manual resistance. Stretching of this
muscle is accomplished by extreme hip adduction
with slight external rotation.

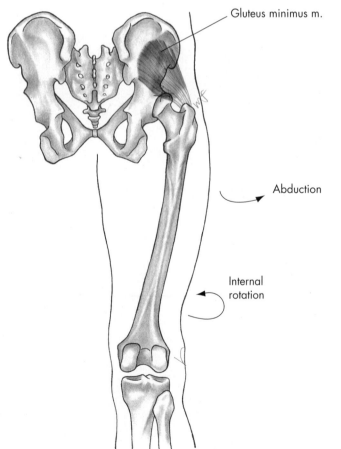
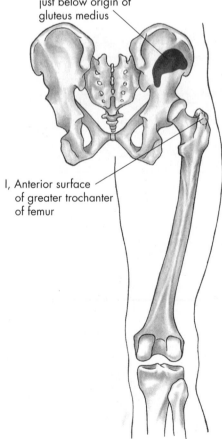

FIG. 9.36 ● Gluteus minimus muscle, posterior view. O, Origin; I, Insertion.

Tensor fasciae latae muscle   FIG. 9.37

(ten´sor fas´i-e la´te)

Origin

Anterior iliac crest and surface of the ilium just
below the crest

Insertion

One-fourth of the way down the thigh into the
iliotibial tract, which in turn inserts onto Gerdy's
tubercle of the anterolateral tibial condyle

Action

Abduction of the hip
Flexion of the hip
Tendency to rotate the hip internally as it flexes
Anterior pelvic rotation

Palpation

Anterolaterally, between anterior iliac crest and
greater trochanter during internal rotation, flexion,
and abduction

Innervation

Superior gluteal nerve (L4, L5, S1)

Application, strengthening, and flexibility

The tensor fasciae latae muscle aids in preventing
external rotation of the hip as it is flexed by other
flexor muscles.

The tensor fasciae latae muscle is used when
flexion and internal rotation take place. This is a
weak movement but is important in helping direct
the leg forward so that the foot is placed straight
forward in walking and running. Thus, from the
supine position, raising the leg with definite internal rotation of the femur will call it into action.

The tensor fasciae latae may be developed by
performing hip abduction exercises against gravity
and resistance while in a side-lying position. This
is done simply by abducting the hip that is up and
then slowly lowering it back to rest against the
other leg. Stretch may be applied by remaining on
the side and having a partner passively move the
downside hip into full extension, adduction, and
external rotation.

Hip
abduction

Hip flexion

Hip
internal
rotation

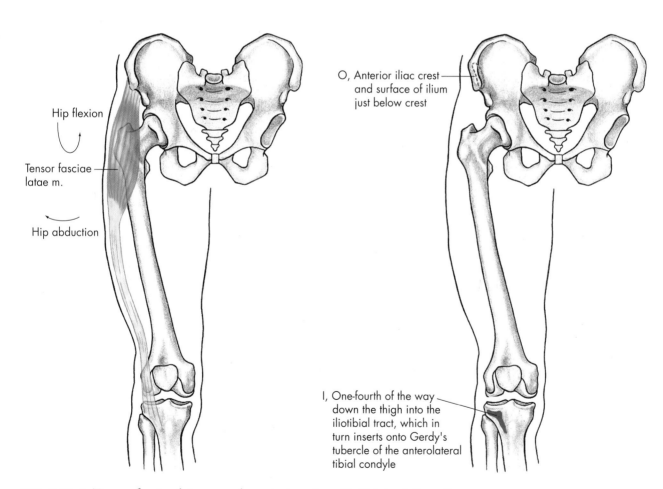

Hip flexion

Tensor fasciae
latae m.

Hip abduction

O, Anterior iliac crest
and surface of ilium
just below crest

I, One-fourth of the way
down the thigh into the
iliotibial tract, which in
turn inserts onto Gerdy's
tubercle of the anterolateral
tibial condyle

Chapter

9

FIG. 9.37 ● Tensor fasciae latae muscle, anterior view. O, Origin; I, Insertion.

The six deep lateral rotator muscles—piriformis, gemellus superior, gemellus inferior, obturator externus, obturator internus, quadratus femoris   FIG. 9.38

(pi-ri-for´mis)
(je-mel´us su-pe´ri-or)
(je-mel´us in-fe´ri-or)
(ob-tu-ra´tor eks-ter´nus)
(ob-tu-ra´tor in-ter´nus)
(kwad-ra´tus fem´or-is)

Origin

Anterior sacrum, posterior portions of the ischium, and obturator foramen

Insertion

Superior and posterior aspect of the greater trochanter

Action

External rotation of the hip

Palpation

Although not directly palpable, deep palpation is possible between the posterior superior greater trochanter and obturator foramen with the subject prone during relaxation of the gluteus maximus while passively using the lower leg flexed at the knee to passively internally and externally rotate the femur or alternately contract/relax the external rotators slightly

Innervation

Piriformis: first or second sacral nerve (S1, S2)

Gemellus superior: sacral nerve (L5, S1, S2)
Gemellus inferior: branches from sacral plexus (L4, L5, S1, S2)
Obturator externus: obturator nerve (L3, L4)
Obturator internus: branches from sacral plexus (L4, L5, S1, S2)
Quadratus femoris: branches from sacral plexus (L4, L5, S1)

Application, strengthening, and flexibility

The six lateral rotators are used powerfully in movements of external rotation of the femur, as in sports in which the individual takes off on one leg from preliminary internal rotation. Throwing a baseball and swinging a baseball bat, in which there is rotation of the hip, are typical examples.

Standing on one leg and forcefully turning the body away from that leg is accomplished by contraction of these muscles, and this may be repeated for strengthening purposes. A partner may provide resistance as development progresses. The six deep lateral rotators may be stretched in the supine position with a partner passively internally rotating and slightly flexing the hip.

Of special note, the sciatic nerve usually passes just inferior to the piriformis muscle but may pass through it. As a result, tightness in the piriformis muscle may contribute to compression on the sciatic nerve. The piriformis may be stretched by having the subject lie on the uninvolved side with a partner passively taking the hip into full internal rotation combined with hip adduction and slight to moderate hip flexion.

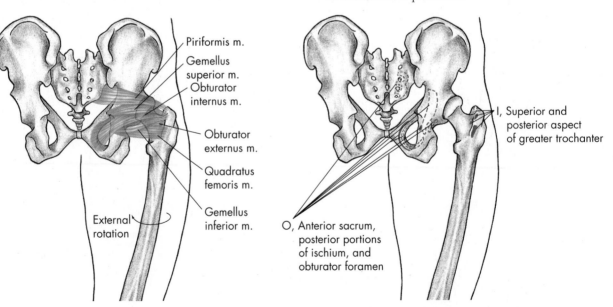

FIG. 9.38 ● The six deep lateral rotator muscles, posterior view: piriformis, gemellus superior, gemellus inferior, obturator externus, obturator internus, and quadratus femoris. O, Origin; I, Insertion.

REVIEW EXERCISES

1. List the planes in which each of the following hip joint movements occurs. List the respective axis of rotation for each movement in each plane.
   a. Flexion
   b. Extension
   c. Adduction
   d. Abduction
   e. External rotation
   f. Internal rotation

2. How is walking different from running in terms of the use of the hip joint muscle actions and the range of motion?

3. Research common hip disorders such as osteoarthritis, groin strains, hamstring strains, greater trochanteric bursitis, and slipped capital femoral epiphysis. Report your findings in class.

4. **Muscle analysis chart** • Hip joint

Complete the chart by listing the muscles primarily involved in each movement.	
Flexion	Extension
Abduction	Adduction
External rotation	Internal rotation

5. **Antagonistic muscle action chart** • Hip joint and pelvic girdle

Complete the chart by listing the muscle(s) or parts of muscles that are antagonist in their actions to the muscles in the left column.

Agonist	Antagonist
Gluteus maximus	
Gluteus medius	
Gluteus minimus	
Biceps femoris	
Semimembranosus/Semitendinosus	
Adductor magnus/Adductor brevis	
Adductor longus	
Gracilis	
Lateral rotators	
Rectus femoris	
Sartorius	
Pectineus	
Iliopsoas	
Tensor fasciae latae	

Chapter

9

LABORATORY EXERCISES

1. Locate the following parts of the pelvic girdle and hip joint on a human skeleton and on a subject:
   a. Skeleton
      1. Ilium
      2. Ischium
      3. Pubis
      4. Symphysis pubis
      5. Acetabulum
      6. Rami (ascending and descending)
      7. Obturator foramen
      8. Ischial tuberosity
      9. Anterior superior iliac spine
      10. Greater trochanter
      11. Lesser trochanter
   b. Subject
      1. Crest of ilium
      2. Anterior superior iliac spine
      3. Ischial tuberosity
      4. Greater trochanter

2. How and where can the following muscles be palpated on a human subject?
   a. Gracilis
   b. Sartorius
   c. Gluteus maximus
   d. Gluteus medius
   e. Gluteus minimus
   f. Biceps femoris
   g. Rectus femoris
   h. Semimembranosus
   i. Semitendinosus
   j. Adductor magnus
   k. Adductor longus
   l. Adductor brevis

3. Be prepared to indicate on a human skeleton, using a long rubber band, where each muscle has its origin and insertion.

4. Distinguish between hip flexion and trunk flexion by performing each individually and then both together.

5. Demonstrate the movement and list the muscles primarily responsible for the following hip movements:
   a. Flexion
   b. Extension
   c. Adduction
   d. Abduction
   e. External rotation
   f. Internal rotation

6. How may the walking gait be affected by a weakness in the gluteus medius muscle? Have a laboratory partner demonstrate the gait pattern associated with gluteus medius weakness. What is the name of this dysfunctional gait?

7. How might bilateral iliopsoas tightness affect the posture and movement of the lumbar spine in the standing position? Demonstrate and discuss this effect with a laboratory partner.

8. How might bilateral hamstring tightness affect the posture and movement of the lumbar spine in the standing position? Demonstrate and discuss this effect with a laboratory partner.

9. **Hip joint exercise movement analysis chart**

After analyzing each exercise in the chart, break each into two primary movement phases, such as a lifting phase and a lowering phase. For each phase, determine what hip joint movements occur, and then list the hip joint muscles primarily responsible for causing/controlling those movements. Beside each muscle in each movement, indicate the type of contraction as follows: I—isometric; C—concentric; E—eccentric.

Exercise	Initial movement (lifting) phase		Secondary movement (lowering) phase	
	Movement(s)	Agonist(s)—(contraction type)	Movement(s)	Agonist(s)—(contraction type)
Push-up				
Squat				
Dead lift				
Hip sled				
Forward lunge				
Rowing exercise				
Stair machine				

10. Hip joint sport skill analysis chart

Analyze each skill in the chart, and list the movements of the right and left hip joints in each phase of the skill. You may prefer to list the initial position the hip joint is in for the stance phase. After each movement, list the hip joint muscle(s) primarily responsible for causing/controlling the movement. Beside each muscle in each movement, indicate the type of contraction as follows: I—isometric, C—concentric; E—eccentric. It may be desirable to review the concepts for analysis in Chapter 8 for the various phases.

Exercise		Stance phase	Preparatory phase	Movement phase	Follow-through phase
Baseball pitch	(R)				
	(L)				
Football punting	(R)				
	(L)				
Walking	(R)				
	(L)				
Softball pitch	(R)				
	(L)				
Soccer pass	(R)				
	(L)				
Batting	(R)				
	(L)				
Bowling	(R)				
	(L)				
Basketball jump shot	(R)				
	(L)				

Chapter

9

References

Field D: *Anatomy: palpation and surface markings,* ed 3, Oxford, 2001, Butterworth-Heinemann.

Hamilton N, Weimer W, Luttgens K: *Kinesiology: scientific basis of human motion,* ed 12, New York, 2012, McGraw-Hill.

Hislop HJ, Montgomery J: *Daniels and Worthingham's muscle testing: techniques of manual examination,* ed 8, Philadelphia, 2007, Saunders.

Kendall FP, McCreary EK, Provance, PG, Rodgers MM, Romani WA: *Muscles: testing and function, with posture and pain,* ed 5, Baltimore, 2005, Lippincott Williams & Wilkins.

Lindsay DT: *Functional human anatomy,* St. Louis, 1996, Mosby.

Lysholm J, Wikland J: Injuries in runners, *American Journal of Sports Medicine* 15:168, September–October 1986.

Magee DJ: *Orthopedic physical assessment,* ed 5, Philadelphia, 2008, Saunders.

Muscolino JE: *The muscular system manual: the skeletal muscles of the human body,* ed 3, St. Louis, 2010, Elsevier Mosby.

Noahes TD, et al: Pelvic stress fractures in long distance runners, *American Journal of Sports Medicine* 13:120, March–April 1985.

Oatis CA: *Kinesiology: the mechanics and pathomechanics of human movement,* ed 2, Philadelphia, 2008, Lippincott Williams & Wilkins.

Prentice WE: *Principles of athletic training: a competency-based approach,* ed 15, New York, 2014, McGraw-Hill.

Saladin KS: *Anatomy & physiology: the unity of form and function,* ed 5, New York, 2010, McGraw-Hill.

Seeley RR, Stephens TD, Tate P: *Anatomy & physiology,* ed 8, New York, 2008, McGraw-Hill.

Shier D, Butler J, Lewis R: *Hole's human anatomy and physiology,* ed 12, New York, 2010, McGraw-Hill.

Sieg KW, Adams SP: *Illustrated essentials of musculoskeletal anatomy,* ed 4, Gainesville, FL, 2002, Megabooks.

Stone RJ, Stone JA: *Atlas of the skeletal muscles,* ed 6, New York, 2009, McGraw-Hill.

Thibodeau GA, Patton KT: *Anatomy & physiology,* ed 9, St. Louis, 1993, Mosby.

Van De Graaff KM: *Human anatomy,* ed 6, Dubuque, IA, 2002, McGraw-Hill.

For additional resources and a list of related websites, visit **www.mhhe.com/floyd19e**.

Chapter

9

Worksheet Exercises

For in- or out-of-class assignments, or for testing, utilize this tear-out worksheet.

Worksheet 1

Using crayons or colored markers, draw and label on the worksheet the following muscles. Indicate the origin and insertion of each muscle with an "O" and an "I," respectively, and draw in the origin and insertion on the contralateral side of the skeleton.

a. Iliopsoas

b. Rectus femoris

c. Sartorius

d. Pectineus

e. Adductor brevis

f. Adductor longus

g. Adductor magnus

h. Gracilis

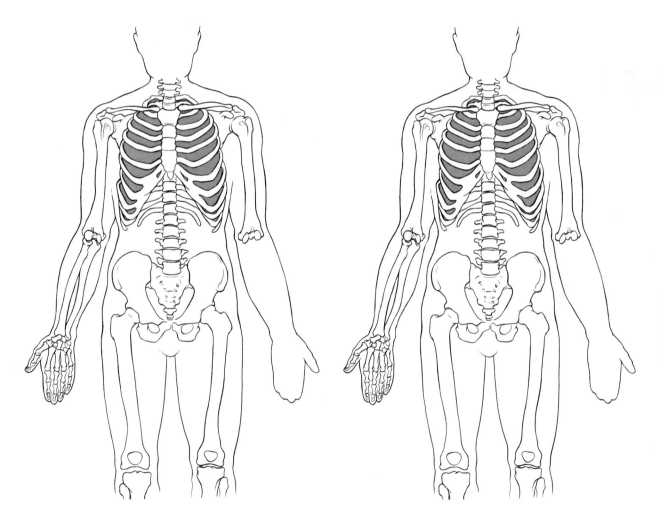

Worksheet Exercises

For in- or out-of-class assignments, or for testing, utilize this tear-out worksheet.

Worksheet 2

Using crayons or colored markers, draw and label on the worksheet the following muscles. Indicate the origin and insertion of each muscle with an "O" and an "I," respectively, and draw in the origin and insertion on the contralateral side of the skeleton.

a. Semitendinosus

b. Semimembranosus

c. Biceps femoris

d. Gluteus maximus

e. Gluteus medius

f. Gluteus minimis

g. Tensor fasciae latae

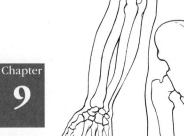

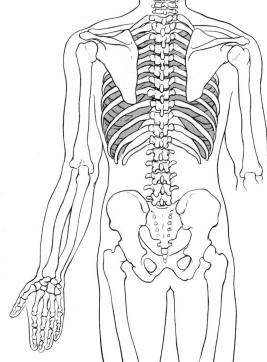

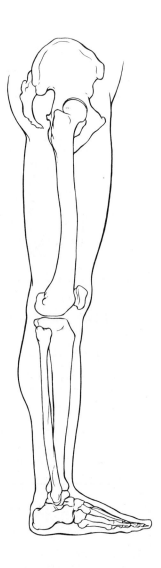

THE KNEE JOINT

Objectives

- To identify on a human skeleton selected bony features of the knee

- To explain the cartilaginous and ligamentous structures of the knee joint

- To draw and label on a skeletal chart muscles and ligaments of the knee joint

- To palpate the superficial knee joint structures and muscles on a human subject

- To demonstrate and palpate with a fellow student all the movements of the knee joint and list their respective planes of motion and axes of rotation

- To name and explain the actions and importance of the quadriceps and hamstring muscles

- To list and organize the muscles that produce the movements of the knee joint and list their antagonists

- To determine, through analysis, the knee movements and muscles involved in selected skills and exercises

Online Learning Center Resources

Visit *Manual of Structural Kinesiology's* Online Learning Center at **www.mhhe.com/floyd19e** for additional information and study material for this chapter, including:

- *Self-grading quizzes*
- *Anatomy flashcards*
- *Animations*
- *Related websites*

The knee joint is the largest diarthrodial joint in the body and is very complex. It is primarily a hinge joint. The combined functions of weight bearing and locomotion place considerable stress, strain, compression, and torsion on the knee joint. Powerful knee joint extensor and flexor muscles, combined with a strong ligamentous structure, provide a strong functioning joint in most instances.

Bones   FIG. 10.1

The enlarged femoral condyles articulate on the enlarged condyles of the tibia, somewhat in a horizontal line. Because the femur projects downward at an oblique angle toward the midline, its medial condyle is slightly larger than the lateral condyle.

The top of the medial and lateral tibial condyles, known as the medial and lateral tibial plateaus, serve as receptacles for the femoral condyles. The tibia is the medial bone in the leg and bears much more of the body's weight than the fibula. The fibula serves as the attachment for some very important knee joint structures, although it does not articulate with the femur or patella and is not part of the knee joint.

The patella is a sesamoid (floating) bone contained within the quadriceps muscle group and the patellar tendon. Its location allows it to serve the quadriceps in a fashion similar to the work of a pulley by creating an improved angle of pull. This results in a greater mechanical advantage when performing knee extension.

Chapter
10

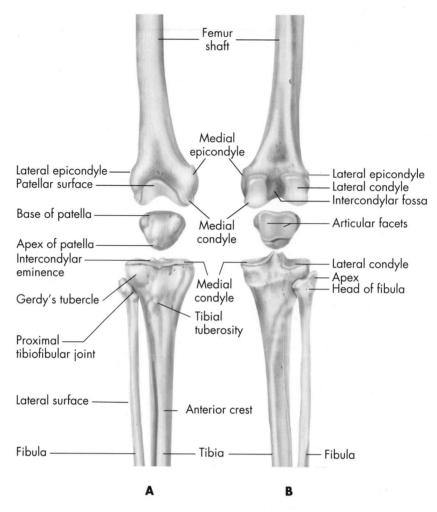

Femur
shaft

Medial
epicondyle

Lateral epicondyle
Patellar surface

Base of patella

Apex of patella

Intercondylar
eminence

Gerdy's tubercle

Proximal
tibiofibular joint

Lateral surface

Fibula

Medial
condyle

Tibial
tuberosity

Anterior crest

Tibia

Lateral epicondyle
Lateral condyle
Intercondylar fossa

Articular facets

Lateral condyle
Apex
Head of fibula

Fibula

A   **B**

FIG. 10.1 • Bones of the right knee—femur, patella, tibia, and fibula. **A,** Anterior view; **B,** Posterior view.

Key bony landmarks of the knee include the superior and inferior poles of the patella, the tibial tuberosity, Gerdy's tubercle, the medial and lateral femoral condyles, the upper anterior medial surface of the tibia, and the head of the fibula. The three vasti muscles of the quadriceps originate on the proximal femur and insert along with the rectus femoris on the superior pole of the patella. Their specific insertion into the patella varies in that the vastus medialis and vastus lateralis insert into the patella from a superomedial and superolateral angle, respectively. The superficial rectus femoris and the vastus intermedius, which lies directly beneath it, both attach to the patella from the superior direction. From here their insertion is ultimately on the tibial tuberosity by way of the large patellar tendon, which runs from the inferior patellar pole to the tibial tuberosity. Gerdy's tubercle, located on the

anterolateral aspect of the lateral tibial condyle, is the insertion point for the iliotibial tract of the tensor fasciae latae.

The upper anteromedial surface of the tibia just below the medial condyle serves as the insertion for the sartorius, gracilis, and semitendinosus. The semimembranosus inserts posteromedially on the medial tibial condyle. The head of the fibula is the primary location of the biceps femoris insertion, although some of its fibers insert on the lateral tibial condyle. The popliteus origin is located on the lateral aspect of the lateral femoral condyle.

Additionally, the tibial collateral ligament originates on the medial aspect of the upper medial femoral condyle and inserts on the medial surface of the tibia. Laterally, the shorter fibula collateral originates on the lateral femoral condyle very close to the popliteus origin and inserts on the head of the fibula.

274   www.mhhe.com/floyd19e

Joints   FIGS. 10.2, 10.3

The knee joint proper, or tibiofemoral joint, is classified as a ginglymus joint because it functions like a hinge. It moves between flexion and extension without side-to-side movement into abduction or adduction. However, it is sometimes referred to as a trochoginglymus joint because of the internal and external rotation movements that can occur during flexion. Some authorities argue that it should be classified as a condyloid or "double condyloid" joint due to its bicondylar structure. The patellofemoral joint is classified as an arthrodial joint due to the gliding nature of the patella on the femoral condyles.

The ligaments provide static stability to the knee joint, and contractions of the quadriceps and hamstrings produce dynamic stability. The surfaces between the femur and tibia are protected by articular cartilage, as is true of all diarthrodial joints. In addition to the articular cartilage covering the ends of the bones, specialized cartilages (see Fig. 10.2), known as the menisci, form cushions between the bones. These menisci are attached to the tibia and deepen the tibial plateaus, thereby enhancing stability.

The medial semilunar cartilage, or, more technically, the medial meniscus, is located on the medial tibial plateau to form a receptacle for the medial femoral condyle. The lateral semilunar cartilage (lateral meniscus) sits on the lateral tibial plateau to receive the lateral femoral condyle. Both of these menisci are thicker on the outside border and taper down to be very thin on the inside border. They can slip about slightly and are held in place by various small ligaments. The medial meniscus is the larger of the two and has a much more open C appearance than the rather closed C configuration of the lateral meniscus. Either or both of the menisci may be torn in several different areas from a variety of mechanisms, resulting in varying degrees of severity and problems. These injuries often occur due to the significant compression and shear forces that develop as the knee rotates while flexing or extending during quick directional changes in running.

Two very important ligaments of the knee are the anterior and posterior cruciate ligaments, so named because they cross within the knee between the tibia and the femur. These ligaments are vital in maintaining the anterior and posterior stability of the knee joint, respectively, as well as its rotatory stability (see Fig. 10.2).

The anterior cruciate ligament (ACL) tear is one of the most common serious injuries to the knee and has been shown to be significantly more common in females than males during similar sports such as basketball and soccer. The mechanism of this injury often involves noncontact rotary forces associated with planting and cutting. Studies have also shown that the ACL may be disrupted in a hyperextension mechanism or solely by a violent contraction of the quadriceps that pulls the tibia forward on the femur. Recent studies suggest that ACL injury prevention programs incorporating detailed conditioning exercises and techniques designed to improve neuromuscular coordination and control among the hamstrings and quadriceps, maintain proper knee alignment, and utilize proper landing techniques may be effective in reducing the likelihood of injury.

Fortunately, the posterior cruciate ligament (PCL) is not often injured. Injuries of the posterior cruciate usually come about through direct contact with an opponent or with the playing surface. Many of the PCL injuries that do occur are partial tears with minimal involvement of other knee structures. In many cases, even with complete tears, athletes may remain fairly competitive at a high level after a brief nonsurgical treatment and rehabilitation program.

On the medial side of the knee is the tibial (medial) collateral ligament (MCL; see Fig. 10.2), which maintains medial stability by resisting valgus forces or preventing the knee joint from being abducted. Injuries to the tibial collateral occur quite commonly, particularly in contact or collision sports in which a teammate or an opponent falls against the lateral aspect of the knee or leg, causing medial opening of the knee joint and stress to the medial ligamentous structures. Its deeper fibers are attached to the medial meniscus, which may be affected with injuries to the ligament.

On the lateral side of the knee, the fibular (lateral) collateral ligament (LCL) joins the fibula and the femur. Injuries to this ligament are infrequent.

In addition to the other intraarticular ligaments detailed in Fig. 10.2, there are numerous other ligaments not shown that are contiguous with the joint capsule. These ligaments are generally of lesser importance and will not be discussed further.*

The knee joint (see Fig. 10.3) is well supplied with synovial fluid from the synovial cavity, which lies under the patella and between the surfaces of the tibia and the femur. Commonly, this synovial cavity is called the capsule of the knee. Just posterior to the patellar tendon is the infrapatellar fat

Chapter
10

*More detailed discussion of the knee is found in anatomy texts and athletic training manuals.

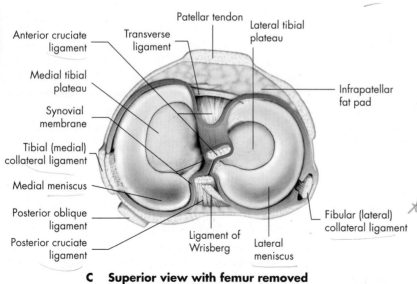

A **Anterior view with patella removed**

- Femur
- Lateral condyle of femur
- Medial condyle of femur
- Posterior cruciate ligament
- Anterior cruciate ligament
- Lateral meniscus
- Medial meniscus
- Fibular (lateral) collateral ligament
- Tibial (medial) collateral ligament
- Gerdy's tubercle
- Superior tibiofibular joint
- Tibial tuberosity
- Fibula
- Tibia

B **Posterior view**

- Femur
- Ligament of Wrisberg
- Anterior cruciate ligament
- Medial femoral condyle
- Lateral femoral condyle
- Medial meniscus
- Lateral meniscus
- Medial tibial condyle
- Lateral tibial condyle
- Posterior cruciate ligament
- Fibular (lateral) collateral ligament
- Tibial (medial) collateral ligament
- Superior tibiofibular joint
- Tibia
- Fibula

C **Superior view with femur removed**

- Patellar tendon
- Transverse ligament
- Lateral tibial plateau
- Anterior cruciate ligament
- Medial tibial plateau
- Synovial membrane
- Infrapatellar fat pad
- Tibial (medial) collateral ligament
- Medial meniscus
- Posterior oblique ligament
- Posterior cruciate ligament
- Ligament of Wrisberg
- Lateral meniscus
- Fibular (lateral) collateral ligament

FIG. 10.2 • Ligaments and menisci of the right knee. **A,** Anterior view with patella removed; **B,** Posterior view; **C,** Superior view with femur removed.

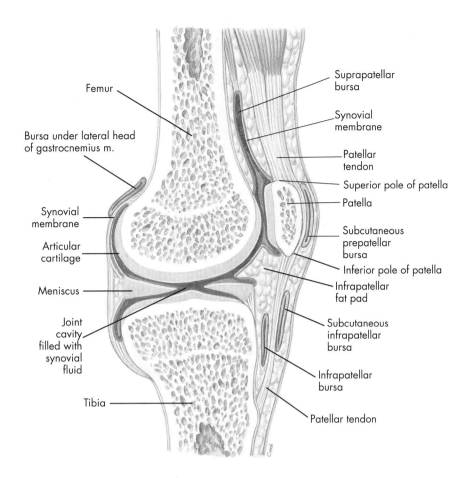

FIG. 10.3 • Knee joint, sagittal view.

pad, which is often an insertion point for synovial folds of tissue known as **plica**. A plica is an anatomical variant among some individuals that may be irritated or inflamed with injuries or overuse of the knee. More than 10 bursae are located around the knee, some of which are connected to the synovial cavity. Bursae are located where they can absorb shock or reduce friction.

The knee can usually extend to 180 degrees, or a straight line, although it is not uncommon for some knees to hyperextend up to 10 degrees or more. When the knee is in full extension, it can move from there to about 150 degrees of flexion. With the knee flexed 30 degrees or more, approximately 30 degrees of internal rotation and 45 degrees of external rotation can occur (Fig. 10.4).

Due to the shape of the medial femoral condyle, the knee must "screw home" to fully extend. As the knee approaches full extension, the tibia must externally rotate approximately 10 degrees to achieve proper alignment of the tibial and femoral condyles. In full extension, due to the close congruency of the articular surfaces, there is no appreciable

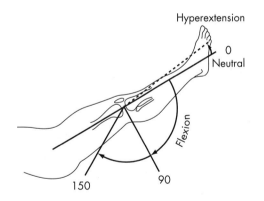

FIG. 10.4 • Active motion of the knee. Flexion is measured in degrees from the zero starting position, which is an extended straight leg with the subject either prone or supine. Hyperextension is measured in degrees opposite the zero starting point.

rotation of the knee. During initial flexion from a fully extended position, the knee "unlocks" by the tibia's rotating internally, to a degree, from its externally rotated position to achieve flexion.

Movements   FIG. 10.5

Flexion and extension of the knee occur in the sagittal plane, whereas internal and external rotation occur in the horizontal plane. The knee will not allow rotation unless it is flexed 20 to 30 degrees or more.

Flexion: bending or decreasing the angle between the femur and the lower leg; characterized by the heel moving toward the buttocks

Extension: straightening or increasing the angle between the femur and the lower leg

External rotation: rotary movement of the lower leg laterally away from the midline

Internal rotation: rotary movement of the lower leg medially toward the midline

Muscles   FIG 9.12

Some of the muscles involved in knee joint movements were discussed in Chapter 9 because of their biarticular arrangement with the hip and the knee joints. These will not be covered fully in this chapter. The knee joint muscles that have already been addressed are

Knee extensor: rectus femoris

Knee flexors: sartorius, biceps femoris, semitendinosus, semimembranosus, and gracilis

The gastrocnemius muscle, discussed in Chapter 11, also assists minimally with knee flexion.

The muscle group that extends the knee is located in the anterior compartment of the thigh and is known as the quadriceps. It consists of four muscles: the rectus femoris, the vastus lateralis, the vastus intermedius, and the vastus medialis. All four muscles work together to pull the patella superiorly, which in turn pulls the leg into extension at the knee by its attachment to the tibial tuberosity via the patellar tendon.

The central line of pull for the entire quadriceps runs from the anterior superior iliac spine (ASIS) to the center of the patella. The line of pull of the patellar tendon runs from the center of the patella to the center of the tibial tuberosity. The angle formed by the intersection of these two lines at the patella is known as the **Q angle** or quadriceps angle (Fig. 10.6). Normally, in the anatomical position, this angle will be 15 degrees or less for males and 20 degrees or less for females. Generally, females have higher angles due to a wider pelvis. Dynamic Q angles vary significantly during planting and cutting activities. Higher Q angles generally predispose people, in varying degrees, to a variety of potential knee problems, including patellar subluxation or dislocation, patellar compression syndrome, chondromalacia, and ligamentous injuries.

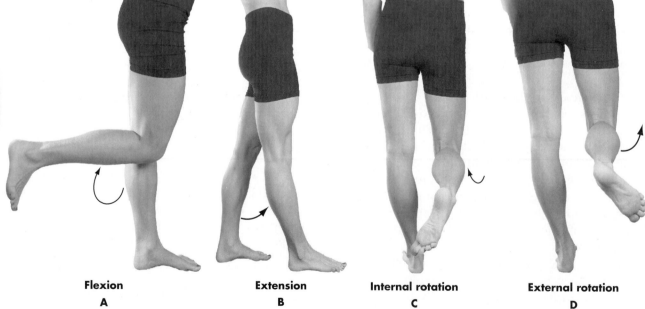

Flexion
A

Extension
B

Internal rotation
C

External rotation
D

FIG. 10.5 • Movements of the right knee.

The hamstring muscle group is located in the posterior compartment of the thigh and is responsible for knee flexion. The hamstrings consist of three muscles: the semitendinosus, the semimembranosus, and the biceps femoris. The semimembranosus and semitendinosus muscles (medial hamstrings) are assisted by the popliteus in internally rotating the knee, whereas the biceps femoris (lateral hamstring) is responsible for knee external rotation.

Two-joint muscles are most effective when either the origin or the insertion is stabilized to prevent movement in the direction of the muscle when it contracts. Additionally, muscles are able to exert greater force when lengthened than when shortened. All the hamstring muscles, as well as the rectus femoris, sartorius, and gracilis, are biarticular (two-joint) muscles.

As an example, the sartorius muscle becomes a better flexor at the knee when the pelvis is rotated posteriorly and stabilized by the abdominal muscles, thus increasing its total length by moving its origin farther from its insertion. This is exemplified by trying to flex the knee and cross the legs in the sitting position. One usually leans backward to flex the legs at the knees. This is also illustrated by kicking a football. The kicker invariably leans well backward to raise and fix the origin of the rectus femoris muscle to make it more effective as an extensor of the leg at the knee. And when youngsters hang by the knees, they flex the hips to fix or raise the origin of the hamstrings to make the latter more effective flexors of the knees.

The sartorius, gracilis, and semitendinosus all join together distally to form a tendinous expansion known as the **pes anserinus**, which attaches to the anteromedial aspect of the proximal tibia below the level of the tibial tuberosity. This attachment and the line of pull these muscles have posteromedially to the knee enable them to assist with knee flexion, particularly once the knee is flexed and the hip is externally rotated. The medial and lateral heads of the gastrocnemius attach posteriorly on the medial and lateral femoral condyles, respectively. This relationship to the knee provides the gastrocnemius with a line of pull to assist with knee flexion.

Knee joint muscles—location

Muscle location is closely related to muscle function with the knee. While viewing the muscles, in Fig. 9.12, correlate them with Table 10.1.

Anterior
    Primarily knee extension
        Rectus femoris*
        Vastus medialis
        Vastus intermedius
        Vastus lateralis
Posterior
    Primarily knee flexion
        Biceps femoris*
        Semimembranosus*
        Semitendinosus*
        Sartorius*
        Gracilis*
        Popliteus
        Gastrocnemius*

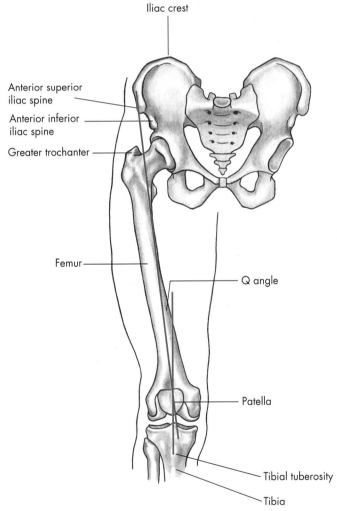

FIG. 10.6 • Q angle, represented by the angle between the line from the anterior superior iliac spine to the central patella and the line from the central patella to the tibial tuberosity.

Labels on figure: Iliac crest; Anterior superior iliac spine; Anterior inferior iliac spine; Greater trochanter; Femur; Q angle; Patella; Tibial tuberosity; Tibia

*Two-joint muscles; hip actions are discussed in Chapter 9, and ankle actions are discussed in Chapter 11.

TABLE 10.1 • Agonist muscles of the knee joint

	Muscle	Origin	Insertion	Action	Plane of motion	Palpation	Innervation
Anterior muscles	Rectus femoris	Anterior inferior iliac spine of the ilium and groove (posterior) above the acetabulum	Superior aspect of the patella and patellar tendon to tibial tuberosity	Extension of the knee / Flexion of the hip / Anterior pelvic rotation	Sagittal	Straight down anterior thigh from anterior inferior iliac spine to patella with resisted hip flexion/knee extension	Femoral nerve (L2–L4)
	Vastus inter-medius	Upper 2/3 of anterior surface of femur	Upper border of the patella and patellar tendon to tibial tuberosity	Extension of the knee	Sagittal	Anteromedial distal 1/3 of thigh just above the superomedial patella and deep to the rectus femoris, with extension of the knee, particularly full extension against resistance	Femoral nerve (L2–L4)
	Vastus lateralis (externus)	Intertrochanteric line, anterior and inferior borders of the greater trochanter, gluteal tuberosity, upper half of the linea aspera, and entire lateral intermuscular septum	Lateral border of the patella and patellar tendon to tibial tuberosity	Extension of the knee	Sagittal	Slightly distal to greater trochanter down the anterolateral aspect of the thigh to the superolateral patella, with extension of the knee, particularly full extension against resistance	Femoral nerve (L2–L4)
	Vastus medialis (internus)	Whole length of linea aspera and medial condyloid ridge	Medial half of upper border of patella and patellar tendon to tibial tuberosity	Extension of the knee	Sagittal	Anterior medial side of the thigh just above the superomedial patella, with extension of the knee, particularly full extension against resistance	Femoral nerve (L2–L4)

TABLE 10.1 (continued) • Agonist muscles of the knee joint

	Muscle	Origin	Insertion	Action	Plane of motion	Palpation	Innervation
Posterior muscles	Biceps femoris	Long head: ischial tuberosity. Short head: lower half of the linea aspera, and lateral condyloid ridge	Head of the fibula and lateral condyle of the tibia	Flexion of the knee	Sagittal	Posterolateral aspect of distal thigh with combined knee flexion and external rotation against resistance; just distal to the ischial tuberosity in a prone position with hip internally rotated during active knee flexion	Long head: sciatic nerve—tibial division (S1–S3) Short head: sciatic nerve—peroneal division (L5, S1, S2)
				Extension of the hip			
				Posterior pelvic rotation			
				External rotation of the knee	Transverse		
				External rotation of the hip			
	Popliteus	Posterior surface of lateral condyle of femur	Upper posterior medial surface of tibia	Internal rotation of the knee as it flexes	Transverse	With subject sitting, knee flexed 90 degrees, palpate deep to the gastrocnemius medially on the posterior proximal tibia and proceed superolaterally toward lateral epicondyle of tibia just deep to fibular collateral ligament, while subject internally rotates knee	Tibial nerve (L5, S1)
				Flexion of the knee	Sagittal		
	Semi-membranosus	Ischial tuberosity	Postero-medial surface of the medial tibial condyle	Extension of the hip	Sagittal	Largely covered by other muscles, tendon can be felt at posteromedial aspect of knee just deep to semitendinosus tendon with combined knee flexion and internal rotation against resistance	Sciatic nerve—tibial division (L5, S1, S2)
				Flexion of the knee			
				Posterior pelvic rotation			
				Internal rotation of the hip	Transverse		
				Internal rotation of the knee			
	Semi-tendinosus	Ischial tuberosity	Upper anterior medial surface of the tibia just below the condyle	Extension of the hip	Sagittal	Posteromedial aspect of the distal thigh with combined knee flexion and internal rotation against resistance; just distal to ischial tuberosity in a prone position with hip internally rotated during active knee flexion	Sciatic nerve—tibial division (L5, S1, S2)
				Flexion of the knee			
				Posterior pelvic rotation			
				Internal rotation of the hip	Transverse		
				Internal rotation of the knee			

Note: The sartorius and gracilis assist, although not primarily, with knee flexion and internal rotation and are discussed in detail in Chapter 9. The gastrocnemius, discussed in Chapter 11, assists to some degree with knee flexion.

Chapter
10

Nerves

The femoral nerve (Fig. 9.20) innervates the knee extensors—rectus femoris, vastus medialis, vastus intermedius, and vastus lateralis. The knee flexors, consisting of the semitendinosus, semimembranosus, biceps femoris (long head), and popliteus, are innervated by the tibial division of the sciatic nerve (Fig. 9.22). The biceps femoris short head is supplied by the peroneal nerve (Fig. 11.10).

Quadriceps muscles FIG. 10.7

(kwod´ri-seps)

Knee
extension

Hip flexion

The ability to jump is essential in nearly all sports. Individuals who have good jumping ability always have strong quadriceps muscles that extend the leg at the knee. The quadriceps function as a decelerator when it is necessary to decrease speed for changing direction or to prevent falling when landing. This deceleration function is also evident in stopping the body when coming down from a jump. The contraction that occurs in the quadriceps during braking or decelerating actions is eccentric. This eccentric action of the quadriceps controls the slowing of movements initiated in previous phases of the sport skill.

The muscles are the rectus femoris (the only two-joint muscle of the group), vastus lateralis (the largest muscle of the group), vastus intermedius, and vastus medialis. All attach to the patella and by the patellar tendon to the tuberosity of the tibia. All are superficial and palpable, except the vastus intermedius, which is under the rectus femoris. The vertical jump is a simple test that may be used to indicate the strength or power of the quadriceps. It is generally desired that this muscle group be 25% to 33% stronger than the hamstring muscle group (knee flexors).

Development of the strength and endurance of the quadriceps, or "quads," is essential for maintenance of patellofemoral stability, which is often a problem in many physically active individuals. This problem is exacerbated by the quads' being particularly prone to atrophy when injuries occur. The muscles of the quadriceps may be developed by resisted knee extension activities from a seated position; however, full-range knee extensions may be contraindicated with certain patellofemoral conditions. Performing functional weight-bearing activities such as step-ups or squats is particularly useful for strengthening and endurance.

Rectus femoris muscle FIG. 9.24

(rek´tus fem´o-ris)

Origin

Anterior inferior iliac spine of the ilium and superior margin of the acetabulum

Insertion

Superior aspect of the patella and patellar tendon to the tibial tuberosity

Action

Flexion of the hip
Extension of the knee
Anterior pelvic rotation

Palpation

Straight down anterior thigh from anterior inferior iliac spine to patella, with resisted knee extension and hip flexion

Innervation

Femoral nerve (L2–L4)

Application, strengthening, and flexibility

When the hip is flexed, the rectus femoris becomes shorter, which reduces its effectiveness as an extensor of the knee. The work is then done primarily by the three vasti muscles.

Also see the rectus femoris discussion in Chapter 9, p. 252 (Fig. 9.24).

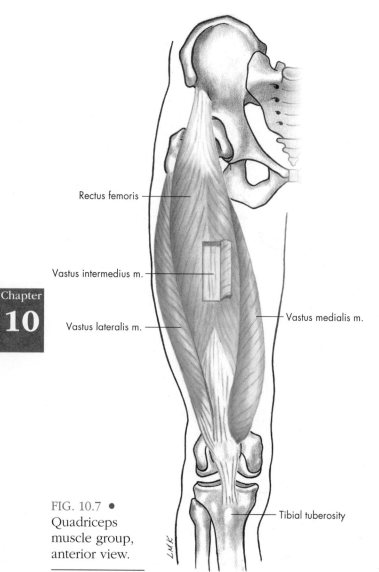

Rectus femoris

Vastus intermedius m.

Vastus lateralis m.

Vastus medialis m.

Tibial tuberosity

FIG. 10.7 ●
Quadriceps
muscle group,
anterior view.

Chapter
10

Vastus lateralis (externus) muscle

FIG. 10.8

  (vas´tus lat-er-a´lis)

Origin

Intertrochanteric line, anterior and inferior borders
  of the greater trochanter, gluteal tuberosity, upper
  half of the linea aspera, and entire lateral intermuscular septum

Insertion

Lateral border of the patella and patellar tendon to
  the tibial tuberosity

Action

Extension of the knee

Palpation

Slightly distal to the greater trochanter down the
  anterolateral aspect of the thigh to the superolateral patella, with extension of the knee, particularly
  full extension against resistance

Innervation

Femoral nerve (L2–L4)

Application, strengthening, and flexibility

All three of the vasti muscles function with the
rectus femoris in knee extension. They are typically used in walking and running and must be
used to keep the knee straight, as in standing.
The vastus lateralis has a slightly superior lateral
pull on the patella and, as a result, is occasionally
blamed in part for common lateral patellar subluxation and dislocation problems.

  The vastus lateralis is strengthened through
knee extension activities against resistance. See
Appendix 3 for more commonly used exercises
for the vastus lateralis and other muscles in this
chapter. Stretching occurs by pulling the knee
into maximum flexion, such as by standing on
one leg and pulling the heel of the other leg up
to the buttocks.

Knee
extension

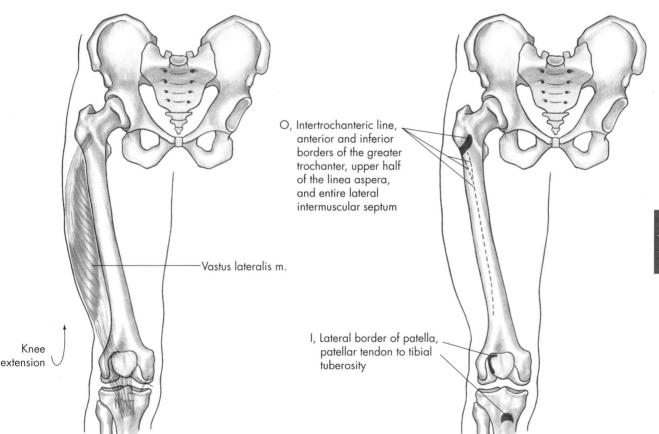

O, Intertrochanteric line,
anterior and inferior
borders of the greater
trochanter, upper half
of the linea aspera,
and entire lateral
intermuscular septum

Vastus lateralis m.

Knee
extension

I, Lateral border of patella,
patellar tendon to tibial
tuberosity

Chapter
10

FIG. 10.8 ● Vastus lateralis muscle, anterior view. O, Origin; I, Insertion.

Vastus intermedius muscle   FIG. 10.9

(vas´tus in´ter-me´di-us)

Origin
Upper two-thirds of the anterior surface of the femur

Insertion
Upper border of the patella and patellar tendon to the tibial tuberosity

Action
Extension of the knee

Palpation
Anteromedial distal one-third of thigh just above the superomedial patella and deep to the rectus femoris, with extension of the knee, particularly full extension against resistance

Innervation
Femoral nerve (L2–L4)

Application, strengthening, and flexibility

The three vasti muscles all contract in knee extension. They are used together with the rectus femoris in running, jumping, hopping, skipping, and walking. The vasti muscles are primarily responsible for extending the knee while the hip is flexed or being flexed. Thus, in doing a knee bend with the trunk bent forward at the hip, the vasti are exercised with little involvement of the rectus femoris. The natural activities mentioned above develop the quadriceps.

If done properly, squats with a barbell of varying weights on the shoulders, depending on strength, are an excellent exercise for developing the quadriceps. Caution should be used, along with strict attention to proper technique, to avoid injuries to the knees and lower back. Leg press exercises and knee extensions with weight machines are other good exercises. Full knee flexion stretches all of the quadriceps musculature.

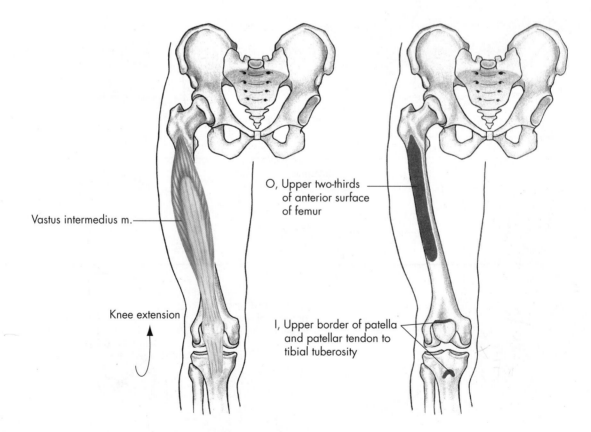

Vastus intermedius m.

Knee extension

O, Upper two-thirds of anterior surface of femur

I, Upper border of patella and patellar tendon to tibial tuberosity

FIG. 10.9 ● Vastus intermedius muscle, anterior view. O, Origin; I, Insertion.

Vastus medialis (internus) muscle

FIG. 10.10

(vas´tus me-di-a´lis)

Origin

Whole length of the linea aspera and the medial
condyloid ridge

Insertion

Medial half of the upper border of the patella and
patellar tendon to the tibial tuberosity

Action

Extension of the knee

Palpation

Anterior medial side of the thigh just above the
superomedial patella, with extension of the knee,
particularly full extension against resistance

Innervation

Femoral nerve (L2–L4)

Knee
extension

Application, strengthening, and flexibility

The vastus medialis is thought to be very important
in maintaining patellofemoral stability because of
the oblique attachment of its distal fibers to the
superior medial patella. This portion of the vastus medialis is referred to as the vastus medialis
obliquus (VMO). The vastus medialis is strengthened similarly to the other quadriceps muscles by
squats, knee extensions, and leg presses, but the
VMO is not really emphasized until the last 10 to
20 degrees of knee extension. Full knee flexion
stretches all the quadriceps muscles.

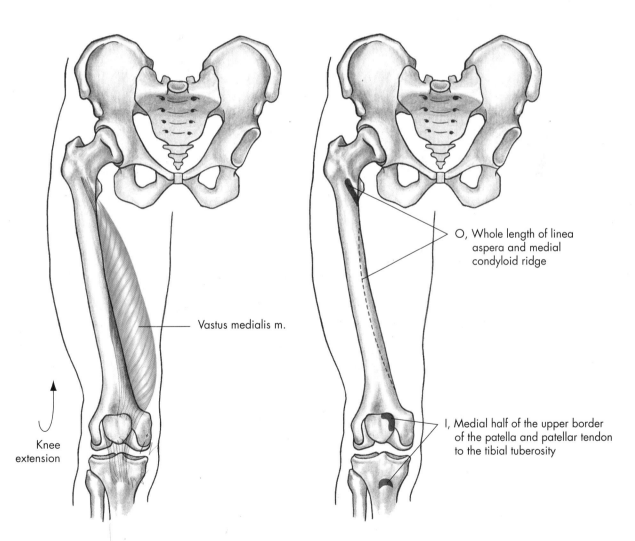

Vastus medialis m.

Knee
extension

O, Whole length of linea
aspera and medial
condyloid ridge

I, Medial half of the upper border
of the patella and patellar tendon
to the tibial tuberosity

FIG. 10.10 • Vastus medialis muscle, anterior view. O, Origin; I, Insertion.

Knee
flexion

Hip
extension

Knee
internal
rotation

Knee
external
rotation

Hip
internal
rotation

Chapter
10

Hip
external
rotation

Hamstring muscles   FIG. 10.11

The hamstring muscle group, consisting of the biceps femoris, semimembranosus, and semitendinosus, is covered in complete detail in Chapter 9, but further discussion is included here because of its importance in knee function.

Muscle strains involving the hamstrings are very common in football and other sports that require explosive running. This muscle group is often referred to as the running muscle because of its function in acceleration. The hamstring muscles are antagonists to the quadriceps muscles at the knee and are named for their cordlike attachments at the knee. All the hamstring muscles originate on the ischial tuberosity of the pelvic bone, with the exception of the short head of the biceps femoris, which originates on the lower half of the linea aspera and lateral condyloid ridge. The semitendinosus and semimembranosus insert on the anteromedial and posteromedial sides of the tibia, respectively. The biceps femoris inserts on the lateral tibial condyle and head of the fibula—hence the saying "Two to the inside and one to the outside." The short head of the biceps femoris originates on the linea aspera of the femur.

Special exercises to improve the strength and flexibility of this muscle group are important in decreasing knee injuries. Inability to touch the floor with the fingers when the knees are straight is largely a result of a lack of flexibility of the hamstrings. The hamstrings may be strengthened by performing knee or hamstring curls on a knee table against resistance. Tight or inflexible hamstrings are also contributing factors in painful conditions involving the lower back and knee. The flexibility of these muscles may be improved by performing slow, static stretching exercises, such as flexing the hip slowly while maintaining knee extension in a long sitting position.

The hamstrings are primarily knee flexors in addition to serving as hip extensors. Rotation of the knee can occur when it is in a flexed position. Once the knee is flexed approximately 20 degrees, it may rotate by actions of the hamstring muscles. The biceps femoris externally rotates the lower leg at the knee. The semitendinosus and semimembranosus perform internal rotation. Rotation of the knee permits pivoting movements and change in direction of the body. This rotation of the knee is vital in accommodating to forces developing at the hip or ankle during directional changes in order to make the total movement more functional as well as more fluid in appearance. See Figures 9.31, 9.32, and 9.33 on pages 259, 260, and 261 for the semitendinosus, semimembranosus, and biceps femoris, respectively.

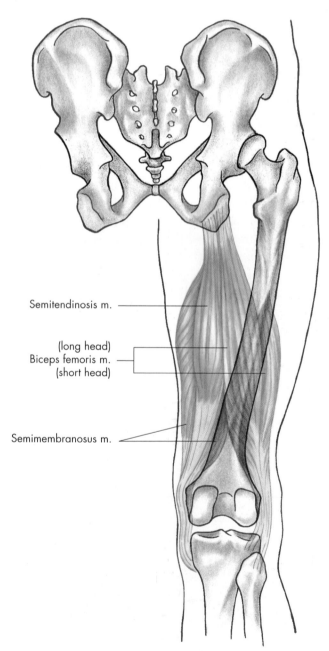

Semitendinosis m.

(long head)
Biceps femoris m.
(short head)

Semimembranosus m.

FIG. 10.11 • The hamstring muscle group, posterior view.

Popliteus muscle   FIG. 10.12

(pop´li-te´us)

Origin

Posterior surface of the lateral condyle of the femur

Insertion

Upper posterior medial surface of the tibia

Action

Flexion of the knee
Internal rotation of the knee as it flexes

Palpation

With subject sitting, knee flexed 90 degrees, palpate deep to the gastrocnemius medially on the posterior proximal tibia and proceed superolaterally toward lateral epicondyle of tibia just deep to fibular collateral ligament, while subject internally rotates knee.

Innervation

Tibial nerve (L5, S1)

Application, strengthening, and flexibility

The popliteus muscle is the only true flexor of the leg at the knee. All other flexors are two-joint muscles. The popliteus is vital in providing posterolateral stability to the knee. It assists the medial hamstrings in internal rotation of the lower leg at the knee and is crucial in internally rotating the knee to unlock it from the "screwed home" full extension position.

Knee internal rotation

Hanging from a bar with the legs flexed at the knee strenuously exercises the popliteus muscle. Also, the less strenuous activities of walking and running exercise this muscle. Specific efforts to strengthen this muscle combine knee internal rotation and flexion exercises against resistance. Stretching of the popliteus is difficult but may be done through passive full knee extension without flexing the hip. Passive maximum external rotation with the knee flexed approximately 20 to 30 degrees also stretches the popliteus.

Knee flexion

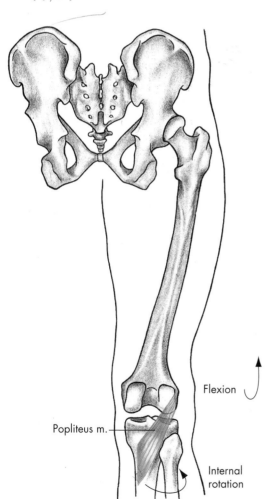

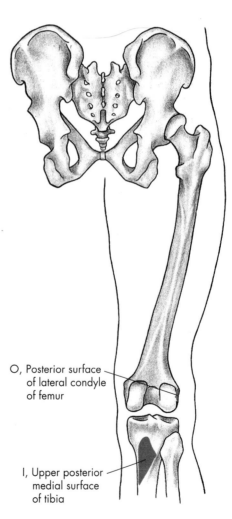

FIG. 10.12 • Popliteus muscle, posterior view. O, Origin; I, Insertion.

REVIEW EXERCISES

1. List the planes in which each of the following movements occurs. List the axis of rotation for each movement in each plane.
   a. Extension of the leg at the knee
   b. Flexion of the leg at the knee
   c. Internal rotation of the leg at the knee
   d. External rotation of the leg at the knee
2. Research the acceptability of deep knee bends and duck-walk activities in a physical education program, and report your findings in class.
3. Prepare a report on the knee on one of the following topics: anterior cruciate ligament injuries, meniscal injuries, medial collateral ligament injuries, patellar tendon tendonitis, plica syndrome, anterior knee pain, osteochondritis dissecans, patella subluxation/dislocation, knee bracing, quadriceps rehabilitation.
4. Research preventive and rehabilitative exercises to strengthen the knee joint, and report your findings in class.
5. Which muscle group about the knee would be most important for an athlete with a torn anterior cruciate ligament to develop? Why? For an athlete with a torn posterior cruciate ligament? Why?

6. **Muscle analysis chart** • Knee joint

Complete the chart by listing the muscles primarily involved in each movement.	
Flexion	Extension
Internal rotation	External rotation

7. **Antagonistic muscle action chart** • Knee joint

Complete the antagonistic muscle action chart by listing the muscle(s) or parts of muscles that are antagonist in their actions to the muscles in the left column.	
Agonist	Antagonist
Biceps femoris	
Semitendinosus	
Semimembranosus	
Popliteus	
Rectus femoris	
Vastus lateralis	
Vastus intermedius	
Vastus medialis	

LABORATORY EXERCISES

1. Locate the following bony landmarks on a human skeleton and on a subject:
   a. Skeleton
      1. Head and neck of femur
      2. Greater trochanter
      3. Shaft of femur
      4. Lesser trochanter
      5. Linea aspera
      6. Adductor tubercle
      7. Medial femoral condyle
      8. Lateral femoral condyle
      9. Patella
      10. Fibula head
      11. Medial tibial condyle
      12. Lateral tibial condyle
      13. Tibial tuberosity
      14. Gerdy's tubercle
   b. Subject
      1. Greater trochanter
      2. Adductor tubercle
      3. Medial femoral condyle
      4. Lateral femoral condyle
      5. Patella
      6. Fibula head
      7. Medial tibial condyle
      8. Lateral tibial condyle
      9. Tibial tuberosity
      10. Gerdy's tubercle
2. How and where can the following muscles be palpated on a human subject?

Note: Palpate the previously studied hip joint muscles as they are performing actions at the knee.
   a. Gracilis   f. Rectus femoris
   b. Sartorius   g. Vastus lateralis
   c. Biceps femoris   h. Vastus intermedius
   d. Semitendinosus   i. Vastus medialis
   e. Semimembranosus   j. Popliteus

3. Be prepared to indicate on a human skeleton, by using a long rubber band, the origin and insertion of the muscles listed in Question 2.
4. Demonstrate the following movements, and list the muscles primarily responsible for each.
   a. Extension of the leg at the knee
   b. Flexion of the leg at the knee
   c. Internal rotation of the leg at the knee
   d. External rotation of the leg at the knee
5. With a laboratory partner, determine how and why maintaining the position of full knee extension limits the ability to maximally flex the hip both actively and passively. Does maintaining excessive hip flexion limit the ability to accomplish full knee extension?
6. With a laboratory partner, determine how and why maintaining the position of full knee flexion limits the ability to maximally extend the hip both actively and passively. Does maintaining excessive hip extension limit the ability to accomplish full knee flexion?
7. Compare and contrast the bony, ligamentous, articular, and cartilaginous aspects of the medial knee joint with those of the lateral knee joint.

8. ## Knee joint exercise movement analysis chart

After analyzing each exercise in the chart, break each into two primary movement phases, such as a lifting phase and a lowering phase. For each phase, determine what knee joint movements occur, and then list the knee joint muscles primarily responsible for causing/controlling those movements. Beside each muscle in each movement, indicate the type of contraction as follows: I—isometric; C—concentric; E—eccentric.

Chapter
10

Exercise	Initial movement (lifting) phase		Secondary movement (lowering) phase	
	Movement(s)	Agonist(s)—(contraction type)	Movement(s)	Agonist(s)—(contraction type)
Push-up				
Squat				
Dead lift				
Hip sled				
Forward lunge				
Rowing exercise				
Stair machine				

9. Knee joint sport skill analysis chart

Analyze each skill in the chart, and list the movements of the right and left knee joints in each phase of the skill. You may prefer to list the initial position the knee joint is in for the stance phase. After each movement, list the knee joint muscle(s) primarily responsible for causing/controlling the movement. Beside each muscle in each movement, indicate the type of contraction as follows: I—isometric; C—concentric; E—eccentric. It may be desirable to review the concepts for analysis in Chapter 8 for the various phases.

Exercise		Stance phase	Preparatory phase	Movement phase	Follow-through phase
Baseball pitch	(R)				
	(L)				
Football punt	(R)				
	(L)				
Walking	(R)				
	(L)				
Softball pitch	(R)				
	(L)				
Soccer pass	(R)				
	(L)				
Batting	(R)				
	(L)				
Bowling	(R)				
	(L)				
Basketball jump shot	(R)				
	(L)				

References

Baker BE, et al: Review of meniscal injury and associated sports, *American Journal of Sports Medicine* 13:1, January–February 1985.

Field D: *Anatomy: palpation and surface markings,* ed 3, Oxford, 2001, Butterworth-Heinemann.

Garrick JG, Regna RK: Prophylactic knee bracing, *American Journal of Sports Medicine* 15:471, September–October 1987.

Hamilton N, Weimer W, Luttgens K: *Kinesiology: scientific basis of human motion,* ed 12, New York, 2012, McGraw-Hill.

Hislop HJ, Montgomery J: *Daniels and Worthingham's muscle testing: techniques of manual examination,* ed 8, Philadelphia, 2007, Saunders.

Kelly DW, et al: Patellar and quadriceps tendon ruptures—jumping knee, *American Journal of Sports Medicine* 12:375, September–October 1984.

Lysholm J, Wikland J: Injuries in runners, *American Journal of Sports Medicine* 15:168, September–October 1986.

Magee DJ: *Orthopedic physical assessment,* ed 5, Philadelphia, 2008, Saunders.

Muscolino JE: *The muscular system manual: the skeletal muscles of the human body,* ed 3, St. Louis, 2010, Elsevier Mosby.

Oatis CA: *Kinesiology: the mechanics and pathomechanics of human movement,* ed 2, Philadelphia, 2008, Lippincott Williams & Wilkins.

Prentice WE: *Principles of athletic training: a competency based approach,* ed 15, New York, 2014, McGraw-Hill.

Seeley RR, Stephens TD, Tate P: *Anatomy & physiology,* ed 8, New York, 2008, McGraw-Hill.

Shier D, Butler J, Lewis R: *Hole's human anatomy and physiology,* ed 12, New York, 2010, McGraw-Hill.

Sieg KW, Adams SP: *Illustrated essentials of musculoskeletal anatomy,* ed 4, Gainesville, FL, 2002, Megabooks.

Stone RJ, Stone JA; *Atlas of the skeletal muscles,* ed 6, New York, 2009, McGraw-Hill

Van De Graaff KM: *Human anatomy,* ed 6, Dubuque, IA, 2002, McGraw-Hill.

Wroble RR, et al: Pattern of knee injuries in wrestling, a six-year study, *American Journal of Sports Medicine* 14:55, January–February 1986.

For additional resources and a list of related websites, visit **www.mhhe.com/floyd19e**.

Worksheet Exercises

For in- or out-of-class assignments, or for testing, utilize this tear-out worksheet.

Worksheet 1

Using crayons or colored markers, draw and label on the worksheet the following muscles. Indicate the origin and insertion of each muscle with an "O" and an "I," respectively, and draw in the origin and insertion on the contralateral side of the skeleton.

a. Rectus femoris
b. Vastus lateralis
c. Vastus intermedius
d. Vastus medialis
e. Biceps femoris
f. Semitendinosus
g. Semimembranosus
h. Popliteus

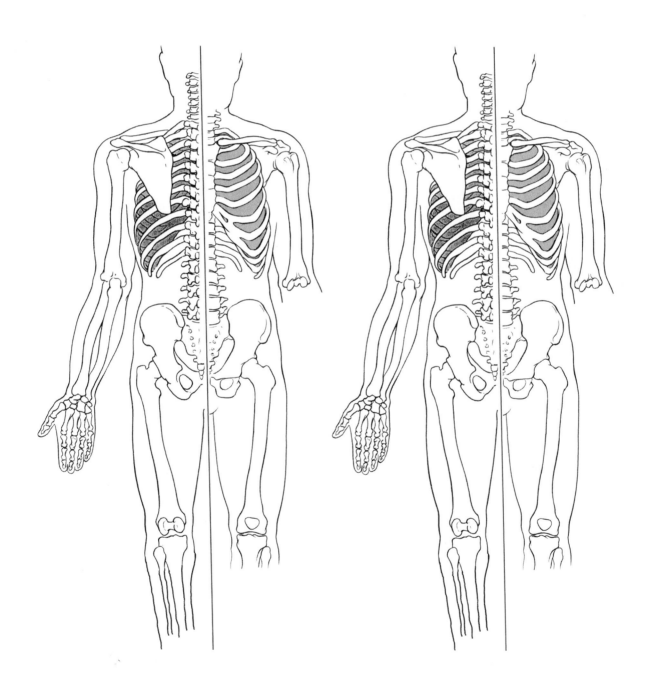

Worksheet Exercises

For in- or out-of-class assignments, or for testing, utilize this tear-out worksheet.

Worksheet 2

Label and indicate with arrows the following movements of the knee joint. For each motion, complete the sentence by supplying the plane in which it occurs and the axis of rotation, as well as the muscles causing the motion.

a. Flexion occurs in the _____ plane about the _____ axis and is accomplished by concentric contractions of the _____
_____ muscles.

b. Extension occurs in the _____ plane about the _____ axis and is accomplished by concentric contractions of the _____
_____ muscles.

c. Internal rotation occurs in the _____ plane about the _____ axis and is accomplished by concentric contractions of the _____
_____ muscles.

d. External rotation occurs in the _____ plane about the _____ axis and is accomplished by concentric contractions of the _____
_____ muscles.

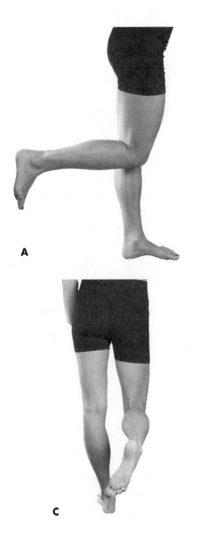

A

B

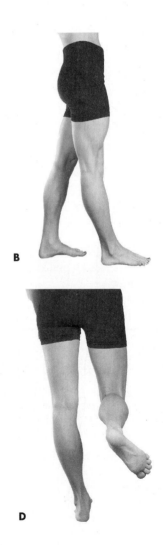

C

D

CHAPTER **11**

THE ANKLE AND FOOT JOINTS

Objectives

- To identify on a human skeleton the most important bone features, ligaments, and arches of the ankle and foot

- To draw and label on a skeletal chart the muscles of the ankle and foot

- To demonstrate and palpate with a fellow student the movements of the ankle and foot and list their respective planes of motion and axes of rotation

- To palpate the superficial joint structures and muscles of the ankle and foot on a human subject

- To list and organize the muscles that produce movement of the ankle and foot and list their antagonists

- To determine, through analysis, the ankle and foot movements and muscles involved in selected skills and exercises.

Online Learning Center Resources

Visit *Manual of Structural Kinesiology*'s **Online Learning Center** at **www.mhhe.com/floyd19e** for additional information and study material for this chapter, including:

- *Self-grading quizzes*
- *Anatomy flashcards*
- *Animations*
- *Related websites*

The complexity of the foot is evidenced by the 26 bones, 19 large muscles, many small (intrinsic) muscles, and more than 100 ligaments that make up its structure.

Support and propulsion are the two functions of the foot. Proper functioning and adequate development of the muscles of the foot and practice of proper foot mechanics are essential for everyone. In our modern society, foot trouble is one of the most common ailments. Quite often, people develop poor foot mechanics or gait abnormalities secondary to improper footwear or other relatively minor problems. Poor foot mechanics early in life inevitably leads to foot discomfort in later years.

Walking may be divided into stance and swing phases (Fig. 11.1). The **stance** phase is further divided into three components—heel-strike, midstance, and toe-off. Midstance may be further separated into loading response, midstance, and terminal stance. Normally, **heel-strike** is characterized by landing on the heel with the foot in supination and the leg in external rotation, followed immediately by pronation and internal rotation of the foot and leg, respectively, during **midstance**. The foot returns to supination and the leg returns to external rotation immediately prior to and during **toe-off**. The **swing** phase occurs when the foot leaves the ground and the leg moves forward to another point of contact. The swing phase may be divided into initial swing, midswing, and terminal swing. Problems often arise when the foot is too rigid and does not pronate adequately or when the foot remains in pronation past midstance. If the foot remains too rigid and does not pronate adequately, then impact forces will not be absorbed through the gait, resulting in shock being transmitted up the

Chapter

11

kinetic chain. If the foot overpronates or remains in pronation too much past midstance, then propulsive forces are diminished and additional stresses are placed on the kinetic chain. Walking differs from running in that one foot is always in contact with the ground and there is a point at which both feet contact the ground whereas in running there is a point at which neither foot is in contact with the ground, and both feet are never in contact with the ground at the same time.

The fitness revolution that has occurred during the past four decades has resulted in great improvements in shoes available for sports and recreational activities. In the past, a pair of sneakers would suffice for most activities. Now there are basketball, baseball, football, jogging, soccer, tennis, walking, and cross-training shoes. Good shoes are important, but there is no substitute for adequate muscle development, strength, and proper foot mechanics.

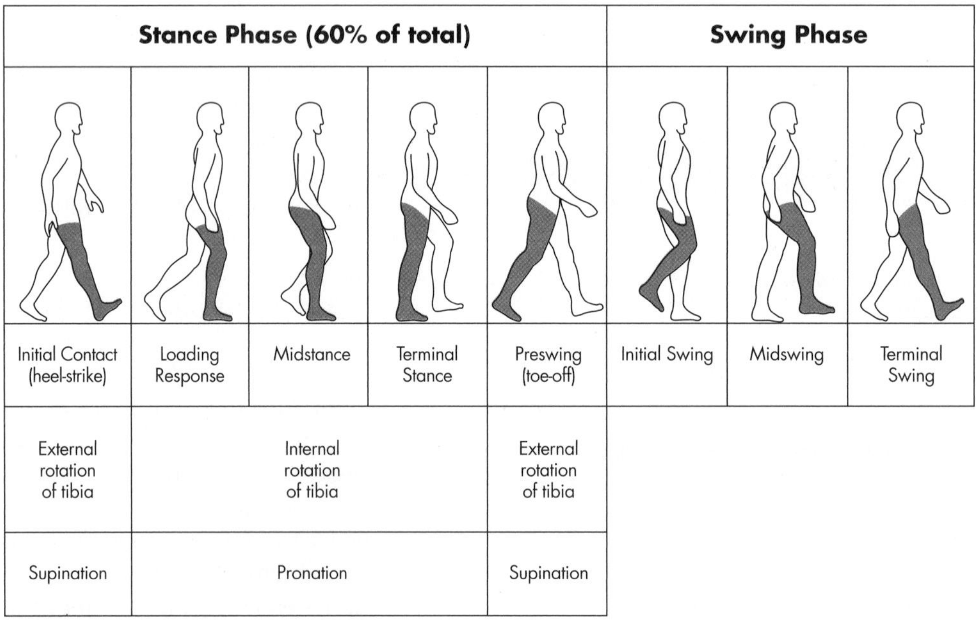

Stance Phase (60% of total)					Swing Phase		
Initial Contact (heel-strike)	Loading Response	Midstance	Terminal Stance	Preswing (toe-off)	Initial Swing	Midswing	Terminal Swing
External rotation of tibia		Internal rotation of tibia		External rotation of tibia			
Supination		Pronation		Supination			

FIG. 11.1 • Walking gait cycle.

Chapter
11

Bones

Each foot has 26 bones, which collectively form the shape of an arch. They connect with the thigh and the remainder of the body through the fibula and tibia (Figs. 11.2 and 11.3). Body weight is transferred from the tibia to the talus and the calcaneus. It should be noted that the talus is one of the few bones involved in locomotion that has no muscle attachments.

The anterior portion of the talus is wider than its posterior portion, and this is a factor in making the ankle stabler in dorsiflexion than in plantar flexion.

In addition to the talus and calcaneus, there are five other bones in the rear foot and midfoot, known as the tarsals. Between the talus and the three cuneiform bones lies the navicular. The cuboid is located between the calcaneus and the fourth and fifth metatarsals. Distal to the tarsals are the five metatarsals, which in turn correspond to each of the five toes. The toes are known as the phalanges. There are three individual bones in each phalange, except for the great toe, which has only two. Each of these bones is known as a phalanx. Finally, there are two sesamoid bones located beneath the first metatarsophalangeal joint and contained within the flexor hallucis longus tendons.

The distal ends of the tibia and fibula are enlarged and protrude horizontally and inferiorly. These bony protrusions, known as malleoli, serve as a sort of pulley for the tendons of the muscles that run directly posterior to them. Specifically, the peroneus brevis and peroneus longus are immediately behind the lateral malleolus. The muscles immediately posterior to the medial malleolus may be remembered by the phrase "Tom, Dick, and Harry" with the "T" standing for the tibialis posterior, the "D" for the flexor digitorum longus, and the "H" for the flexor hallucis longus. This bony arrangement increases the mechanical

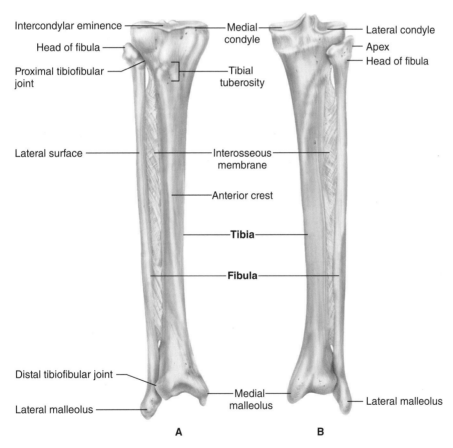

FIG. 11.2 • Right tibia and fibula. **A,** Anterior view; **B,** Posterior view.

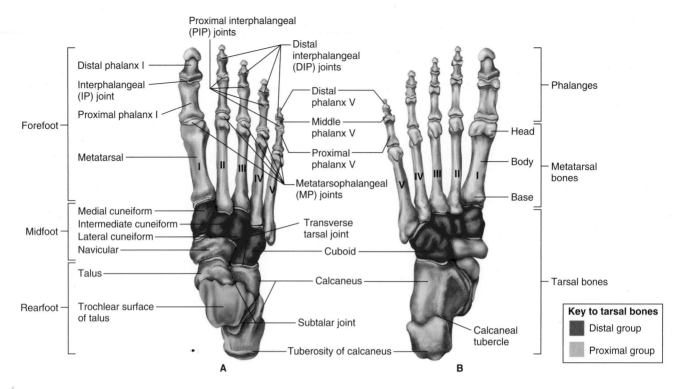

Proximal interphalangeal
(PIP) joints

Distal
interphalangeal
(DIP) joints

Distal phalanx I

Interphalangeal
(IP) joint

Proximal phalanx I

Distal
phalanx V

Middle
phalanx V

Proximal
phalanx V

Metatarsal

Metatarsophalangeal
(MP) joints

Forefoot

Phalanges

Head

Body

Metatarsal
bones

Base

Medial cuneiform

Intermediate cuneiform

Lateral cuneiform

Navicular

Midfoot

Transverse
tarsal joint

Cuboid

Talus

Trochlear surface
of talus

Rearfoot

Calcaneus

Tarsal bones

Subtalar joint

Calcaneal
tubercle

Tuberosity of calcaneus

A

B

Key to tarsal bones

Distal group

Proximal group

FIG. 11.3 • Right foot. **A,** Superior (dorsal) view; **B,** Inferior (plantar) view.

advantage of these muscles in performing their actions of inversion and eversion. The base of the fifth metatarsal is enlarged and prominent to serve as an attachment point for the peroneus brevis and tertius.

The inner surface of the medial cuneiform and the base of the first metatarsal provide insertion points for the tibialis anterior, while the undersurfaces of the same bones serve as the insertion for the peroneus longus. The tibialis posterior has multiple insertions on the lower inner surfaces of the navicular, cuneiform, and second through fifth metatarsal bases. The tops and undersurfaces of the bases of the second through fifth distal phalanxes are the insertion points for the extensor digitorum longus and the flexor digitorum longus, respectively. Similarly, the top and undersurface of the base of the first distal phalanx provide insertions for the extensor hallucis longus and flexor hallucis longus, respectively.

The posterior surface of the calcaneus is very prominent and serves as the attachment point for the Achilles tendon of the gastrocnemius–soleus complex.

Joints

The tibia and fibula form the tibiofibular joint, a syndesmotic amphiarthrodial joint (see Fig. 11.2). The bones are joined at both the proximal and distal tibiofibular joints. In addition to the ligaments supporting both of these joints, there is a strong, dense interosseus membrane between the shafts of these two bones. Although only minimal movement is possible between these bones, the distal joint does become sprained occasionally in heavy contact sports such as football. A common component of this injury involves the ankle, or talocrural, joint being in dorsiflexion, which, by making the ankle more stable, allows the ligamentous stress to be transferred to the syndesmosis joint when the dorsiflexed ankle is forced into external rotation. This injury, a sprain of the syndesmosis joint, is commonly referred to as a high ankle sprain and primarily involves the anterior inferior tibiofibular ligament. Secondarily, and with more severe injuries, the posterior tibiofibular ligament, interosseus ligament, and interosseus membrane may be involved.

The ankle joint, technically known as the talocrural joint, is a hinge or ginglymus-type joint (Fig. 11.4). Specifically, it is the joint made up of the talus, the distal tibia, and the distal fibula. The ankle joint allows approximately 50 degrees of plantar flexion and 15 to 20 degrees of dorsiflexion (Fig. 11.5). Greater range of dorsiflexion, particularly in weight bearing, is possible when the knee is flexed, which reduces the tension of the biarticular gastrocnemius muscle. The fibula rotates on its axis 3 to 5 degrees externally with dorsiflexion of the ankle and 3 to 5 degrees internally during plantar flexion. The syndesmosis joint widens by approximately 1 to 2 millimeters during full dorsiflexion.

Inversion and eversion, though commonly thought to be ankle joint movements, technically occur in the subtalar and transverse tarsal joints. These joints, classified as gliding or arthrodial, combine to allow approximately 20 to 30 degrees of inversion and 5 to 15 degrees of eversion. There is minimal movement within the remainder of the intertarsal and tarsometatarsal arthrodial joints.

The phalanges join the metatarsals to form the metatarsophalangeal joints, which are classified as condyloid-type joints. The metatarsophalangeal (MP) joint of the great toe flexes 45 degrees and extends 70 degrees, whereas the interphalangeal (IP) joint can flex from 0 degrees of full extension to 90 degrees of flexion. The MP joints of the four lesser toes allow approximately 40 degrees of flexion and 40 degrees of extension. The MP joints also abduct and adduct minimally. The proximal interphalangeal (PIP) joints in the lesser toes flex from 0 degrees of extension to 35 degrees of flexion. The distal interphalangeal (DIP) joints flex 60 degrees and extend 30 degrees. There is much variation from joint to joint and from person to person in all these joints.

Ankle sprains are one of the most common injuries among physically active people. Sprains involve the stretching or tearing of one or more ligaments. There are far too many ligaments in the foot and ankle to discuss in this text, but a few of the ankle ligaments are shown in Fig. 11.4. Far and away the most common ankle sprain results from excessive inversion, usually while in some degree of plantar flexion. This most commonly results in damage to the anterior talofibular ligament, particularly when in greater amounts of plantar flexion. When closer to neutral plantar flexion/dorsiflexion, the true inversion mechanism places more stress on the calcaneofibular ligament. Less common are excessive eversion forces causing injury to the deltoid ligament on the medial aspect of the ankle.

Ligaments in the foot and the ankle maintain the position of an arch. All 26 bones in the foot are connected with ligaments. This brief discussion focuses on the longitudinal and transverse arches.

There are two longitudinal arches (Fig. 11.6). The medial longitudinal arch, important for shock absorption, is located on the medial side of the foot and extends from the calcaneus bone to the talus, the navicular, the three cuneiforms, and the distal ends of the three medial metatarsals. The medial longitudinal arch, often implicated in a variety of foot problems, is primarily supported dynamically by the tibialis posterior and tibialis anterior muscles. The lateral longitudinal arch, important in balance, is located on the lateral side of the foot and extends from the calcaneus to the cuboid and the distal ends of the fourth and fifth metatarsals. Individual long arches can be high, medium, or low, but a low arch is not necessarily a weak arch.

The transverse arch (see Fig. 11.6) assists in adapting the foot to the ground and extends across the foot from the first metatarsal to the fifth metatarsal. The distal transverse arch, also a common source of foot problems, is supported by the intrinsic muscles of the foot such as the lumbricals, adductor hallucis, and flexor digiti minimi. These muscles may be strengthened through a towel crunch exercise in which the metatarsophalangeal joints are flexed to grab a towel. The long plantar ligament, longest of all the ligaments of the tarsals, originates on the plantar surface of the calcaneus anterior to the calcaneal tuberosity and inserts on the plantar surface of the cuboid with superficial fibers continuing forward to the bases of the second, third, and fourth metatarsal bones (see Fig. 11.4, *A* and *B*). The plantar fascia, sometimes referred to as the plantar aponeurosis, is a broad structure extending from the medial calcaneal tuberosity to the proximal phalanges of the toes. It assists in stabilizing the medial longitudinal arch and in propelling the body forward during the latter part of the stance phase. A common painful condition involving the plantar fascia is known as plantar fasciitis.

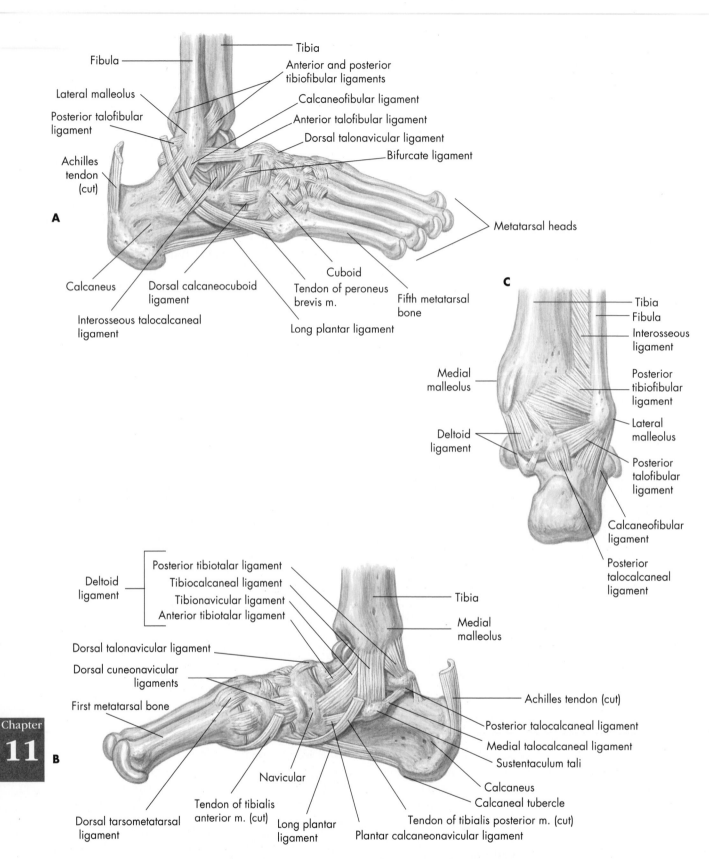

FIG. 11.4 • Right ankle joint. **A,** Lateral view; **B,** Medial view; **C,** Posterior view.

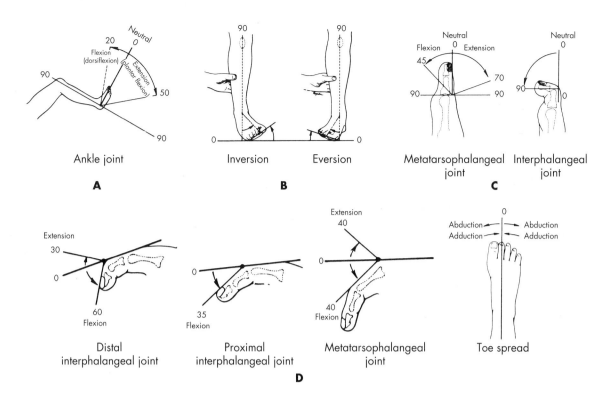

FIG. 11.5 • Active motion of the ankle, foot, and toes. **A,** Dorsiflexion and plantar flexion are measured in degrees from the right-angle neutral position or in percentages of motion as compared to the opposite ankle; **B,** Inversion and eversion normally are estimated in degrees or expressed in percentages as compared to the opposite foot; **C,** Flexion and extension of the great toe; **D,** ROM for the lateral four toes.

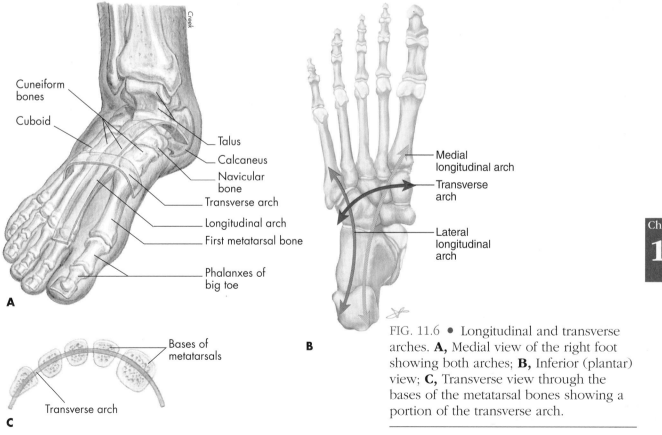

FIG. 11.6 • Longitudinal and transverse arches. **A,** Medial view of the right foot showing both arches; **B,** Inferior (plantar) view; **C,** Transverse view through the bases of the metatarsal bones showing a portion of the transverse arch.

Movements FIG. 11.7

Dorsiflexion (flexion): dorsal flexion; movement of the top of the ankle and foot toward the anterior tibia

Plantar flexion (extension): movement of the ankle and foot away from the tibia

Eversion: turning the ankle and foot outward; abduction, away from the midline; weight is on the medial edge of the foot

Inversion: turning the ankle and foot inward; adduction, toward the midline; weight is on the lateral edge of the foot

Toe flexion: movement of the toes toward the plantar surface of the foot

Toe extension: movement of the toes away from the plantar surface of the foot

Pronation: a combination of ankle dorsiflexion, subtalar eversion, and forefoot abduction (toe-out)

Supination: a combination of ankle plantar flexion, subtalar inversion, and forefoot adduction (toe-in)

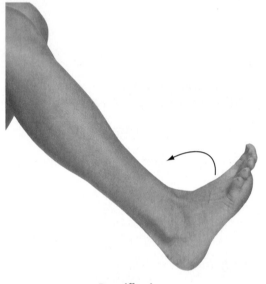

Dorsiflexion
A

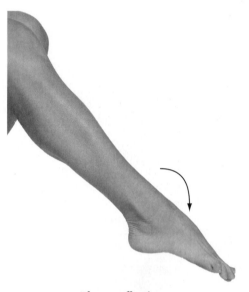

Plantar flexion
B

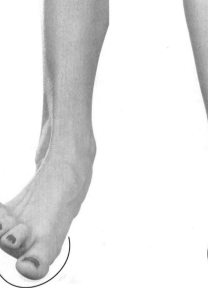
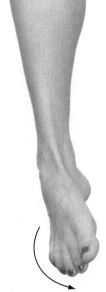

Transverse tarsal and subtalar eversion
C

Transverse tarsal and subtalar inversion
D

FIG. 11.7 • Movements of the right ankle and foot.

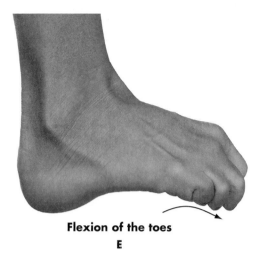

Flexion of the toes
E

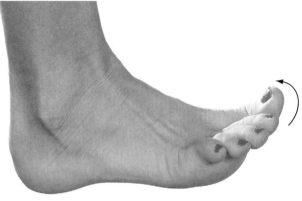

Extension of the toes
F

Great toe
MTP and IP
flexion

Great toe
MTP and IP
extension

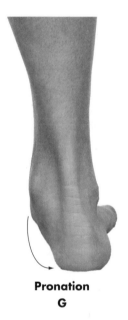

Pronation
G

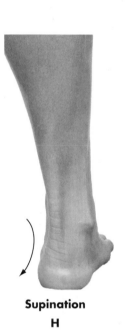

Supination
H

FIG. 11.7 (continued) • Movements of the right ankle and foot.

Chapter
11

Ankle and foot muscles FIGS 11.8, 11.9

The large number of muscles in the ankle and foot may be easier to learn if grouped according to location and function. In general, the muscles located on the anterior aspect of the ankle and foot are the dorsal flexors and/or toe extensors. Those on the posterior aspect are plantar flexors and/or toe flexors. Specifically, the gastrocnemius and the soleus collectively are known as the **triceps surae**, due to their three heads, which together join to the Achilles tendon. Muscles that are evertors are located more to the lateral side, whereas the invertors are located medially.

The lower leg is divided into four compartments, each containing specific muscles (Fig. 11.9). Tightly surrounding and binding each compartment is a dense fascia, which facilitates venous return and prevents excessive swelling of the muscles during exercise. The anterior compartment contains the dorsiflexor group, consisting of the tibialis anterior, peroneus tertius, extensor digitorum longus,

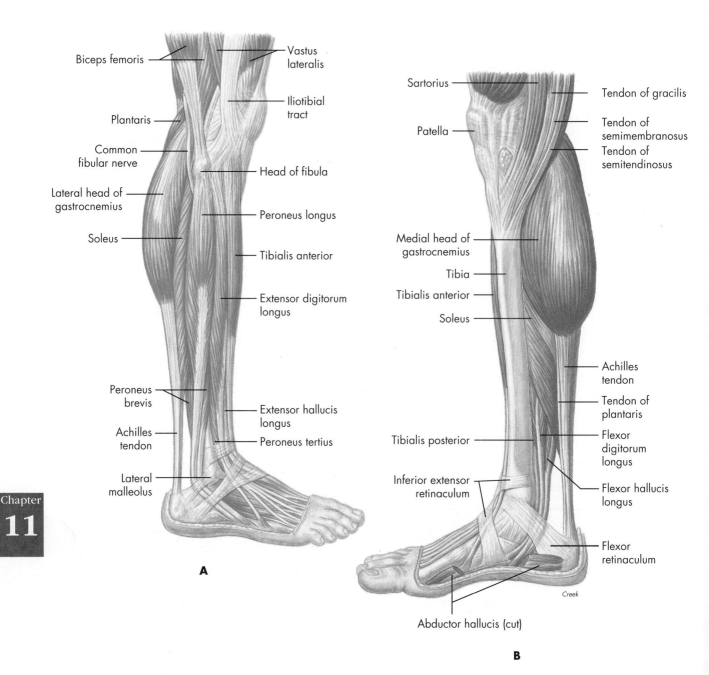

A

B

FIG. 11.8 • Right Lower leg, ankle, and foot muscles. **A,** Lateral view; **B,** Medial view.

and extensor hallucis longus. The lateral compartment contains the peroneus longus and peroneus brevis—the two most powerful evertors. The posterior compartment is divided into deep and superficial compartments. The gastrocnemius, soleus, and plantaris are located in the superficial posterior compartment, while the deep posterior compartment is composed of the flexor digitorum longus, flexor hallucis longus, popliteus, and tibialis posterior. All the muscles of the superficial posterior compartment are primarily plantar flexors. The plantaris, absent in some humans, is a vestigial biarticular muscle that contributes minimally to ankle plantar flexion and knee flexion. It originates on the inferior aspect of the lateral supracondylar line of the distal femur posteriorly, runs just medial to the lateral head of the gastrocnemius

and then deep to it, but superficial to the soleus, to insert on the middle one-third of the posterior calcaneal surface just medial to the Achilles tendon. The deep posterior compartment muscles, except for the popliteus, are plantar flexors but also function as invertors. Although most common with the anterior compartment, any of these components are subject to a condition known as compartment syndrome. This condition may be acute or chronic and may occur secondarily to injury, trauma, or overuse. Symptoms include sharp pain, particularly with increased movement actively or passively, swelling, and weakness in the muscles of the involved compartment. Depending on the severity, emergency surgery may be indicated to release the fascia in order to prevent permanent tissue damage, although many compartment

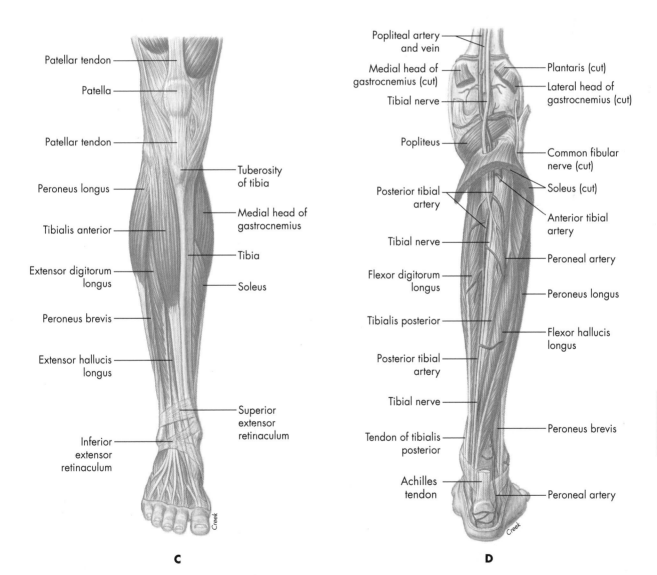

C

D

FIG. 11.8 (continued) • Right Lower leg, ankle, and foot muscles. **C,** Anterior view; **D,** Deep posterior view.

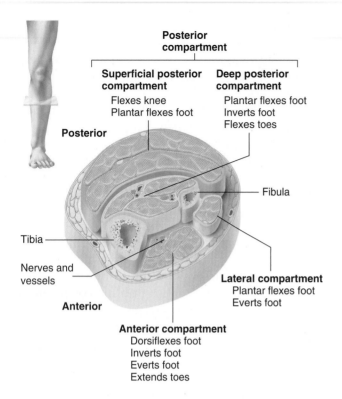

Posterior
compartment

Superficial posterior **Deep posterior**
compartment **compartment**
Flexes knee   Plantar flexes foot
Plantar flexes foot   Inverts foot
                      Flexes toes

Posterior

Fibula

Tibia

Nerves and
vessels

Anterior

Lateral compartment
Plantar flexes foot
Everts foot

Anterior compartment
Dorsiflexes foot
Inverts foot
Everts foot
Extends toes

FIG. 11.9 ● Cross section of the left leg,
demonstrating the muscular compartments.

Ankle and foot muscles by function

Note: A number of the ankle and foot muscles
are capable of helping produce more than one
movement.

Plantar flexors
    Gastrocnemius
    Flexor digitorum longus
    Flexor hallucis longus
    Peroneus (fibularis) longus
    Peroneus (fibularis) brevis
    Plantaris
    Soleus
    Tibialis posterior
Evertors
    Peroneus (fibularis) longus
    Peroneus (fibularis) brevis
    Peroneus (fibularis) tertius
    Extensor digitorum longus
Dorsiflexors
    Tibialis anterior
    Peroneus (fibularis) tertius
    Extensor digitorum longus (extensor of the
    lesser toes)
    Extensor hallucis longus (extensor of the
    great toe)
Invertors
    Tibialis anterior
    Tibialis posterior
    Flexor digitorum longus (flexor of the
    lesser toes)
    Flexor hallucis longus (flexor of the great toe)

Ankle and foot muscles by compartment

Anterior compartment
    Tibialis anterior
    Extensor hallucis longus
    Extensor digitorum longus
    Peroneus (fibularis) tertius
Lateral compartment
    Peroneus (fibularis) longus
    Peroneus (fibularis) brevis
Deep posterior compartment
    Flexor digitorum longus
    Flexor hallucis longus
    Tibialis posterior
    Popliteus
Superficial posterior compartment
    Gastrocnemius (medial head)
    Gastrocnemius (lateral head)
    Soleus
    Plantaris

    While viewing the muscles in Figs. 11.8 and
11.9, correlate them with Table 11.1.

syndromes may be adequately addressed with
proper acute management.

    Due to the heavy demands placed on the mus-
culature of the legs in the running activities of
most sports, both acute and chronic injuries are
common. "Shin splints" is a common term used to
describe a painful condition of the leg that is often
associated with running activities. This condition
is not a specific diagnosis but rather is attributed
to a number of specific musculotendinous injuries.
Most often the tibialis posterior, medial soleus, or
tibialis anterior is involved, but the extensor digitorum longus may also be involved. Shin splints
often occur as a result of an inappropriate level of
flexibility, strength, and endurance for the specific
demands of the activity and may be prevented in
part by stretching the plantar flexors and strengthening the dorsiflexors.

    Additionally, painful cramps caused by acute
muscle spasm in the gastrocnemius and soleus
occur somewhat commonly and may be relieved
through active and passive dorsiflexion. Also, a
very disabling injury involves the complete rupture of the strong Achilles tendon, which connects these two plantar flexors to the calcaneus.

TABLE 11.1 • Agonist muscles of the ankle and foot joints

	Muscle	Origin	Insertion	Action	Plane of motion	Palpation	Innervation
Superficial posterior compartment	Gastroc-nemius	Medial head: posterior surface of the medial femoral condyle Lateral head: posterior surface of the lateral femoral condyle	Posterior surface of the calcaneus (Achilles tendon)	Plantar flexion of the ankle	Sagittal	Upper half of the posterior aspect of the lower leg	Tibial nerve (S1, S2)
				Flexion of the knee			
	Soleus	Posterior surface of the proximal fibula and proximal 2/3 of the posterior tibial surface	Posterior surface of the calcaneus (Achilles tendon)	Plantar flexion of the ankle	Sagittal	Posteriorly under the gastrocnemius muscle on the medial and lateral sides of the lower leg, particularly while prone with knee flexed approximately 90 degrees and actively plantarflexing ankle	Tibial nerve (S1, S2)
Deep posterior compartment	Tibialis posterior	Posterior surface of the upper half of the interosseus membrane and the adjacent surfaces of the tibia and fibula	Inferior surfaces of the navicular, cuneiform, and cuboid bones and bases of the 2nd, 3rd, and 4th metatarsal bones	Inversion of the foot	Frontal	The tendon may be palpated both proximally and distally immediately behind medial malleolus with inversion and plantar flexion and is better distinguished from flexor digitorum longus and flexor hallucis longus if toes can be maintained in slight extension	Tibial nerve (L5, S1)
				Plantar flexion of the ankle	Sagittal		
	Flexor digitorum longus	Middle 1/3 of the posterior surface of the tibia	Base of the distal phalanx of each of the four lesser toes	Flexion of the four lesser toes at the metatarso-phalangeal and the proximal and distal interphalangeal joints	Sagittal	The tendon may be palpated immediately posterior to the medial malleolus and tibialis posterior and immediately anterior to the flexor hallucis longus with flexion of the lesser toes while maintaining great toe extension, ankle dorsiflexion, and foot eversion	Tibial nerve (L5, S1)
				Plantar flexion of the ankle			
				Inversion of the foot	Frontal		
	Flexor hallucis longus	Middle 2/3 of the posterior surface of the fibula	Base of the distal phalanx of the great toe; plantar surface	Flexion of the great toe at the metatarso-phalangeal and interphalangeal joints	Sagittal	Most posterior of three tendons immediately behind medial malleolus; between medial soleus and tibia with active great toe flexion while maintaining extension of four lesser toes, ankle dorsiflexion, and foot eversion	Tibial nerve (L5, S1, S2)
				Plantar flexion of the ankle			
				Inversion of the foot	Frontal		

Chapter

11

TABLE 11.1 (continued) • Agonist muscles of the ankle and foot joints

	Muscle	Origin	Insertion	Action	Plane of motion	Palpation	Innervation
Lateral compartment	Peroneus (fibularis) longus	Head and upper 2/3 of the lateral surface of the fibula	Undersurfaces of the medial cuneiform and 1st metatarsal bone	Eversion of the foot	Frontal	Upper lateral side of tibia just distal to fibular head and down to immediately posterior to lateral malleolus; just posterolateral from tibialis anterior and extensor digitorum longus with active eversion	Superficial peroneal nerve (L4, L5, S1)
				Plantar flexion of the ankle	Sagittal		
	Peroneus (fibularis) brevis	Mid to lower 2/3 of the lateral surface of the fibula	Tuberosity of the 5th metatarsal bone	Eversion of the foot	Frontal	Tendon of muscle at proximal end of 5th metatarsal; just proximal and posterior to lateral malleolus; immediately deep anteriorly and posteriorly to peroneus longus with active eversion	Superficial peroneal nerve (L4, L5, S1)
				Plantar flexion of the ankle	Sagittal		
Anterior compartment	Peroneus (fibularis) tertius	Distal 1/3 of the anterior fibula	Superior aspect of the base of the 5th metatarsal	Dorsiflexion of the ankle	Sagittal	Just medial to distal fibula; lateral to extensor digitorum longus tendon on anterolateral aspect of foot, down to medial side of base of 5th metatarsal with dorsiflexion and eversion	Deep peroneal nerve (L4, L5, S1)
				Eversion of the foot	Frontal		
	Extensor digitorum longus	Lateral condyle of the tibia, head of the fibula, and upper 2/3 of the anterior surface of the fibula	Tops of the middle and distal phalanges of the four lesser toes	Extension of the four lesser toes at the metatarsophalangeal and the proximal and distal interphalangeal joints	Sagittal	Second muscle to lateral side of anterior tibial border; upper lateral side of tibia between tibialis anterior medially and fibula laterally; divides into four tendons just distal to anterior ankle with active toe extension	Deep peroneal nerve (L4, L5, S1)
				Dorsiflexion of the ankle			
				Eversion of the foot	Frontal		
	Extensor hallucis longus	Middle 2/3 of the medial surface of the anterior fibula	Base of the distal phalanx of the great toe	Extension of great toe at the metatarsophalangeal and interphalangeal joints	Sagittal	From dorsal aspect of great toe to just lateral to tibialis anterior and medial to extensor digitorum longus at anterior ankle joint	Deep peroneal nerve (L4, L5, S1)
				Dorsiflexion of the ankle			
				Weak inversion of the foot	Frontal		
	Tibialis anterior	Upper 2/3 of the lateral surface of the tibia	Inner surface of the medial cuneiform and the base of the 1st metatarsal bone	Dorsiflexion of the ankle	Sagittal	First muscle to the lateral side of the anterior tibial border, particularly palpable with fully active ankle dorsiflexion; most prominent tendon crossing the ankle anteromedially	Deep peroneal nerve (L4, L5, S1)
				Inversion of the foot	Frontal		

Note: The plantaris is not included because its contribution to knee flexion and plantar flexion is relatively minimal.

Chapter
11

Nerves

As described in Chapter 9, the sciatic nerve originates from the sacral plexus and becomes the tibial nerve and peroneal nerve. The tibial division of the sciatic nerve (Fig. 9.22) continues down to the posterior aspect of the lower leg to innervate the gastrocnemius (medial head), soleus, tibialis posterior, flexor digitorum longus, and flexor hallucis longus. Just before reaching the ankle, the tibial nerve branches to become the medial and lateral plantar nerves, which innervate the intrinsic muscles of the foot. The medial plantar nerve innervates the abductor hallucis, flexor hallucis brevis, first lumbrical, and flexor digitorum brevis. The lateral plantar nerve supplies the adductor hallucis, quadratus plantae, lumbricals (2, 3, and 4), dorsal interossei, plantar interossei, abductor digiti minimi, and flexor digiti minimi.

The peroneal, or fibular, nerve (Fig. 11.10) divides just below the head of the fibula to become the superficial and deep peroneal nerves. The superficial branch innervates the peroneus longus and peroneus brevis, while the deep branch innervates the tibialis anterior, extensor digitorum longus, extensor hallucis longus, peroneus tertius, and extensor digitorum brevis.

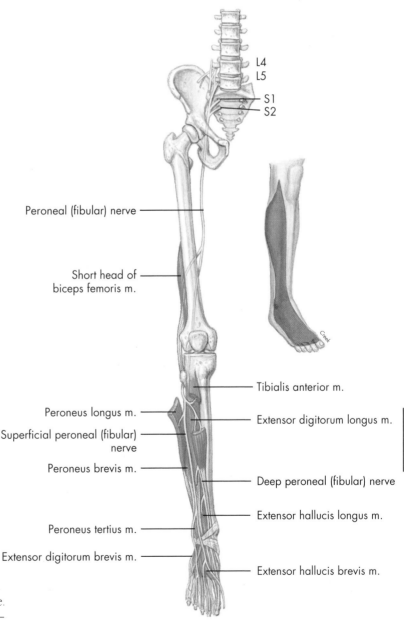

FIG. 11.10 • Muscular and cutaneous distribution of the peroneal (fibular) nerve.

Gastrocnemius muscle   FIG. 11.11

(gas-trok-ne´mi-us)

Ankle plantar flexion

Knee flexion

Origin

Medial head: posterior surface of the medial femoral condyle

Lateral head: posterior surface of the lateral femoral condyle

Insertion

Posterior surface of the calcaneus (Achilles tendon)

Action

Plantar flexion of the ankle

Flexion of the knee

Palpation

Easiest muscle in the lower extremity to palpate; upper one-half of posterior aspect of lower leg

Innervation

Tibial nerve (S1, S2)

Application, strengthening, and flexibility

The gastrocnemius and soleus together are known as the triceps surae with triceps referring to the heads of the medial and lateral gastrocnemius and the soleus and surae referring to the calf. Because the gastrocnemius is a biarticular muscle, it is more effective as a knee flexor if the ankle is dorsiflexed and more effective as a plantar flexor of the foot if the knee is held in extension. This is observed when one sits too close to the steering wheel in driving a car, which significantly shortens the entire muscle, reducing its effectiveness. When the knees are bent, the muscle becomes an ineffective plantar flexor, and it is more difficult to depress the brakes. Running, jumping, hopping, and skipping exercises all depend significantly on the gastrocnemius and soleus to propel the body upward and forward. Heel-raising exercises with the knees in full extension and the toes resting on a block of wood are an excellent way to strengthen the muscle through the full range of motion. Holding a barbell on the shoulders can increase the resistance. See Appendix 3 for more commonly used exercises for the gastrocnemius and other muscles in this chapter.

The gastrocnemius may be stretched by standing and placing both palms on a wall about 3 feet away and leaning into the wall. The feet should be pointed straight ahead, and the heels should remain on the floor. The knees should remain fully extended throughout the exercise to accentuate the stretch on the gastrocnemius.

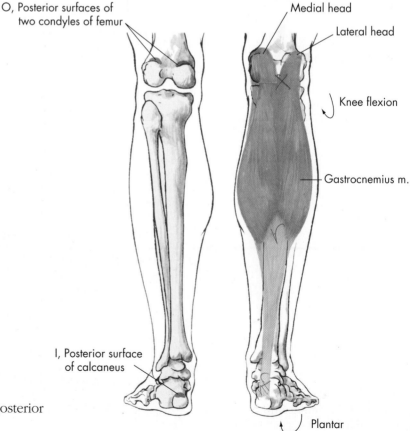

O, Posterior surfaces of two condyles of femur

Medial head

Lateral head

Knee flexion

Gastrocnemius m.

I, Posterior surface of calcaneus

Plantar flexion

FIG. 11.11 • Gastrocnemius muscle, posterior view. O, Origin; I, Insertion.

Soleus muscle   FIG. 11.12

(so′le-us)

Origin

Posterior surface of the proximal fibula and proximal two-thirds of the posterior tibial surface

Insertion

Posterior surface of the calcaneus (Achilles tendon)

Action

Plantar flexion of the ankle

Palpation

Posteriorly under the gastrocnemius muscle on the medial and lateral sides of the lower leg, particularly prone with knee flexed approximately 90 degrees and actively plantarflexing ankle

Innervation

Tibial nerve (S1, S2)

Application, strengthening, and flexibility

The soleus muscle is one of the most important plantar flexors of the ankle. Some anatomists believe that it is nearly as important in this movement as the gastrocnemius. This is especially true when the knee is flexed. When one rises up on the toes, the soleus muscle can plainly be seen on the outside of the lower leg if one has exercised the legs extensively, as in running and walking.

The soleus muscle is used whenever the ankle plantar flexes. Any movement with body weight on the foot with the knee flexed or extended calls it into action. When the knee is flexed slightly, the effect of the gastrocnemius is reduced, placing more work on the soleus. Running, jumping, hopping, skipping, and dancing on the toes are all exercises that depend heavily on the soleus. It may be strengthened through any plantar flexion exercise against resistance, particularly if the knee is flexed slightly to deemphasize the gastrocnemius. Heel-raising exercises as described for the gastrocnemius, except with the knees flexed slightly, are one way to isolate this muscle for strengthening. Resistance may be increased by holding a barbell on the shoulders.

The soleus is stretched in the same manner as the gastrocnemius, except that the knees must be flexed slightly, which releases the stretch on the gastrocnemius and places it on the soleus. Again, it is important to attempt to keep the heels on the floor.

Ankle plantar flexion

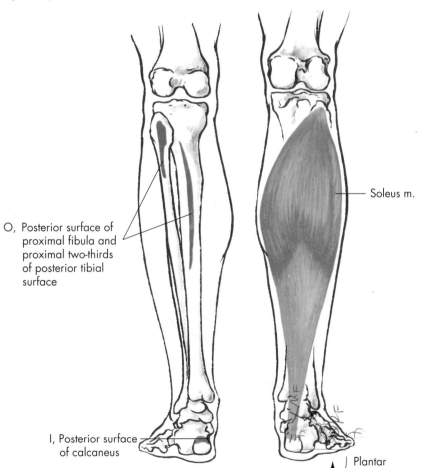

O, Posterior surface of proximal fibula and proximal two-thirds of posterior tibial surface

Soleus m.

I, Posterior surface of calcaneus

Plantar flexion

FIG. 11.12 ● Soleus muscle, posterior view. O, Origin; I, Insertion.

Chapter

11

Transverse
tarsal and
subtalar
eversion

Ankle
plantar
flexion

Peroneus (fibularis) longus muscle FIG. 11.13

(per-o-ne´us lon´gus)

Origin

Head and upper two-thirds of the lateral surface of
the fibula

Insertion

Undersurfaces of the medial cuneiform and first
metatarsal bones

Action

Eversion of the foot
Plantar flexion of the ankle

Palpation

Upper lateral side of the tibia; just distal to fibular
head and down to immediately posterior to lateral
malleolus; just posterolateral from the tibialis anterior
and extensor digitorum longus with active eversion

Innervation

Superficial peroneal nerve (L4, L5, S1)

Application, strengthening, and flexibility

The peroneus longus muscle passes posteroinferiorly to the lateral malleolus and under the
foot from the outside to under the inner surface.
Because of its line of pull, it is a strong evertor
and assists in plantar flexion.

When the peroneus longus muscle is used
effectively with the other ankle flexors, it helps
bind the transverse arch as it contracts. Developed
without the other plantar flexors, it would produce a weak, everted foot. In running, jumping,
hopping, and skipping, the foot should be placed
so that it is pointing forward to ensure proper
development of the group. Walking barefoot or
in stocking feet on the inside of the foot (everted
position) is the best exercise for this muscle.

Eversion exercises to strengthen this muscle
may be performed by turning the sole of the foot
outward while resistance is applied in the opposite direction.

The peroneus longus may be stretched by passively taking the foot into extreme inversion and
dorsiflexion while the knee is flexed.

Chapter
11

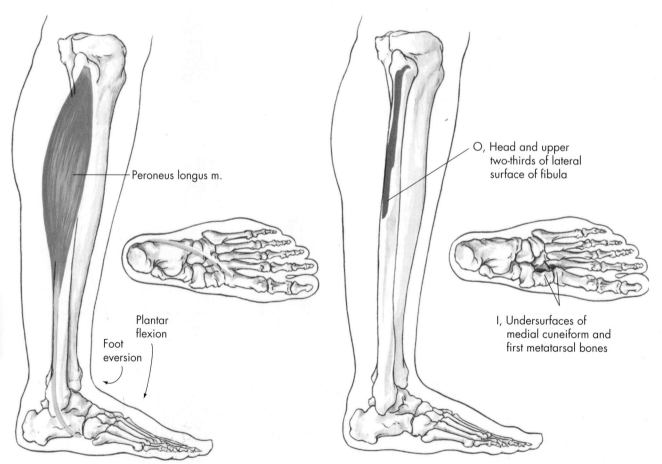

FIG. 11.13 • Peroneus longus muscle, lateral plantar views, right leg and foot. O, Origin; I, Insertion.

Peroneus (fibularis) brevis muscle   FIG. 11.14

(per-o-ne´us bre´vis)

Origin

Mid to lower two-thirds of the lateral surface of the fibula

Insertion

Tuberosity of the fifth metatarsal

Action

Eversion of the foot
Plantar flexion of the ankle

Palpation

Tendon of the muscle at the proximal end of the fifth metatarsal just proximal and posterior to the lateral malleolus; immediately deep anteriorly and posteriorly to the peroneus longus with active eversion

Innervation

Superficial peroneal nerve (L4, L5, S1)

Application, strengthening, and flexibility

The peroneus brevis muscle passes posteroinferiorly to the lateral malleolus to pull on the base of the fifth metatarsal. It is a primary evertor of the foot and assists in plantar flexion. In addition, it aids in maintaining the lateral longitudinal arch as it depresses the foot.

The peroneus brevis muscle is exercised with other plantar flexors in the powerful movements of running, jumping, hopping, and skipping. It may be strengthened in a fashion similar to that for the peroneus longus by performing eversion exercises, such as turning the sole of the foot outward against resistance.

The peroneus brevis is stretched in the same manner as the peroneus longus.

Transverse tarsal and subtalar eversion

Ankle plantar flexion

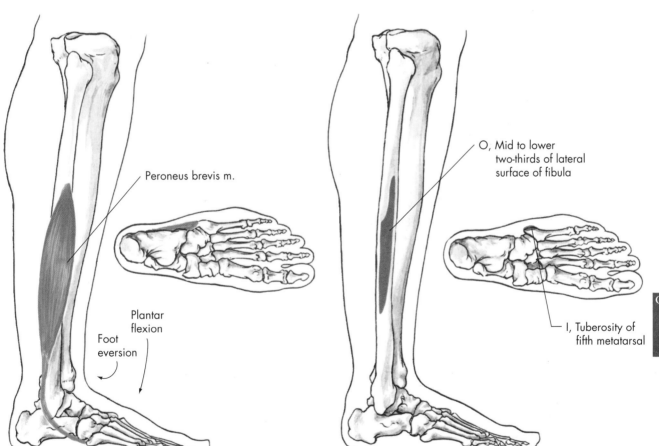

Peroneus brevis m.

Plantar flexion

Foot eversion

O, Mid to lower two-thirds of lateral surface of fibula

I, Tuberosity of fifth metatarsal

Chapter
11

FIG. 11.14 • Peroneus brevis muscle, lateral and plantar views, right leg and foot. O, Origin; I, Insertion.

Transverse
tarsal and
subtalar
eversion

Ankle
dorsal
flexion

Peroneus (fibularis) tertius muscle FIG. 11.15

(per-o-ne´us ter´shi-us)

Origin

Distal third of the anterior fibula

Insertion

Superior aspect of the base of the fifth metatarsal

Action

Eversion of the foot
Dorsiflexion of the ankle

Palpation

Just medial to distal fibula; lateral to the extensor
digitorum longus tendon on the anterolateral aspect
of the foot, down to the medial side of the base of
the fifth metatarsal with dorsiflexion and eversion

Innervation

Deep peroneal nerve (L4, L5, S1)

Application, strengthening, and flexibility

The peroneus tertius, absent in some humans, is a
small muscle that assists in dorsiflexion and eversion. Some authorities refer to it as the fifth tendon of the extensor digitorum longus. It may be
strengthened by pulling the foot up toward the
shin against a weight or resistance. Everting the
foot against resistance, such as weighted eversion towel drags, can also be used for strength
development.

The peroneus tertius may be stretched by passively taking the foot into extreme inversion and
plantar flexion.

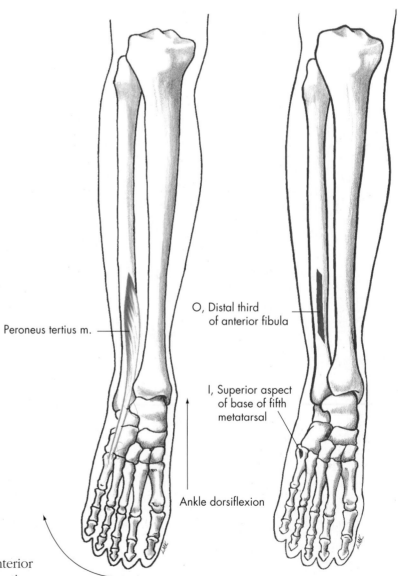

Peroneus tertius m.

O, Distal third
of anterior fibula

I, Superior aspect
of base of fifth
metatarsal

Ankle dorsiflexion

Foot eversion

FIG. 11.15 ● Peroneus tertius muscle, anterior
view, right leg and foot. O, Origin; I, Insertion.

Extensor digitorum longus muscle   FIG. 11.16

(eks-ten´sor dij-i-to´rum lon´gus)

Origin

Lateral condyle of the tibia, head of the fibula, and upper two-thirds of the anterior surface of the fibula

Insertion

Tops of the middle and distal phalanxes of the four lesser toes

Action

Extension of the four lesser toes at the metatarso-phalangeal and the proximal and distal interphalangeal joints
Dorsiflexion of the ankle
Eversion of the foot

Palpation

Second muscle to the lateral side of the anterior tibial border; upper lateral side of the tibia between the tibialis anterior medially and the fibula laterally; divides into four tendons just distal to anterior ankle with active toe extension

Innervation

Deep peroneal nerve (L4, L5, S1)

Application, strengthening, and flexibility

Strength is necessary in the extensor digitorum longus muscle to maintain balance between the plantar and dorsal flexors.

Action that involves dorsal flexion of the ankle and extension of the toes against resistance strengthens both the extensor digitorum longus and the extensor hallucis longus muscles. This may be accomplished by manually applying a downward force on the toes while attempting to extend them up.

The extensor digitorum longus may be stretched by passively taking the four lesser toes into full flexion while the foot is inverted and plantarflexed.

2nd–5th MTP, PIP, and DIP extension

Ankle dorsal flexion

Transverse tarsal and subtalar eversion

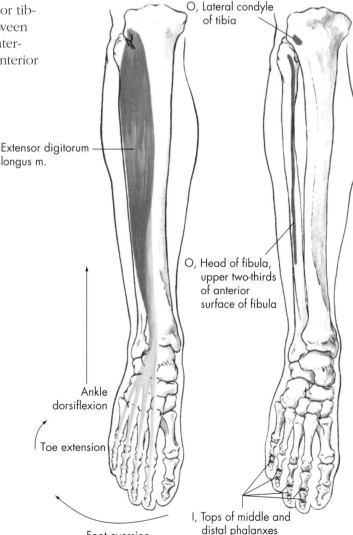

O, Lateral condyle of tibia

Extensor digitorum longus m.

O, Head of fibula, upper two-thirds of anterior surface of fibula

Ankle dorsiflexion

Toe extension

Foot eversion

I, Tops of middle and distal phalanxes of four lesser toes

FIG. 11.16 ● Extensor digitorum longus muscle, anterior view, right leg and foot. O, Origin; I, Insertion.

Extensor hallucis longus muscle

Great toe
MTP and IP
extension

FIG. 11.17

(eks-ten´sor hal-u´sis lon´gus)

Origin

Middle two-thirds of the medial surface of the anterior fibula

Insertion

Top of the base of the distal phalanx of the great toe

Ankle
dorsal
flexion

Action

Dorsiflexion of the ankle

Extension of the great toe at the metatarsophalangeal and interphalangeal joints

Weak inversion of the foot

Transverse
tarsal and
subtalar
inversion

Palpation

From the dorsal aspect of the great toe to just lateral to the tibialis anterior and medial to the extensor digitorum longus at the anterior ankle joint

Innervation

Deep peroneal nerve (L4, L5, S1)

Application, strengthening, and flexibility

The four dorsiflexors of the foot—tibialis anterior, extensor digitorum longus, extensor hallucis longus, and peroneus tertius—may be exercised by attempting to walk on the heels with the ankle flexed dorsally and toes extended. Extension of the great toe, as well as ankle dorsiflexion against resistance, will provide strengthening for this muscle.

The extensor hallucis longus may be stretched by passively taking the great toe into full flexion while the foot is everted and plantarflexed.

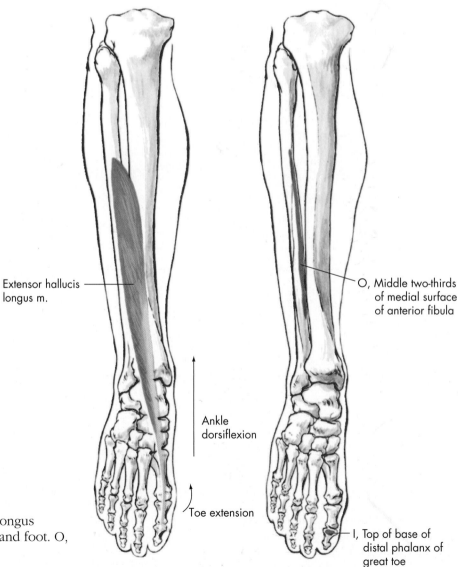

Extensor hallucis longus m.

O, Middle two-thirds of medial surface of anterior fibula

Ankle dorsiflexion

Toe extension

I, Top of base of distal phalanx of great toe

FIG. 11.17 ● Extensor hallucis longus muscle, anterior view, right leg and foot. O, Origin; I, Insertion.

Tibialis anterior muscle   FIG. 11.18

(tib-i-a´lis an-te´ri-or)

Origin

Upper two-thirds of the lateral surface of the tibia

Insertion

Inner surface of the medial cuneiform and the base of the first metatarsal bone

Action

Dorsiflexion of the ankle
Inversion of the foot

Palpation

First muscle to the lateral side of the anterior tibial border, particularly palpable with fully active ankle dorsiflexion; most prominent tendon crossing the ankle anteromedially

Innervation

Deep peroneal nerve (L4, L5, S1)

Application, strengthening, and flexibility

By its insertion, the tibialis anterior muscle is in a fine position to hold up the inner margin of the foot. However, as it contracts concentrically, it dorsiflexes the ankle and is used as an antagonist to the plantar flexors of the ankle. The tibialis anterior is forced to contract strongly when a person ice skates or walks on the outside of the foot. It strongly supports the medial longitudinal arch in inversion.

Turning the sole of the foot to the inside against resistance to perform inversion exercises is one way to strengthen this muscle. Dorsal flexion exercises against resistance may also be used for this purpose. Walking barefoot or in stocking feet on the outside of the foot (inversion) is an excellent exercise for the tibialis anterior muscle.

The tibialis anterior may be stretched by passively taking the foot into extreme eversion and plantar flexion.

Ankle dorsal flexion

Transverse tarsal and subtalar inversion

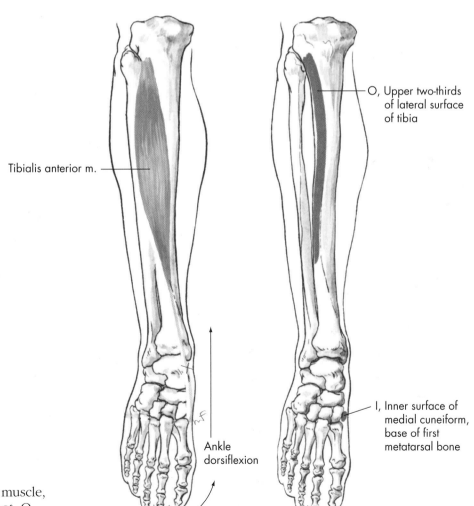

Tibialis anterior m.

O, Upper two-thirds of lateral surface of tibia

I, Inner surface of medial cuneiform, base of first metatarsal bone

Ankle dorsiflexion

Foot inversion

FIG. 11.18 ● Tibialis anterior muscle, anterior view, right leg and foot. O, Origin; I, Insertion.

Tibialis posterior muscle FIG. 11.19

(tib-i-a'lis pos-te'ri-or)

Ankle
plantar
flexion

Origin

Posterior surface of the upper half of the interosseus
membrane and adjacent surfaces of the tibia and
fibula

Insertion

Inferior surfaces of the navicular, cuneiform, and
cuboid bones and bases of the second, third, and
fourth metatarsal bones

Transverse
tarsal and
subtalar
inversion

Action

Plantar flexion of the ankle
Inversion of the foot

Palpation

The tendon may be palpated both proximally and
distally immediately behind the medial malleolus
with inversion and plantar flexion and is better
distinguished from the flexor digitorum longus and
flexor hallucis longus if the toes can be maintained
in slight extension

Innervation

Tibial nerve (L5, S1)

Application, strengthening, and flexibility

Passing down the back of the leg, under the
medial malleolus, then forward to the navicular
and medial cuneiform bones, the tibialis posterior
muscle pulls down from the underside and, when
contracted concentrically, inverts and plantar
flexes the foot. As a result, it is in position to support the medial longitudinal arch. **Shin splints** is
a slang term frequently used to describe an often
chronic condition in which the tibialis posterior,
tibialis anterior, and extensor digitorum longus
muscles are inflamed. This inflammation is usually a tendinitis of one or more of these structures
but may be a result of stress fracture, periostitis,
tibial stress syndrome, or compartment syndrome.
Sprints and long-distance running are common
causes, particularly if the athlete has not developed appropriate strength, flexibility, and endurance in the lower-leg musculature.

Use of the tibialis posterior muscle in plantar
flexion and inversion gives support to the longitudinal arch of the foot. This muscle is generally strengthened by performing heel raises, as
described for the gastrocnemius and soleus, as
well as inversion exercises against resistance.

The tibialis posterior may be stretched by passively taking the foot into extreme eversion and
dorsiflexion while the knee and toes are passively
flexed.

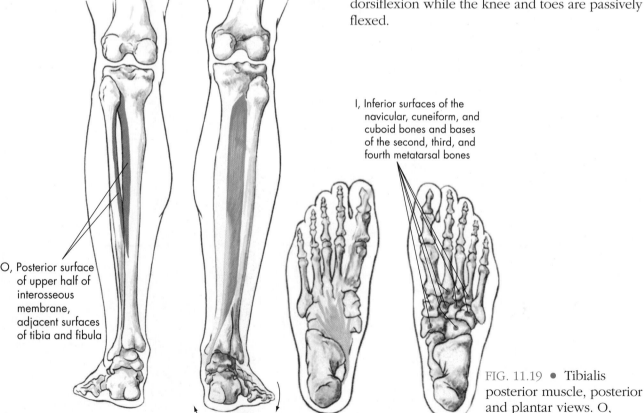

I, Inferior surfaces of the
navicular, cuneiform, and
cuboid bones and bases
of the second, third, and
fourth metatarsal bones

O, Posterior surface
of upper half of
interosseous
membrane,
adjacent surfaces
of tibia and fibula

Foot inversion   Plantar flexion of ankle

FIG. 11.19 • Tibialis
posterior muscle, posterior
and plantar views. O,
Origin; I, Insertion.

Flexor digitorum longus muscle

FIG. 11.20

(fleks´or dij-i-to´rum lon´gus)

Origin

Middle third of the posterior surface of the tibia

Insertion

Base of the distal phalanx of each of the four lesser toes

Action

Flexion of the four lesser toes at the metatarsophalangeal and the proximal and distal interphalangeal joints
Inversion of the foot
Plantar flexion of the ankle

Palpation

The tendon may be palpated immediately posterior to the medial malleolus and tibialis posterior and immediately anterior to the flexor hallucis longus with flexion of the four lesser toes while maintaining great toe extension, ankle dorsiflexion, and foot eversion

Innervation

Tibial nerve (L5, S1)

Application, strengthening, and flexibility

Passing down the back of the lower leg under the medial malleolus and then forward, the flexor digitorum longus muscle draws the four lesser toes down into flexion toward the heel as it plantar flexes the ankle. It is very important in helping other foot muscles maintain the longitudinal arch. Some of the weak foot and ankle conditions result from ineffective use of the flexor digitorum longus. Walking barefoot with the toes curled downward toward the heel and with the foot inverted will exercise this muscle. It may be strengthened by performing towel grabs against resistance in which the heel rests on the floor while the toes extend to grab a flat towel and then flex to pull the towel under the foot. This may be repeated numerous times, with a small weight placed on the opposite end of the towel for added resistance.

The flexor digitorum longus may be stretched by passively taking the four lesser toes into extreme extension while the foot is everted and dorsiflexed. The knee should be flexed.

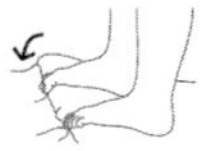

2nd–5th MTP, PIP, and DIP flexion

Transverse tarsal and subtalar inversion

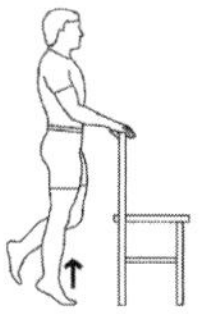

Ankle plantar flexion

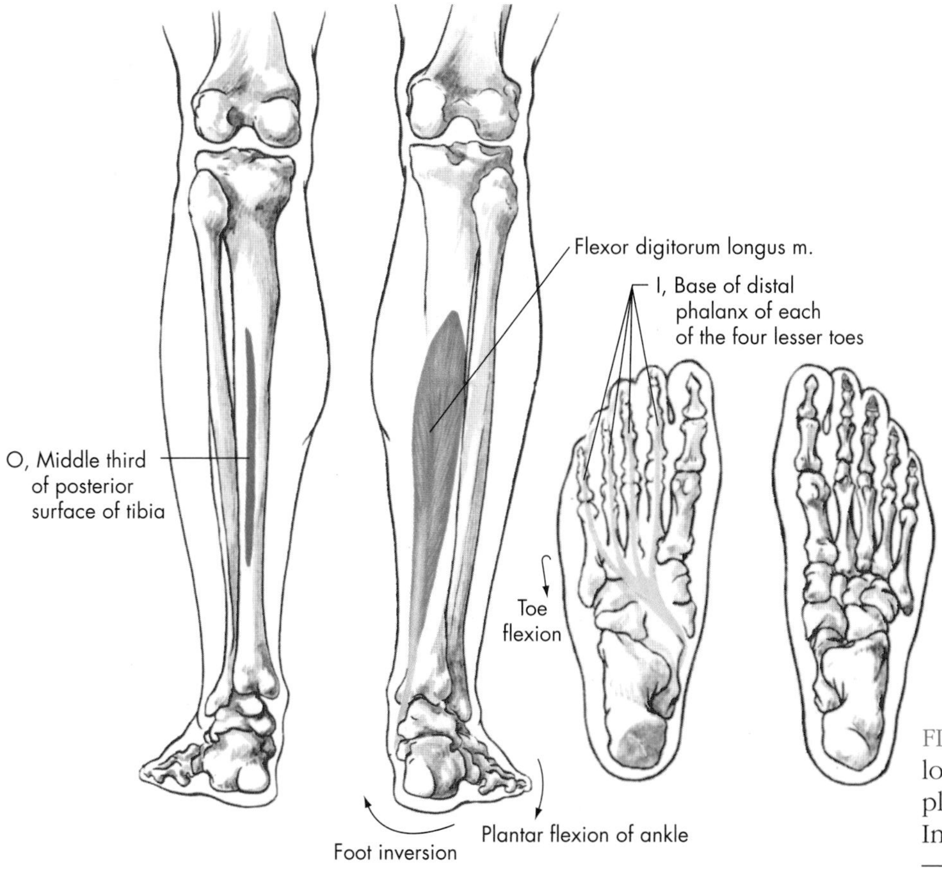

O, Middle third of posterior surface of tibia

Flexor digitorum longus m.

I, Base of distal phalanx of each of the four lesser toes

Toe flexion

Foot inversion

Plantar flexion of ankle

Chapter

11

FIG. 11.20 • Flexor digitorum longus muscle, posterior and plantar views. O, Origin; I, Insertion.

Flexor hallucis longus muscle FIG. 11.21

(fleks′or hal-u′sis lon′gus)

Great toe
MTP and
IP flexion

Transverse
tarsal and
subtalar
inversion

Ankle
plantar
flexion

Origin

Middle two-thirds of the posterior surface of the fibula

Insertion

Base of the distal phalanx of the great toe, plantar
    surface

Action

Flexion of the great toe at the metatarsophalangeal
    and interphalangeal joints
Inversion of the foot
Plantar flexion of the ankle

Palpation

Most posterior of the three tendons immediately
    behind the medial malleolus; between the medial
    soleus and the tibia with active great toe flexion
    while maintaining extension of the four lesser toes,
    ankle dorsiflexion, and foot eversion

Innervation

Tibial nerve (L5, S1, S2)

Application, strengthening, and flexibility

Pulling from the underside of the great toe, the flexor
hallucis longus muscle may work independently
of the flexor digitorum longus muscle or together
with it. If these two muscles are poorly developed,
they cramp easily when they are called on to do
activities to which they are unaccustomed.

These muscles are used effectively in walking if
the toes are used (as they should be) in maintaining balance as each step is taken. Walking "with
the toes" rather than "over" them is an important
action for them.

When the gastrocnemius, soleus, tibialis posterior, peroneus longus, peroneus brevis, flexor digitorum longus, flexor digitorum brevis, and flexor
hallucis longus muscles are all used effectively in
walking, the strength of the ankle is evident. If
an ankle and a foot are weak, in most cases it is
because of lack of use of all the muscles just mentioned. Running, walking, jumping, hopping, and
skipping provide exercise for this muscle group.
The flexor hallucis longus muscle may be specifically strengthened by performing towel grabs as
described for the flexor digitorum longus muscle.

The flexor hallucis longus may be stretched by
passively taking the great toe into extreme extension while the foot is everted and dorsiflexed. The
knee should be flexed.

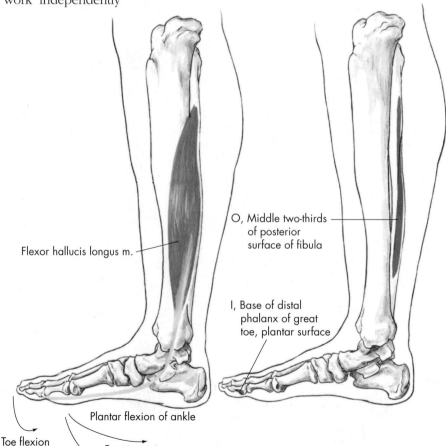

O, Middle two-thirds
of posterior
surface of fibula

Flexor hallucis longus m.

I, Base of distal
phalanx of great
toe, plantar surface

Plantar flexion of ankle

Toe flexion

Foot inversion

FIG. 11.21 • Flexor hallucis
longus muscle, medial
view, right leg and foot. O,
Origin; I, Insertion.

Intrinsic muscles of the foot

FIGS. 11.22, 11.23

The intrinsic muscles of the foot have their origins and insertions on the bones within the foot (Figs. 11.22 and 11.23). One of these muscles, the extensor digitorum brevis, is found on the dorsum of the foot. This muscle includes a band that attaches to the base of the first proximal phalanx and is sometimes labeled the extensor hallucis brevis. The remainder of the muscles are found in a plantar compartment in four layers on the plantar surface of the foot, as follows:

First (superficial) layer: abductor hallucis, flexor digitorum brevis, abductor digiti minimi (quinti)
Second layer: quadratus plantae, lumbricals (four)
Third layer: flexor hallucis brevis, adductor hallucis, flexor digiti minimi (quinti) brevis
Fourth (deep) layer: dorsal interossei (four), plantar interossei (three)

The intrinsic foot muscles may be grouped by location as well as by the parts of the foot on which they act. The abductor hallucis, flexor hallucis brevis, and adductor hallucis all insert either medially or laterally on the proximal phalanx of the great toe. The abductor hallucis and flexor hallucis brevis are located somewhat medially, whereas the adductor hallucis is more centrally located beneath the metatarsals.

The quadratus plantae, four lumbricals, four dorsal interossei, three plantar interossei, flexor digitorum brevis, and extensor digitorum brevis are all located somewhat centrally. All are beneath the foot except the extensor digitorum brevis, which is the only intrinsic muscle in the foot located in the dorsal compartment. Although the entire extensor digitorum brevis has its origin on the anterior and lateral calcaneus, some anatomists refer to its first tendon as the extensor hallucis brevis in order to maintain consistency in naming according to function and location.

Located laterally beneath the foot are the abductor digiti minimi and the flexor digiti minimi brevis, which both insert on the lateral aspect of the base of the proximal phalanx of the fifth phalange. Because of these two muscles' insertion and action on the fifth toe, the name "quinti" is sometimes used instead of "minimi."

Four muscles act on the great toe. The abductor hallucis is solely responsible for abduction of the great toe but assists the flexor hallucis brevis in flexing the great toe at the metatarsophalangeal joint. The adductor hallucis is the sole adductor of the great toe, while the extensor digitorum brevis is the only intrinsic extensor of the great toe at the metatarsophalangeal joint.

The four lumbricals are flexors of the second, third, fourth, and fifth phalanges at their metatarsophalangeal joints, while the quadratus plantae muscles are flexors of these phalanges at their distal interphalangeal joints. The three plantar interossei are adductors and flexors of the proximal phalanxes of the third, fourth, and fifth phalanges, while the four dorsal interossei are abductors and flexors of the second, third, and fourth phalanges, also at their metatarsophalangeal joints. The flexor digitorum brevis flexes the middle phalanxes of the second, third, fourth, and fifth phalanges. The extensor digitorum brevis, as previously mentioned, is an extensor of the great toe but also extends the second, third, and fourth phalanges at their metatarsophalangeal joints.

There are two muscles that act solely on the fifth toe. The proximal phalanx of the fifth phalange is abducted by the abductor digiti minimi and is flexed by the flexor digiti minimi brevis.

Refer to Table 11.2 for further details regarding the intrinsic muscles of the foot.

Muscles are developed and maintain their strength only when they are used. One factor in the great increase in weak foot conditions is the lack of exercise to develop these muscles. Walking is one of the best activities for maintaining and developing the many small muscles that help support the arch of the foot. Some authorities advocate walking without shoes or with shoes designed to enhance proper mechanics. Additionally, towel exercises such as those described for the flexor digitorum longus and flexor hallucis longus are helpful in strengthening the intrinsic muscles of the foot.

Chapter 11

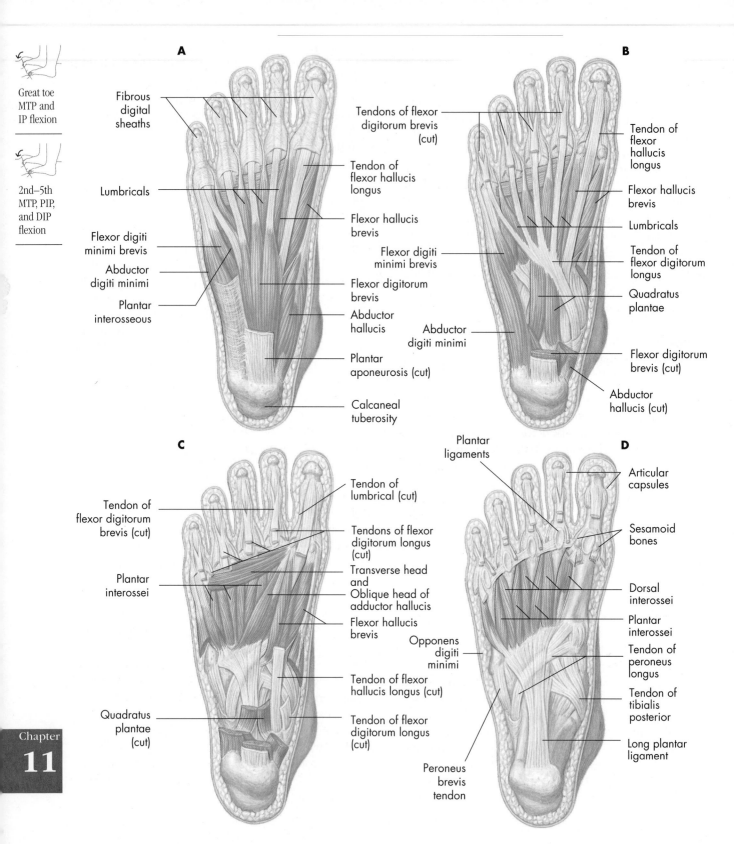

Great toe
MTP and
IP flexion

2nd–5th
MTP, PIP,
and DIP
flexion

A

Fibrous
digital
sheaths

Lumbricals

Flexor digiti
minimi brevis

Abductor
digiti minimi

Plantar
interosseous

Tendon of
flexor hallucis
longus

Flexor hallucis
brevis

Flexor digitorum
brevis

Abductor
hallucis

Plantar
aponeurosis (cut)

Calcaneal
tuberosity

B

Tendons of flexor
digitorum brevis
(cut)

Tendon of
flexor
hallucis
longus

Flexor hallucis
brevis

Lumbricals

Tendon of
flexor digitorum
longus

Quadratus
plantae

Flexor digiti
minimi brevis

Abductor
digiti minimi

Flexor digitorum
brevis (cut)

Abductor
hallucis (cut)

C

Tendon of
flexor digitorum
brevis (cut)

Plantar
interossei

Quadratus
plantae
(cut)

Tendon of
lumbrical (cut)

Tendons of flexor
digitorum longus
(cut)

Transverse head
and
Oblique head of
adductor hallucis

Flexor hallucis
brevis

Tendon of flexor
hallucis longus (cut)

Tendon of flexor
digitorum longus
(cut)

Plantar
ligaments

D

Opponens
digiti
minimi

Peroneus
brevis
tendon

Articular
capsules

Sesamoid
bones

Dorsal
interossei

Plantar
interossei

Tendon of
peroneus
longus

Tendon of
tibialis
posterior

Long plantar
ligament

FIG. 11.22 • The four musculotendinous layers of the plantar aspect of the right foot, detailing the intrinsic muscles. **A,** First (superficial) layer; **B,** Second layer; **C,** Third layer; **D,** Fourth (deep) layer.

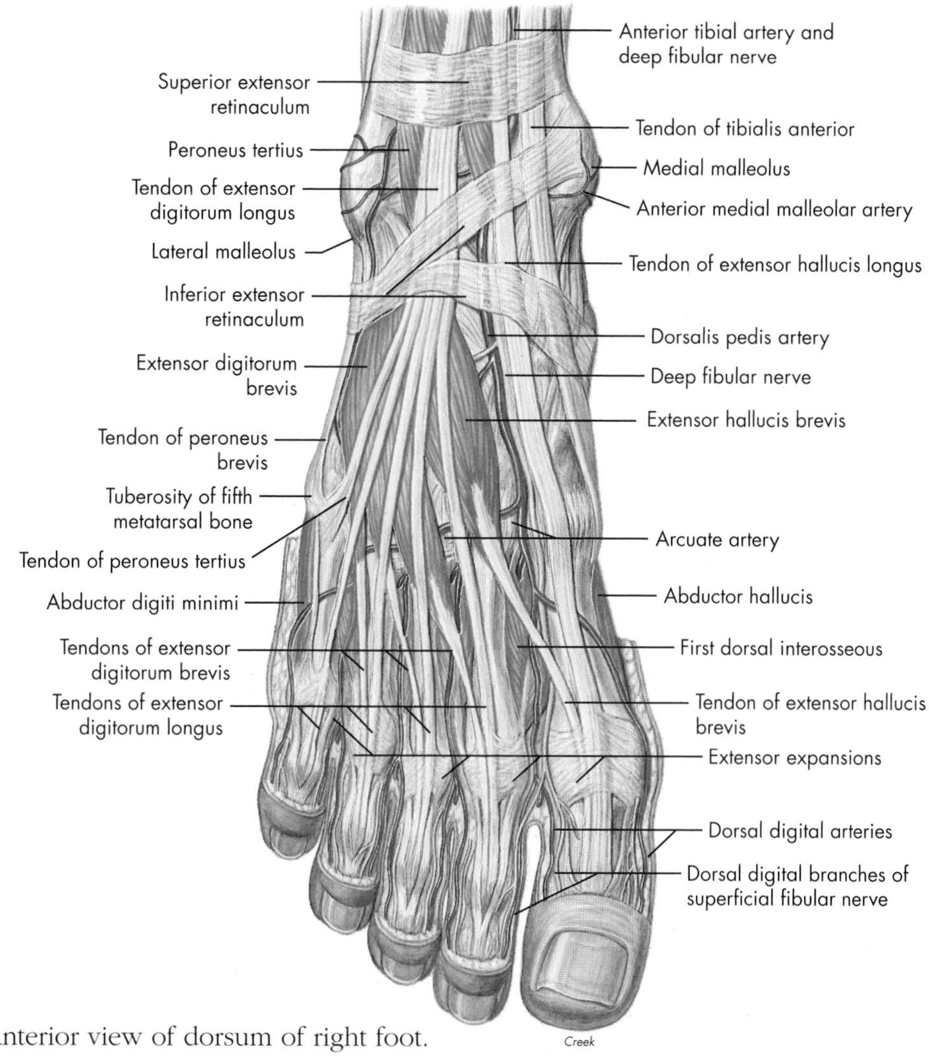
Anterior tibial artery and deep fibular nerve
Superior extensor retinaculum
Tendon of tibialis anterior
Peroneus tertius
Medial malleolus
Tendon of extensor digitorum longus
Anterior medial malleolar artery
Lateral malleolus
Tendon of extensor hallucis longus
Inferior extensor retinaculum
Dorsalis pedis artery
Extensor digitorum brevis
Deep fibular nerve
Extensor hallucis brevis
Tendon of peroneus brevis
Tuberosity of fifth metatarsal bone
Arcuate artery
Tendon of peroneus tertius
Abductor digiti minimi
Abductor hallucis
First dorsal interosseous
Tendons of extensor digitorum brevis
Tendons of extensor digitorum longus
Tendon of extensor hallucis brevis
Extensor expansions
Dorsal digital arteries
Dorsal digital branches of superficial fibular nerve
Creek

FIG. 11.23 • Anterior view of dorsum of right foot.

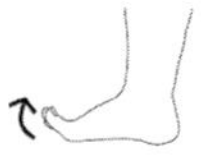

Great toe MTP and IP extension

TABLE 11.2 • **Intrinsic muscles of the foot**

	Muscle	Origin	Insertion	Action	Palpation	Innervation
Superficial layer	Flexor digitorum brevis	Tuberosity of calcaneus, plantar aponeurosis	Medial and lateral aspects of 2nd, 3rd, 4th, and 5th middle phalanxes	MP and PIP flexion of 2nd, 3rd, 4th, and 5th phalanges	Cannot be palpated	Medial plantar nerve (L4, L5)
Superficial layer	Abductor digiti minimi (quinti)	Tuberosity of calcaneus, plantar aponeurosis	Lateral aspect of 5th proximal phalanx	MP abduction of 5th phalange	Cannot be palpated	Lateral plantar nerve (S1, S2)
Superficial layer	Abductor hallucis	Tuberosity of calcaneus, flexor retinaculum, plantar aponeurosis	Medial aspect of base of 1st proximal phalanx	MP flexion, abduction of 1st phalange	On the plantar aspect of the foot from medial tubercle of calcaneus to medial side of great toe proximal phalanx with great toe abduction	Medial plantar nerve (L4, L5)

Chapter 11

TABLE 11.2 (continued) • **Intrinsic muscles of the foot**

	Muscle	Origin	Insertion	Action	Palpation	Innervation
Second layer	Quadratus plantae	Medial head: medial surface of calcaneus Lateral head: lateral border of inferior surface of calcaneus	Lateral margin of flexor digitorum longus tendon	DIP flexion of 2nd, 3rd, 4th, and 5th phalanges	Cannot be palpated	Lateral plantar nerve (S1, S2)
	Lumbricals (4)	Tendons of flexor digitorum longus	Dorsal surface of 2nd, 3rd, 4th, and 5th proximal phalanxes	MP flexion of 2nd, 3rd, 4th, and 5th phalanges	Cannot be palpated	1st lumbricals: medial plantar nerve (L4, L5) 2nd, 3rd, 4th lumbricals: lateral plantar nerve (S1, S2)
Third layer	Adductor hallucis	Oblique head: 2nd, 3rd, and 4th metatarsals and sheath of peroneus longus tendon Transverse head: plantar metatarso-phalangeal ligaments of 3rd, 4th, and 5th phalanges and transverse metatarsal ligaments	Lateral aspect of base of 1st proximal phalanx	MP adduction of 1st phalange	Cannot be palpated	Lateral plantar nerve (S1, S2)
	Flexor hallucis brevis	Cuboid, lateral cuneiform	Medial head: medial aspect of 1st proximal phalanx Lateral head: lateral aspect of 1st proximal phalanx	MP flexion of 1st phalange	Cannot be palpated	Medial plantar nerve (L4, L5, S1)
	Flexor digiti minimi (quinti) brevis	Base of 5th meta-tarsal, sheath of peroneus longus tendon	Lateral aspect of base of 5th proximal phalanx	MP flexion of 5th phalange	Cannot be palpated	Lateral plantar nerve (S2, S3)
Fourth layer	Plantar interossei (3)	Bases and medial shafts of 3rd, 4th, and 5th metatarsals	Medial aspects of bases of 3rd, 4th, and 5th proximal phalanxes	MP adduction and flexion of 3rd, 4th, and 5th phalanges	Cannot be palpated	Lateral plantar nerve (S1, S2)

Chapter

11

TABLE 11.2 (continued) • Intrinsic muscles of the foot

	Muscle	Origin	Insertion	Action	Palpation	Innervation
Fourth layer	Dorsal interossei (4)	Two heads on shafts of adjacent metatarsals	1st interosseus: medial aspect of 2nd proximal phalanx 2nd, 3rd, and 4th interossei: lateral aspects of 2nd, 3rd, and 4th proximal phalanxes	MP abduction and flexion of 2nd, 3rd, and 4th phalanges	Cannot be palpated	Lateral plantar nerve (S1, S2)
	Extensor digitorum brevis (including extensor hallucis brevis)	Anterior and lateral calcaneus, lateral talocalcaneal ligament, inferior extensor retinaculum	Base of proximal phalanx of 1st phalange, lateral sides of extensor digitorum longus tendons of 2nd, 3rd, and 4th phalanges	Assists in MP extension of 1st phalange and extension of middle three phalanges	As a mass anterior to and slightly below lateral malleolus on dorsum of foot	Deep peroneal nerve (L5, S1)

REVIEW EXERCISES

1. List the planes in which each of the following movements occurs. List the axis of rotation for each movement in each plane.
   a. Plantar flexion
   b. Dorsiflexion
   c. Inversion
   d. Eversion
   e. Flexion of the toes
   f. Extension of the toes
2. Why are "low arches" and "flat feet" not synonymous terms?

3. Discuss the value of proper footwear in various sports and activities.
4. What are orthotics and how do they function?
5. Research common foot and ankle disorders, such as flat feet, lateral ankle sprains, high ankle sprains, bunions, plantar fasciitis, and hammertoes. Report your findings in class.
6. Research the anatomical factors related to the prevalence of inversion versus eversion ankle sprains and report your findings in class.
7. Report orally or in writing on magazine articles that rate running and walking shoes.

8. **Muscle analysis chart** • Ankle, transverse tarsal and subtalar joints, and toes

Complete the chart by listing the muscles primarily involved in each movement.	
Ankle	
Dorsiflexion	Plantar flexion
Transverse tarsal and subtalar joints	
Eversion	Inversion
Toes	
Flexion	Extension

Chapter

11

9. **Antagonistic muscle action chart** • Ankle, transverse tarsal and subtalar joints, and toes

Complete the chart by listing the muscle(s) or parts of muscles that are antagonist in their actions to the muscles in the left column.

Agonist	Antagonist
Gastrocnemius	
Soleus	
Tibialis posterior	
Flexor digitorum longus	
Flexor hallucis longus	
Peroneus longus/Peroneus brevis	
Peroneus tertius	
Extensor digitorum longus	
Extensor hallucis longus	

LABORATORY EXERCISES

1. Locate the following bony landmarks of the ankle and foot on a human skeleton and on a subject:
   a. Lateral malleolus
   b. Medial malleolus
   c. Calcaneus
   d. Navicular
   e. Three cuneiform bones
   f. Metatarsal bones
   g. Phalanges and individual phalanxes
2. How and where can the following muscles be palpated on a human subject?
   a. Tibialis anterior
   b. Extensor digitorum longus
   c. Peroneus longus
   d. Peroneus brevis
   e. Soleus
   f. Gastrocnemius
   g. Extensor hallucis longus

   h. Flexor digitorum longus
   i. Flexor hallucis longus
   j. Tibialis posterior
3. Demonstrate and palpate the following movements:
   a. Plantar flexion of ankle
   b. Dorsiflexion of ankle
   c. Inversion of the foot
   d. Eversion of the foot
   e. Flexion of the toes
   f. Extension of the toes
4. Have a laboratory partner rise up on the toes (heel raise) with the knees fully extended and then repeat with the knees flexed approximately 20 degrees. Which exercise position appears to be more difficult to maintain for an extended period of time and why? What are the implications for strengthening these muscles? For stretching these muscles?

5. **Ankle and foot joint exercise movement analysis chart**

After analyzing each exercise in the chart, break each into two primary movement phases, such as a lifting phase and a lowering phase. For each phase, determine what ankle and foot joint movements occur, and then list the ankle and foot joint muscles primarily responsible for causing/controlling those movements. Beside each muscle in each movement, indicate the type of contraction as follows: I—isometric; C—concentric; E—eccentric.

Exercise	Initial movement (lifting) phase		Secondary movement (lowering) phase	
	Movement(s)	Agonist(s)—(contraction type)	Movement(s)	Agonist(s)—(contraction type)
Push-up				
Squat				
Dead lift				
Hip sled				
Forward lunge				
Rowing exercise				
Stair machine				

6. **Ankle and foot joint sport skill analysis chart**

Analyze each skill in the chart, and list the movements of the right and left ankle and foot joints in each phase of the skill. You may prefer to list the initial positions that the ankle and foot joints are in for the stance phase. After each movement, list the ankle and foot joint muscle(s) primarily responsible for causing/controlling the movement. Beside each muscle in each movement, indicate the type of contraction as follows: I—isometric; C—concentric; E—eccentric. It may be desirable to review the concepts for analysis in Chapter 8 for the various phases. Assume right hand/leg dominant where applicable.

Exercise		Stance phase	Preparatory phase	Movement phase	Follow-through phase
Baseball pitch	(R)				
	(L)				
Football punt	(R)				
	(L)				
Walking	(R)				
	(L)				
Softball pitch	(R)				
	(L)				
Soccer pass	(R)				
	(L)				
Batting	(R)				
	(L)				
Bowling	(R)				
	(L)				
Basketball jump shot	(R)				
	(L)				

References

Astrom M, Arvidson T: Alignment and joint motion in the normal foot, *Journal of Orthopaedic and Sports Physical Therapy* 22:5, November 1995.

Booher JM, Thibodeau GA: *Athletic injury assessment,* ed 4, New York, 2000, McGraw-Hill.

Field D: *Anatomy: palpation and surface markings,* ed 3, Oxford, 2001, Butterworth-Heinemann.

Gench BE, Hinson MM, Harvey PT: *Anatomical kinesiology,* Dubuque, IA, 1995, Eddie Bowers.

Hamilton N, Weimer W, Luttgens K: *Kinesiology: scientific basis of human motion,* ed 12, New York, 2012, McGraw-Hill.

Hislop HJ, Montgomery J: *Daniels and Worthingham's muscle testing: techniques of manual examination,* ed 8, Philadelphia, 2007, Saunders.

Lindsay DT: *Functional human anatomy,* St. Louis, 1996, Mosby.

Magee DJ: *Orthopedic physical assessment,* ed 5, Philadelphia, 2008, Saunders.

Muscolino JE: *The muscular system manual: the skeletal muscles of the human body,* ed 3, St. Louis, 2010, Elsevier Mosby.

Oatis CA: *Kinesiology: the mechanics and pathomechanics of human movement,* ed 2, Philadelphia, 2008, Lippincott Williams & Wilkins.

Prentice WE: *Principles of athletic training: a competency based approach,* ed 15, New York, 2014, McGraw-Hill.

Rockar PA: The subtalar joint: anatomy and joint motion, *Journal of Orthopaedic and Sports Physical Therapy* 21:6, June 1995.

Saladin KS: *Anatomy & physiology: the unity of form and function,* ed 5, New York, 2010, McGraw-Hill.

Sammarco GJ: Foot and ankle injuries in sports, *American Journal of Sports Medicine* 14:6, November–December 1986.

Seeley RR, Stephens TD, Tate P: *Anatomy & physiology,* ed 8, New York, 2008, McGraw-Hill.

Sieg KW, Adams SP: *Illustrated essentials of musculoskeletal anatomy,* ed 4, Gainesville, FL, 2002, Megabooks.

Stone RJ, Stone JA: *Atlas of the skeletal muscles,* ed 6, New York, 2009, McGraw-Hill.

Thibodeau GA, Patton KT: *Anatomy & physiology,* ed 9, St. Louis, 1993, Mosby.

Van De Graaff KM: *Human anatomy,* ed 6, Dubuque, IA, 2002, McGraw-Hill.

For additional resources and a list of related websites, visit **www.mhhe.com/floyd19e**.

Worksheet Exercises

For in- or out-of-class assignments, or for testing, utilize this tear-out worksheet.

Worksheet 1

Using crayons or colored markers, draw and label on the worksheet the following muscles. Indicate the origin and insertion of each muscle with an "O" and an "I," respectively, and draw in the origin and insertion on the anterior or posterior view as applicable.

a. Tibialis anterior

b. Extensor digitorum longus

c. Peroneus longus

d. Peroneus brevis

e. Peroneus tertius

f. Soleus

g. Gastrocnemius

h. Extensor hallucis longus

i. Tibialis posterior

j. Flexor digitorum longus

k. Flexor hallucis longus

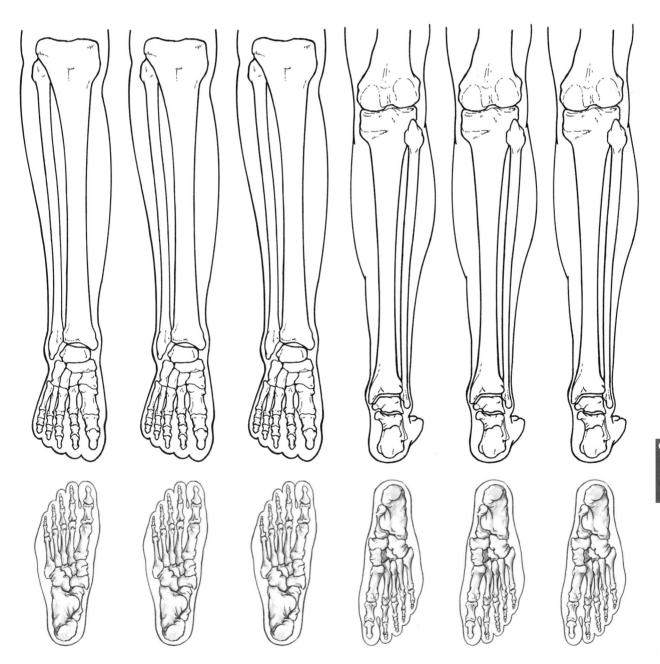

Worksheet Exercises

For in- or out-of-class assignments, or for testing, utilize this tear-out worksheet.

Worksheet 2

Label and indicate with arrows the following movements of the talocrural, transverse tarsal, and subtalar joints. For each motion, complete the sentence by supplying the plane in which it occurs and the axis of rotation as well as the muscles causing the motion.

a. Dorsiflexion occurs in the _____ plane about the _____ axis and is accomplished by concentric contractions of the _____ _____ muscles.

b. Plantar flexion occurs in the _____ plane about the _____ axis and is accomplished by concentric contractions of the _____ _____ muscles.

c. Eversion occurs in the _____ plane about the _____ axis and is accomplished by concentric contractions of the _____ _____ muscles.

d. Inversion occurs in the _____ plane about the _____ axis and is accomplished by concentric contractions of the _____ _____ muscles.

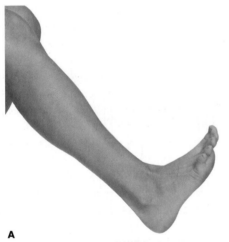

A

B

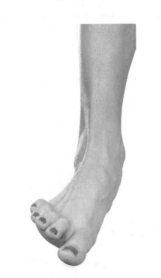

C

D

CHAPTER 12

THE TRUNK AND SPINAL COLUMN

Objectives

- To identify and differentiate the different types of vertebrae in the spinal column

- To label on a skeletal chart the types of vertebrae and their important features

- To draw and label on a skeletal chart some of the muscles of the trunk and the spinal column

- To demonstrate and palpate with a fellow student the movements of the spine and trunk and list their respective planes of motion and axes of rotation

- To palpate on a human subject some of the muscles of the trunk and spinal column

- To list and organize the muscles that produce the primary movements of the trunk and spinal column and their antagonists

- To determine, through analysis, the trunk and spinal column movements and muscles involved in selected skills and exercises

The trunk and spinal column present problems in kinesiology that are not found in the study of other parts of the body. The vertebral column is quite elaborate, consisting of 24 articulating vertebrae with an additional 9 nonmovable vertebrae. These vertebrae contain the spinal column, with its 31 pairs of spinal nerves. Most would agree that it is one of the more complex parts of the human body.

The anterior portion of the trunk contains the abdominal muscles, which are somewhat different from other muscles in that some sections are linked by fascia and tendinous bands and thus do not attach from bone to bone. In addition, there are many small intrinsic muscles acting on the head, vertebral column, and thorax that assist in spinal stabilization or respiration, depending on their location. These muscles are generally too deep to palpate and consequently will not be given the full attention that the larger superficial muscles receive in this chapter.

Bones

Vertebral column

The intricate and complex bony structure of the vertebral column consists of 24 articulating vertebrae (freely movable) and 9 that are fused (Fig. 12.1). The column is further divided into the 7 cervical (neck) vertebrae, 12 thoracic (chest) vertebrae, and 5 lumbar (lower back) vertebrae. The sacrum (posterior pelvic girdle) and the coccyx (tailone) consist of 5 and 4 fused vertebrae, respectively.

Chapter
12

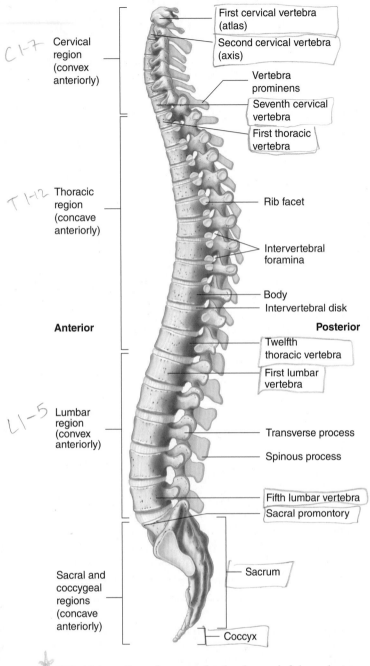

First cervical vertebra (atlas)

Second cervical vertebra (axis)

Vertebra prominens

Seventh cervical vertebra

First thoracic vertebra

Rib facet

Intervertebral foramina

Body

Intervertebral disk

Twelfth thoracic vertebra

First lumbar vertebra

Transverse process

Spinous process

Fifth lumbar vertebra

Sacral promontory

Sacrum

Coccyx

Cervical region (convex anteriorly)

Thoracic region (concave anteriorly)

Anterior

Posterior

Lumbar region (convex anteriorly)

Sacral and coccygeal regions (concave anteriorly)

C 1-7

T 1-12

L 1-5

FIG. 12.1 • Complete vertebral column, left lateral view.

The first two cervical vertebrae are unique in that their shapes allow for extensive rotary movement of the head to the sides, as well as movement forward and backward. The bones in each region of the spine have slightly different sizes and shapes to allow for various functions (Fig. 12.2). The vertebrae increase in size from the cervical region to the lumbar region, primarily because they have to support more weight in the lower back than in the neck. The first two cervical vertebrae are known as the atlas and the axis,

respectively. Vertebrae C3 through L5 have similar architecture: each has a bony block anteriorly, known as the body, a vertebral foramen centrally for the spinal cord to pass through, a transverse process projecting laterally to each side, and a spinous process projecting posteriorly that is easily palpable.

The spine has three normal curves within its movable vertebrae. The primary spinal curve prior to birth and briefly afterward is kyphotic, or C-shaped. As muscle development occurs and activity increases, the secondary curves, which are lordotic, develop in the cervical and lumbar regions. The thoracic curve is concave anteriorly and convex posteriorly, whereas the cervical and lumbar curves are convex anteriorly and concave posteriorly. Finally, the sacral curve, including the coccyx, is concave anteriorly and convex posteriorly. The normal curves of the spine enable it to absorb blows and shocks.

Undesirable deviations from the normal curvatures occur due to a number of factors. Increased posterior concavity of the lumbar and cervical curves is known as **lordosis**, and increased anterior concavity of the normal thoracic curve is known as **kyphosis**. The lumbar spine may have a reduction of its normal lordotic curve, resulting in a flat-back appearance referred to as **lumbar kyphosis**. **Scoliosis** consists of lateral curvatures or sideward deviations of the spine.

Thorax

The skeletal foundation of the thorax is formed by 12 pairs of ribs (Fig. 12.3). Seven pairs are true ribs, in that they attach directly to the sternum via a separate costal cartilage. Five pairs are considered false ribs. Of these, three pairs attach indirectly to the sternum via a shared costal cartilage and two pairs are floating ribs, with their ends free. The manubrium, the body of the sternum, and the xiphoid process are the other bones of the thorax. All the ribs are attached posteriorly to the thoracic vertebrae.

Key bony landmarks for locating the muscles of the neck include the mastoid process, transverse and spinous processes of the cervical spine, spinous processes of the upper four thoracic vertebrae, manubrium of the sternum, and medial clavicle. The spinous and transverse processes of the thoracic spine and the posterior ribs are key areas of attachment for the posterior muscles of the spine. The anterior trunk muscles have attachments on the borders of the lower eight ribs, the costal cartilages of the ribs, the iliac crest, and

Chapter

12

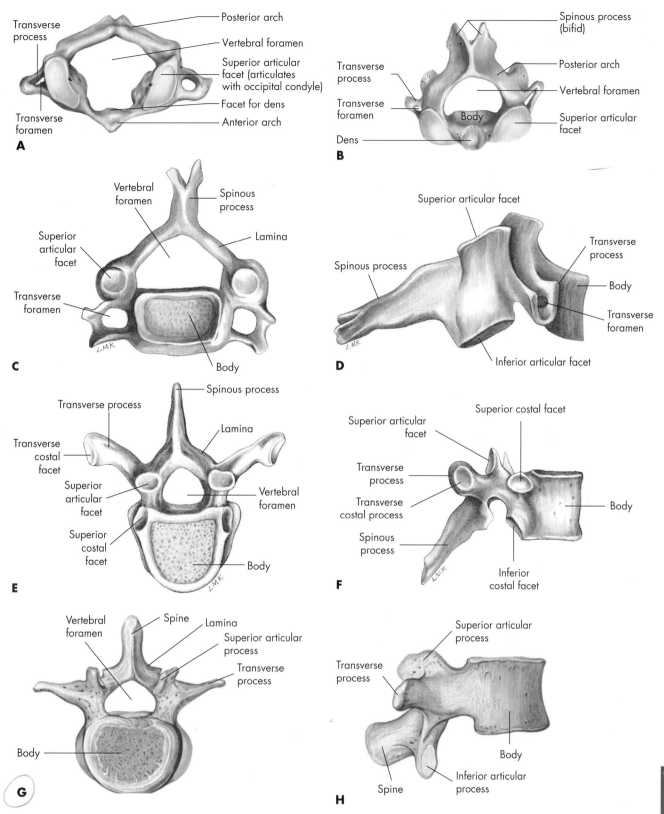

FIG. 12.2 ● Vertebral column. **A,** Atlas (first cervical vertebra), superior view; **B,** Axis (second cervical vertebra), superior view; **C,** Typical cervical vertebra, superior view; **D,** Typical cervical vertebra, lateral view; **E,** Typical thoracic vertebra, superior view; **F,** Typical thoracic vertebra, lateral view; **G,** Third lumbar vertebra, superior view; **H,** Third lumbar vertebra, lateral view.

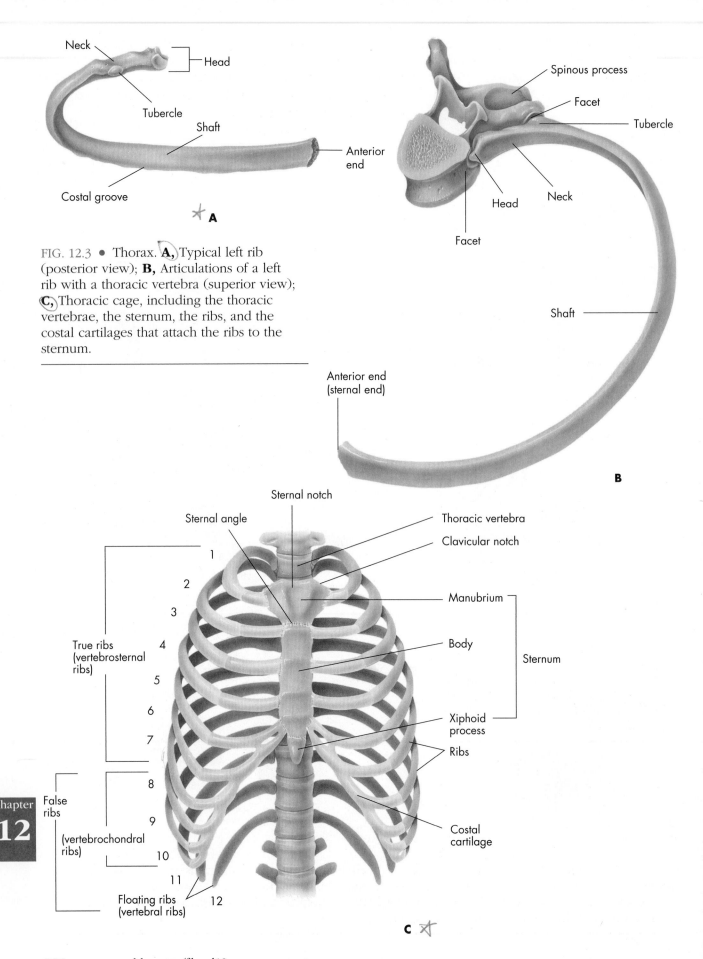

Neck

Head

Tubercle

Shaft

Anterior
end

Costal groove

A

Spinous process

Facet

Tubercle

Head

Neck

Facet

Shaft

Anterior end
(sternal end)

B

FIG. 12.3 • Thorax. **A,** Typical left rib
(posterior view); **B,** Articulations of a left
rib with a thoracic vertebra (superior view);
C, Thoracic cage, including the thoracic
vertebrae, the sternum, the ribs, and the
costal cartilages that attach the ribs to the
sternum.

Sternal notch

Sternal angle

Thoracic vertebra

Clavicular notch

1

2

3

True ribs
(vertebrosternal
ribs)

4

Manubrium

Body

Sternum

5

6

7

Xiphoid
process

Ribs

8

False
ribs

9

(vertebrochondral
ribs)

10

Costal
cartilage

11

Floating ribs
(vertebral ribs)

12

C

the pubic crest. The transverse processes of the upper four lumbar vertebrae also serve as points of insertion for the quadratus lumborum, along with the lower border of the twelfth rib.

Joints

The first joint in the axial skeleton is the atlantooccipital joint, formed by the occipital condyles of the skull sitting on the articular fossa of the first vertebra, which allows capital flexion and extension or flexion and extension of the head on the neck. Although this is a separate articulation, its movements are often grouped with those of the cervical spine. The atlas (C1) in turn sits on the axis (C2) to form the atlantoaxial joint (Fig. 12.4, *A*). Except for the atlantoaxial joint, there is not a great deal of movement possible between any two vertebrae. However, the cumulative effect of combining the movement from several adjacent vertebrae allows for substantial movements within a given area. Most of the rotation within the cervical region occurs in the atlantoaxial joint, which is classified as a trochoid or pivot-type joint. The remainder of the vertebral articulations are

classified as arthrodial or gliding-type joints because of their limited, gliding movements.

As shown in Fig 12.4, *B* and *D,* the anterior longitudinal ligament runs the entire length of the spine from the base of the skull to the sacrum and attaches to the anterior surface of each vertebral body. The posterior longitudinal ligament is located inside the vertebral canal, attaching on the posterior vertebral bodies, and runs from the axis to the sacrum. The ligamentum flavum binds the laminae of adjacent vertebrae together. The interspinous ligaments connect the spinous processes, and the intertransverse ligaments connect the transverse processes. The ligamentum nuchae connects the tips of the cervical spinous processes

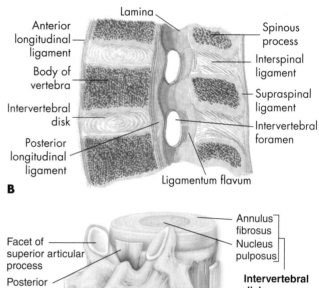

B

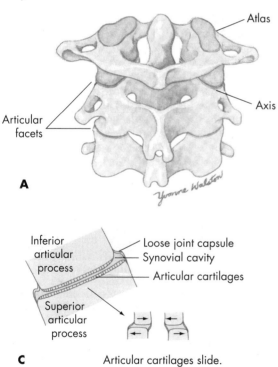

A

C   Articular cartilages slide.

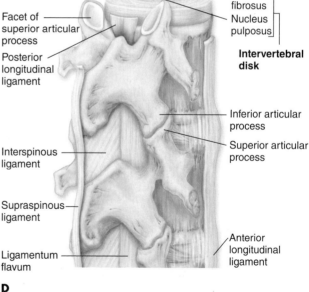

D

FIG. 12.4 • Articular facets of the vertebrae. **A,** The facets of the superior and inferior articular processes articulate between adjacent cervical vertebrae; **B,** Ligaments limit motion between vertebrae, shown in sagittal section through three lumbar vertebrae; **C,** Articular cartilages slide back and forth on each other, a motion allowed by the loose articular capsule; **D,** Intervertebral articulations. Vertebrae articulate with adjacent vertebrae at both their superior and inferior articular processes. Intervertebral disks separate the bodies of adjacent vertebrae.

Chapter
12

from the occipital protuberance to all seven cervical vertebrae, and the supraspinous ligament connects the tips of the spinous processes of the remaining vertebrae.

Gliding movement occurs between the superior and inferior articular processes that form the facet joints of the vertebrae, as depicted in Figs. 12.2 and 12.4, *C* and *D*. Located in between and adhering to the articular cartilage of the vertebral bodies are the intervertebral disks (Fig. 12.4,*B*). These disks are composed of an outer rim of dense fibrocartilage known as the annulus fibrosus and a central gelatinous, pulpy substance known as the nucleus pulposus (Fig. 12.4,*D*). This arrangement of compressed elastic material allows compression in all directions, along with torsion. With age, injury, or improper use of the spine, the intervertebral disks become less resilient, resulting in a weakened annulus fibrosus. Substantial weakening combined with compression can result in the nucleus protruding through the annulus, known as a herniated nucleus pulposus. Commonly referred to as a herniated or "slipped" disk, this protrusion puts pressure on the spinal nerve root, causing a variety of symptoms, including radiating pain, tingling, numbness, and weakness in the dermatomes and myotomes of the extremity supplied by the spinal nerve (Fig. 12.5).

A substantial number of low back problems are caused by improper use of the back over time. These improper mechanics often result in acute strains and muscle spasm of the lumbar extensors and chronic mechanical changes leading to disk herniation. Most problems occur from using the relatively small back muscles to lift objects from a lumbar spine flexed position instead of keeping the lumbar spine in a neutral position while squatting and using the larger, more powerful muscles of the lower extremity. Additionally, our lifestyles chronically place us in lumbar flexion, which over time leads to a gradual loss of lumbar lordosis. This flat-back syndrome results in increased pressure on the lumbar disk and intermittent or chronic low back pain.

Most of the spinal column movement occurs in the cervical and lumbar regions. There is, of course, some thoracic movement, but it is slight in comparison with that of the neck and low back. In discussing movements of the head, it must be remembered that this movement occurs between the cranium and the first cervical vertebra, as well as within the cervical vertebrae. With the

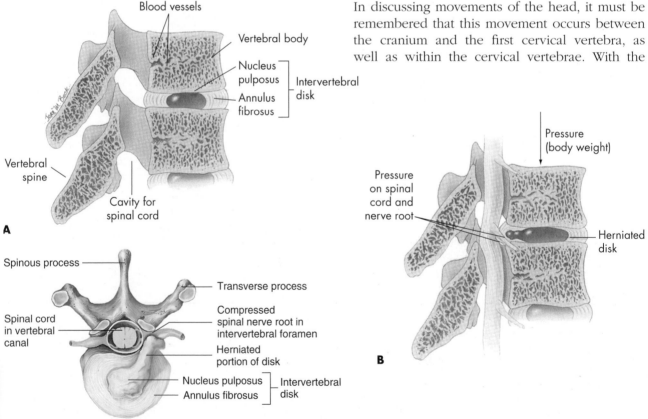

FIG. 12.5 • Intervertebral disks. **A,** Sagittal section of normal disks; **B,** Sagittal section of herniated disks; **C,** Herniated disk, superior view.

understanding that these motions usually occur together, for simplification purposes this text refers to all movements of the head and neck as cervical movements. Similarly, in discussing trunk movements, lumbar motion terminology is used to describe the combined motion that occurs in both the thoracic and the lumbar regions. A closer investigation of specific motion between any two vertebrae is beyond the scope of this text.

The cervical region (Fig. 12.6) can flex 45 degrees and extend 45 degrees. The cervical area laterally flexes 45 degrees and can rotate approximately 60 degrees. The lumbar spine, accounting for most of the trunk movement (Fig. 12.7), flexes approximately 80 degrees and extends 20 to 30 degrees. Lumbar lateral flexion to each side is usually within 35 degrees, and approximately 45 degrees of rotation occurs to the left and right.

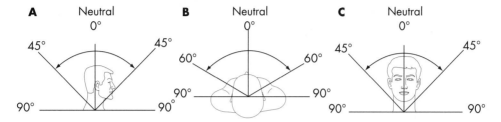

FIG. 12.6 • Active ROM of the cervical spine. **A,** Flexion and extension can be estimated in degrees or indicated by the distance the chin lacks from touching the chest; **B,** Rotation can be estimated in degrees or percentages of motion compared in each direction; **C,** Lateral flexion can be estimated in degrees or indicated by the distance the ear lacks from reaching the shoulders.

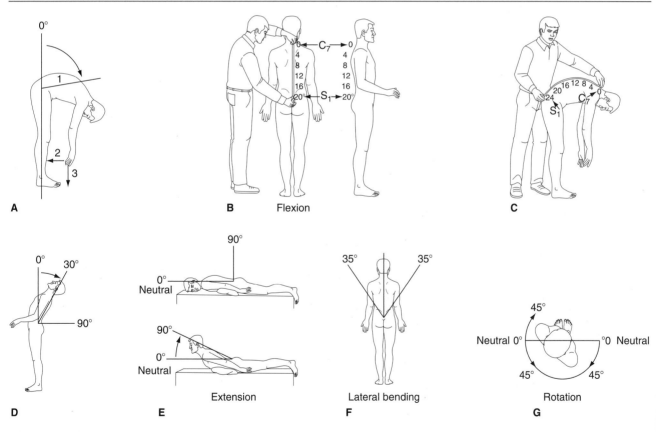

FIG. 12.7 • Active ROM of the thoracic and lumbar spine. **A,** Forward flexion. Motion can be estimated in degrees or by measurement from fingertips to leg or to floor; **B** and **C,** Using a tape measure to compare the increased length of the lumbar and thoracic spine from the anatomical position to the fully flexed position (not shown, the reverse movement may be done to assess the amount of extension); **D,** Extension (hyperextension) with the subject standing; **E,** Extension (hyperextension) with the subject lying prone; **F,** Lateral bending; **G,** Rotation of the spine.

Chapter
12

Cervical
flexion

Cervical
extension

Cervical
lateral
flexion

Cervical
rotation
unilaterally

Movements   FIG. 12.8

Spinal movements are often preceded by the name given to the name given to the region of movement. For example, flexion of the trunk at the lumbar spine is known as lumbar flexion, and extension of the neck is often referred to as cervical extension. Though usually included with cervical flexion and extension, isolated movements of the head or the neck at the atlantoccipital joint are technically known as capital flexion and extension. Additionally, as discussed in Chapter 9, the pelvic girdle rotates as a unit due to movement occurring in the hip joints and the lumbar spine. Refer to Table 9.1.

Spinal flexion: anterior movement of the spine in the sagittal plane; in the cervical region, the head moves toward the chest; in the lumbar region, the thorax moves toward the pelvis

Spinal extension: return from flexion; posterior movement of the spine in the sagittal plane; in the cervical spine, the head moves away from the chest; in the lumbar spine, the thorax moves away from the pelvis; sometimes referred to as hyperextension

Lateral flexion (left or right): sometimes referred to as side bending; the head moves laterally toward the shoulder, and the thorax moves laterally toward the pelvis; both in the frontal plane

Spinal rotation (left or right): rotary movement of the spine in the transverse plane; the chin rotates from neutral toward the shoulder, and the thorax rotates toward one iliac crest

Reduction: return movement from lateral flexion to neutral in the frontal plane

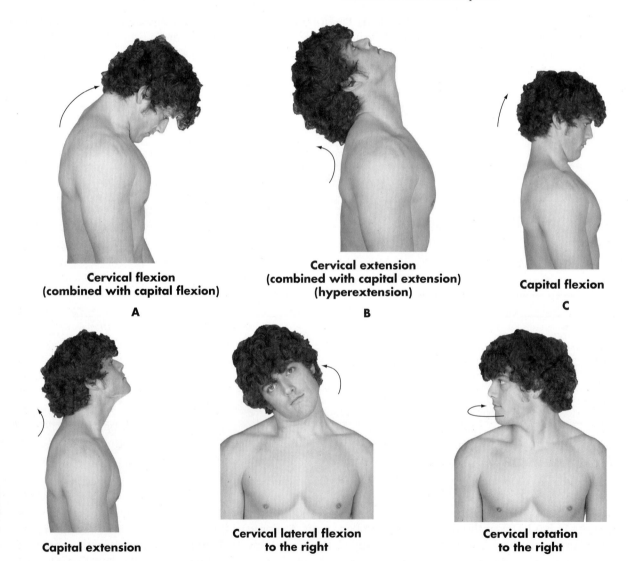

**Cervical flexion
(combined with capital flexion)**

A

**Cervical extension
(combined with capital extension)
(hyperextension)**

B

Capital flexion

C

Capital extension

D

**Cervical lateral flexion
to the right**

E

**Cervical rotation
to the right**

F

FIG. 12.8 ● Movements of the spine.

Chapter
12

| Lumbar flexion
G | Lumbar extension
(hyperextension)
H | Lumbar lateral flexion
to the right
I | Lumbar rotation
to the right
J | Lumbar
flexion

Lumbar
extension

Lumbar
lateral
flexion

Lumbar
rotation
unilaterally |

FIG. 12.8 (continued) • Movements of the spine.

Trunk and spinal column muscles

The largest muscle group in this area is the erector spinae (sacrospinalis), which extends on each side of the spinal column from the pelvic region to the cranium. It is divided into three muscles: the spinalis, the longissimus, and the iliocostalis. From the medial to the lateral side, it has attachments in the lumbar, thoracic, and cervical regions. Thus, the erector spinae group is actually made up of nine muscle segments. Additionally, the sternocleidomastoid and splenius muscles are large muscles involved in cervical and head movements. Large abdominal muscles involved in lumbar movements include the rectus abdominis, external oblique abdominal, and internal oblique abdominal. The quadratus lumborum is involved in the lumbar spine movements of lateral flexion and lumbar extension. See Table 12.1 for further details. Also, the psoas major and minor, as discussed in Chapter 9, are involved in ipsilateral lateral flexion and in flexion of the lumbar spine.

Numerous small muscles are found in the spinal column region. Many of them originate on one vertebra and insert on the next. They are important in the functioning of the spine, but detailed knowledge of these muscles is of limited value to most people who use this text. Consequently, discussion will concentrate on the larger muscles primarily involved in trunk and spinal column movements (see Table 12.1) and will only briefly address the smaller muscles.

So that the muscles of the trunk and spinal column may be better understood, they can be grouped according to both location and function. It should be noted that some muscles have multiple segments. As a result, one segment of a particular muscle may be located and perform movement in one region, while another segment of the same muscle may be located in another region and perform movements there. Many of the muscles of the trunk and spinal column function in moving the spine as well as in aiding respiration. All the muscles of the thorax are primarily involved in respiration. The abdominal wall muscles are different from other muscles that you have studied. They do not go from bone to bone in that they attach from bone to an aponeurosis (fascia) around the rectus abdominis area. They are the external oblique abdominal, internal oblique abdominal, and transversus abdominis.

Chapter
12

TABLE 12.1 • Agonist muscles of the spine

	Muscle	Origin	Insertion	Action	Plane of motion	Palpation	Innervation
Anterior neck	Sternoclei-domastoid (both sides)	Manubrium of sternum, anterior superior surface of medial clavicle	Mastoid process	Extension of head at atlantooccipital joint	Sagittal	Anterolateral neck, diagonally between the origin and insertion, particularly with rotation to contra-lateral side	Spinal accessory (Cr11, C2, C3)
				Flexion of the cervical spine	Sagittal		
				Each side: rotation to contralateral side	Transverse		
				Each side: lateral flexion to ipsilateral side	Frontal		
Posterior neck	Splenius cervicis	Spinous processes of the 3rd through 6th thoracic vertebrae	Transverse processes of the first three cervical vertebrae	Both sides: extension of the cervical spine	Sagittal	Palpate in lower posterior cervical spine just medial to inferior levator scapulae with resisted ipsilateral rotation	Posterior lateral branches of cervical nerves four through eight (C4–C8)
				Each side: rotation to ipsilateral side	Transverse		
				Each side: lateral flexion to ipsilateral side	Frontal		
	Splenius capitis	Lower half of the ligamentum nuchae and the spinous processes of the 7th cervical and upper three or four thoracic vertebrae	Mastoid process and occipital bone	Both sides: extension of the head and cervical spine	Sagittal	Deep to trapezius inferiorly and sternocleidomastoid superiorly; with subject seated, palpate in posterior triangle of neck between upper trapezius and sternocleidomastoid with resisted rotation to ipsilateral side	Posterior lateral branches of cervical nerves four through eight (C4–C8)
				Each side: rotation to ipsilateral side	Transverse		
				Each side: lateral flexion to ipsilateral side	Frontal		
Posterior spine	Erector spinae: Iliocostalis	Medial iliac crest, thoracolumbar aponeurosis from sacrum, posterior ribs 3–12	Posterior ribs 1–12, cervical 4–7 transverse processes	Extension of the spine	Sagittal	Deep and difficult to distinguish from other muscles in the cervical thoracic regions; with subject prone, palpate immediately lateral to spinous processes in lumbar region with active extension	Posterior branches of the spinal nerves
				Anterior pelvic rotation	Sagittal		
				Lateral flexion of the spine	Frontal		
				Lateral pelvic rotation to contralateral side	Frontal		
				Ipsilateral rotation of the spine and head	Transverse		
	Erector spinae: Longissimus	Medial iliac crest, thoracolumbar aponeurosis from sacrum, lumbar 1–5 transverse processes, and thoracic 1–5 transverse processes, cervical 5–7 articular processes	Cervical 2–6 spinous processes, thoracic 1–12 transverse processes, lower 9 ribs, mastoid process	Extension of the spine and head	Sagittal		Posterior branches of the spinal nerves
				Anterior pelvic rotation	Sagittal		
				Lateral flexion of the spine and head	Frontal		
				Lateral pelvic rotation to contralateral side	Frontal		
				Ipsilateral rotation of the spine and head	Transverse		
	Erector spinae: Spinalis	Ligamentum nuchae, 7th cervical spinous process, thoracic 11 and 12 spinous processes, and lumbar 1 and 2 spinous processes	2nd cervical spinous process, thoracic 5–12 spinous processes, occipital bone	Extension of the spine	Sagittal		Posterior branches of the spinal nerves
				Anterior pelvic rotation	Sagittal		
				Lateral flexion of the spine and head	Frontal		
				Lateral pelvic rotation to contralateral side	Frontal		
				Ipsilateral rotation of the spine and head	Transverse		

TABLE 12.1 (continued) • **Agonist muscles of the spine**

	Muscle	Origin	Insertion	Action	Plane of motion	Palpation	Innervation
Lateral lumbar	Quadratus lumborum	Posterior inner lip of the iliac crest	Approximately one-half the length of the lower border of the 12th rib and the transverse process of the upper four lumbar vertebrae	Lateral flexion to the ipsilateral side	Frontal	With subject prone, just superior to iliac crest and lateral to lumbar erector spinae with isometric lateral flexion	Branches of T12, L1 nerves
				Lateral pelvic rotation to contralateral side	Frontal		
				Lumbar extension	Sagittal		
				Anterior pelvic rotation	Sagittal		
				Stabilizes the pelvis and lumbar spine	All planes		
Anterior trunk	Rectus abdominis	Crest of pubis	Cartilage of 5th, 6th, and 7th ribs and xiphoid process	Both sides: lumbar flexion	Sagittal	Anteromedial surface of abdomen, between rib cage and pubic bone with isometric trunk flexion	Intercostal nerves (T7–T12)
				Both sides: posterior pelvic rotation	Sagittal		
				Each side: weak lateral flexion to ipsilateral side	Frontal		
	External oblique abdominal	Borders of lower eight ribs at side of chest dovetailing with serratus anterior	Anterior half of crest of ilium, inguinal ligament, crest of pubis, and fascia of rectus abdominis at lower front	Both sides: lumbar flexion	Sagittal	With subject supine, palpate lateral to the rectus abdominis between iliac crest and lower ribs with active rotation to the contralateral side	Intercostal nerves (T8–T12), iliohypogastric nerve (T12, L1), and ilioinguinal nerve (L1)
				Both sides: posterior pelvic rotation	Sagittal		
				Each side: lumbar lateral flexion to ipsilateral side	Frontal		
				Each side: lateral pelvic rotation to contralateral side	Frontal		
				Each side: lumbar rotation to contralateral side	Transverse		
	Internal oblique abdominal	Upper half of inguinal ligament, anterior 2/3 of crest of ilium, and lumbar fascia	Costal cartilage of 8th, 9th, and 10th ribs and linea alba	Both sides: lumbar flexion	Sagittal	With subject supine, palpate anterolateral abdomen between iliac crest and lower ribs with active rotation to the ipsilateral side	Intercostal nerves (T8–T12), iliohypogastric nerve (T12, L1), and ilioinguinal nerve (L1)
				Both sides: posterior pelvic rotation	Sagittal		
				Each side: lumbar lateral flexion to ipsilateral side	Frontal		
				Each side: lateral pelvic rotation to contralateral side	Frontal		
				Each side: lumbar rotation to ipsilateral side	Transverse		
	Transversus abdominis	Lateral 1/3 of inguinal ligament, inner rim of iliac crest, inner surface of costal cartilages of lower six ribs, lumbar fascia	Crest of the pubis and the iliopectineal line, abdominal aponeurosis to the linea alba	Forced expiration by pulling the abdominal wall inward	Transverse	With subject supine, anterolateral abdomen between iliac crest and lower ribs during forceful exhalation; very difficult to distinguish from abdominal obliques	Intercostal nerves (T7–T12), iliohypogastric nerve (T12, L1), and ilioinguinal nerve (L1)

Chapter

12

Muscles that move the head

Anterior
> Rectus capitis anterior
> Longus capitis

Posterior
> Longissimus capitis
> Obliquus capitis superior
> Obliquus capitis inferior
> Rectus capitis posterior—major and minor
> Trapezius, superior fibers
> Splenius capitis
> Semispinalis capitis

Lateral
> Rectus capitis lateralis
> Sternocleidomastoid

Muscles of the vertebral column

Superficial
> Erector spinae (sacrospinalis)
>> Spinalis—capitis, cervicis, thoracis
>> Longissimus—capitis, cervicis, thoracis
>> Iliocostalis—cervicis, thoracis, lumborum
> Splenius cervicis
> Quadratus lumborum

Deep
> Longus colli—superior oblique, inferior oblique, vertical
> Interspinales—entire spinal column
> Intertransversales—entire spinal column
> Multifidus—entire spinal column
> Psoas minor
> Rotatores—entire spinal column
> Semispinalis—cervicis, thoracis
> Psoas major and minor (see Chapter 9)

Muscles of the thorax

Diaphragm
Intercostalis—external, internal
Levator costarum
Subcostales
Scalenus—anterior, medius, posterior
Serratus posterior—superior, inferior
Transversus thoracis

Muscles of the abdominal wall

Rectus abdominis
External oblique abdominal (obliquus externus abdominis)
Internal oblique abdominal (obliquus internus abdominis)
Transverse abdominis (transversus abdominis)

Core Training

In recent years, a great deal of attention has focused on core training aimed specifically at muscles in and around the abdominal area. Discussions in Chapters 4 and 5 pointed out the importance of the scapula muscles in providing dynamic stability for proper upper extremity function. The same importance and concepts apply to the abdominal core for total body functioning. All too often, there is significant focus on major muscle groups that are strengthened through typical exercises like bench press and squats, with minimal attention given to the link between the upper and lower body—the low back and abdominal core. It is beyond the scope of this text to provide a detailed description of all exercises that may be used to address this area, but it is important to consider core training in designing programs to improve performance and prevent injury.

In addressing the core, some attention is given to the inner core as well as the outer core. The inner core consists of deeper muscles that should be activated as the first step in stabilizing the trunk and pelvis. These muscles consist of the diaphragm, transversus abdominis, lumbar multifidus, and the muscles of the pelvic floor—those that attach to the bony ring of the pelvis. Activating these muscles requires a level of focus and concentration. The outer core consists of the rectus abdominis, external obliques, internal obliques, and erector spinae. These muscles are exercised in a variety of means—including, but not limited to, sit-ups, V-sit-ups, crunches, curl-ups, abdominal twists, prone extensions, superman exercises, and so on.

Nerves

Cranial nerve 11 and the spinal nerves of C2 and C3 innervate the sternocleidomastoid. The posterior lateral branches of C4 through C8 innervate the splenius muscles. The entire erector spinae group is supplied by the posterior branches of the spinal nerves, whereas the intercostal nerves of T7 through T12 innervate the rectus abdominis. Both the internal and external oblique abdominal muscles receive innervation from the intercostal nerves (T8–T12), the iliohypogastric nerve (T12, L1), and the ilioinguinal nerve (L1). The same innervation is provided to the transverse abdominis, except that innervation begins with the T7 intercostal nerve. Branches from T12 and L1 supply the quadratus lumborum. Review Figures 2.6, 4.8, 4.9, and 9.19.

Muscles that move the head

All muscles featured here originate on the cervical vertebrae and insert on the occipital bone of the skull, as implied by their "capitis" name (Figs. 12.9 and 12.10; Table 12.2). Three muscles make up

FIG. 12.9 • Anterior muscles of the neck.

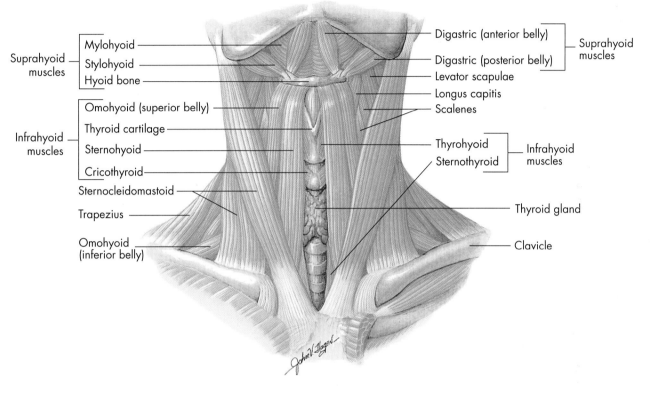

Suprahyoid muscles
- Mylohyoid
- Stylohyoid
- Hyoid bone

Infrahyoid muscles
- Omohyoid (superior belly)
- Thyroid cartilage
- Sternohyoid
- Cricothyroid

Sternocleidomastoid

Trapezius

Omohyoid (inferior belly)

Digastric (anterior belly)
Digastric (posterior belly)
Suprahyoid muscles

Levator scapulae
Longus capitis
Scalenes

Thyrohyoid
Sternothyroid
Infrahyoid muscles

Thyroid gland

Clavicle

Cervical flexion

Cervical extension

Cervical lateral flexion

Cervical rotation unilaterally

FIG. 12.10 • Deep muscles of the posterior neck and upper back regions.

Semispinalis capitis

Splenius capitis

Levator scapulae

Splenius cervicis

Serratus posterior superior

Rhomboideus minor (cut)

Rhomboideus major (cut)

Rectus capitis posterior minor
Rectus capitis posterior major
Obliquus capitis superior
Obliquus capitis inferior
Longissimus capitis
Splenius cervicis (cut)
Levator scapulae (cut)
Scalenus medius
Scalenus posterior

Longissimus cervicis
Iliocostalis cervicis
Longissimus thoracis

Chapter 12

TABLE 12.2 • Muscles that move the head

Muscle	Origin	Insertion	Action	Innervation
Rectus capitis anterior	Anterior surface of lateral mass of atlas	Basilar part of occipital bone anterior to foramen magnum	Flexion of head and stabilization of atlantooccipital joint	C1–C3
Rectus capitis lateralis	Superior surface of transverse processes of atlas	Jugular process of occipital bone	Lateral flexion of head and stabilization of atlantooccipital joint	C1–C3
Rectus capitis posterior (major)	Spinous process of axis	Lateral portion of inferior nuchal line of occipital bone	Extension and rotation of head to ipsilateral side	Posterior rami of C1
Rectus capitis posterior (minor)	Posterior tubercle of posterior arch of atlas	Medial portion of inferior nuchal line of occipital bone	Extension of head	Posterior rami of C1
Longus capitis	Transverse processes of C3–C6	Basilar part of occipital bone	Flexion of head and cervical spine	C1–C3
Obliquus capitis superior	Transverse process of atlas	Occipital bone between inferior and superior nuchal lines	Extension and lateral flexion of head	Posterior rami of C1
Obliquus capitis inferior	Spinous process of axis	Transverse process of atlas	Rotation of atlas	Posterior rami of C1
Semispinalis capitis	Transverse processes of C4–T7	Occipital bone, between superior and inferior nuchal lines	Extension and contralateral rotation of head	Posterior primary divisions on spinal nerves

the anterior vertebral muscles—the longus capitis, the rectus capitis anterior, and the rectus capitis lateralis. All are flexors of the head and upper cervical spine. The rectus capitis lateralis laterally flexes the head, in addition to assisting the rectus capitis anterior in stabilizing the atlantooccipital joint.

The rectus capitis posterior major and minor, obliquus capitis superior and inferior, and semispinalis capitis are located posteriorly. All are extensors of the head, except for the obliquus capitis inferior, which rotates the atlas. The obliquus capitis superior assists the rectus capitis lateralis in lateral flexion of the head. In addition to extension, the rectus capitis posterior major is responsible for rotation of the head to the ipsilateral side. It is assisted by the semispinalis capitis, which rotates the head to the contralateral side. The splenius capitis and the sternocleidomastoid (see Table 12.1) are much larger and more powerful in moving the head and cervical spine; they are covered in detail on the following pages. The remaining muscles that act on the cervical spine are addressed with the muscles of the vertebral column.

Sternocleidomastoid muscle   FIG. 12.11

(ster´no-kli-do-mas´toyd)

Origin

Manubrium of the sternum
Anterior superior surface of the medial clavicle

Insertion

Mastoid process

Action

Extension of the head at the atlantooccipital joint
Flexion of the cervical spine
Right side: rotation to the left and lateral flexion to
the right
Left side: rotation to the right and lateral flexion to
the left

Palpation

Anterolateral side of the neck, diagonally between
the origin and insertion, particularly with rotation
to contralateral side

Innervation

Spinal accessory nerve (Cr11, C2, C3)

Application, strengthening, and flexibility

Cervical
flexion

Cervical
lateral
flexion

Cervical
rotation
unilaterally

The sternocleidomastoid is primarily responsible
for flexion and rotation of the head and neck.
One side of this muscle may be easily visualized
and palpated when the head is rotated to the
opposite side.

The sternocleidomastoid is easily worked for
strength development by placing the hands on
the forehead to apply force posteriorly while
using these muscles to pull the head forward into
flexion. The hand may also be used on one side
of the jaw to apply rotary force in the opposite
direction while the sternocleidomastoid is contracting concentrically to rotate the head in the
direction of the hand.

Cervical hyperextension in combination with
capital flexion provides some bilateral stretching of the sternocleidomastoid. Each side may be
stretched individually. The right side is stretched
by moving into left lateral flexion and right cervical rotation combined with extension. The opposite movements in extension stretch the left side.

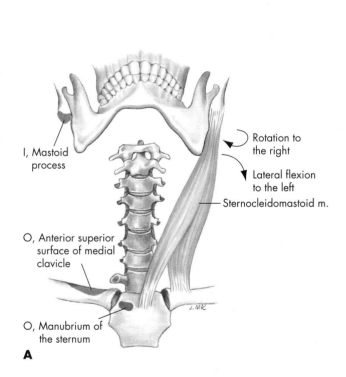

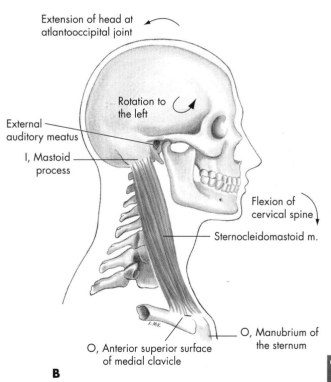

FIG. 12.11 • Sternocleidomastoid muscle. **A,** Anterior view; **B,** Lateral view. O, Origin; I, Insertion.

Chapter
12

Cervical
extension

Cervical
rotation
unilaterally

Cervical
lateral
flexion

Splenius muscles (cervicis, capitis)

FIG. 12.12

(sple´ni-us) (ser´vi-sis) (kap´i-tis)

Origin

Splenius cervicis: spinous processes of the third through sixth thoracic vertebrae

Splenius capitis: lower half of the ligamentum nuchae and spinous processes of the seventh cervical and upper three or four thoracic vertebrae

Insertion

Splenius cervicis: transverse processes of the first three cervical vertebrae

Splenius capitis: mastoid process and occipital bone

Action

Both sides: extension of the head (splenius capitis) and neck (splenius cervicis and capitis)

Right side: rotation and lateral flexion to the right

Left side: rotation and lateral flexion to the left

Palpation

Splenius cervicis: palpate in lower posterior cervical spine just medial to inferior levator scapulae with resisted ipsilateral rotation

Splenius capitis: deep to the trapezius inferiorly and the sternocleidomastoid superiorly; with subject seated, palpate in the posterior triangle of the neck between the upper trapezius and the sternocleidomastoid with resisted rotation to ipsilateral side

Innervation

Posterior lateral branches of cervical nerves four through eight (C4–C8)

Application, strengthening, and flexibility

Any movement of the head and neck into extension, particularly extension and rotation, would bring the splenius muscle strongly into play, together with the erector spinae and the upper trapezius muscles. Tone in the splenius muscle tends to hold the head and neck in proper posture position.

A good exercise for the splenius muscle is to lace the fingers behind the head with it in flexion and then slowly contract the posterior head and neck muscles to move the head and neck into full extension. This exercise may also be performed by using a towel or a partner for resistance.

The entire splenius may be stretched with maximal flexion of the head and cervical spine. The right side can be stretched through combined movements of left rotation, left lateral flexion, and flexion. The same movements to the right apply stretch to the left side.

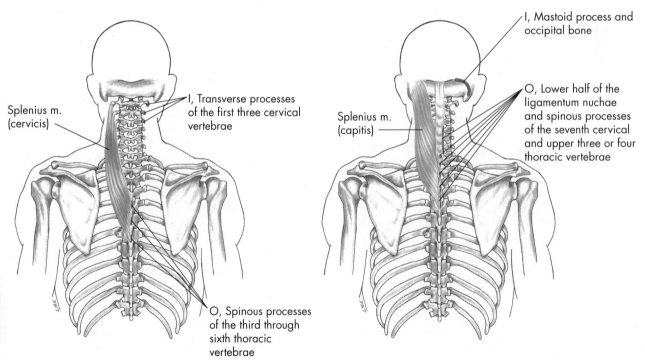

FIG. 12.12 • Splenius muscles (cervicis on the left, capitis on the right), posterior view. O, Origin; I, Insertion.

Chapter
12

Muscles of the vertebral column

In the cervical area, the longus colli muscles are located anteriorly and flex the cervical and upper thoracic vertebrae. Posteriorly, the erector spinae group, the transversospinalis group, the interspinal-intertransverse group, and the splenius all run vertically parallel to the spinal column (Figs. 12.13 and 12.14; Table 12.3). This location enables them to extend the spine as well as assist in rotation and lateral flexion. The splenius and erector spinae group are addressed in detail elsewhere in this chapter. The transversospinalis group consists of the semispinalis, multifidus, and rotatores muscles. These muscles all originate on the transverse processes of their respective vertebrae and generally run posteriorly to attach to the spinous processes on the vertebrae just above their vertebrae of origin. All are extensors of the spine and contract to rotate their respective vertebrae to the contralateral side. The interspinal-intertransverse group lies deep to the rotatores and consists of the interspinalis and the intertransversarii muscles. As a group, they laterally flex and extend but do not rotate the vertebrae. The interspinalis muscles are extensors that connect from the spinous process of one vertebra to the spinous process of the adjacent vertebra. The intertransversarii muscles flex the vertebral column laterally by connecting to the transverse processes of adjacent vertebrae.

Cervical flexion

Cervical extension

Cervical lateral flexion

Cervical rotation unilaterally

Lumbar extension

Lumbar lateral flexion

Lumbar rotation unilaterally

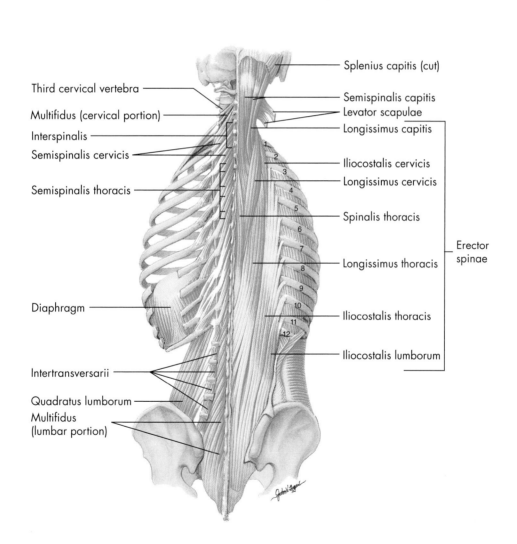

FIG. 12.13 • Deep back muscles, posterior view. **Right,** The erector spinae group of muscles is demonstrated. **Left,** Those muscles have been removed to reveal the deeper back muscles.

Chapter
12

TABLE 12.3 • Muscles of the vertebral column

Muscle	Origin	Insertion	Action	Innervation
Longus colli (superior oblique)	Transverse processes of C3–C5	Anterior arch of atlas	Flexion of cervical spine	C2–C7
Longus colli (inferior oblique)	Bodies of T1–T3	Transverse processes of C5 and C6	Flexion of cervical spine	C2–C7
Longus colli (vertical)	Bodies of C5–C7 and T1–T3	Anterior surface of bodies of C2–C4	Flexion of cervical spine	C2–C7
Interspinalis	Spinous process of each vertebra	Spinous process of next vertebra	Extension of spinal column	Posterior primary ramus of spinal nerves
Intertransversarii	Tubercles of transverse processes of each vertebra	Tubercles of transverse processes of next vertebra	Lateral flexion of spinal column	Anterior primary ramus of spinal nerves
Multifidus	Sacrum, iliac spine, transverse processes of lumbar, thoracic, and lower four cervical vertebrae	Spinous process of 2nd, 3rd, or 4th vertebra above origin	Extension and contralateral rotation of spinal column	Posterior primary ramus of spinal nerves
Rotatores	Transverse processes of each vertebra	Base of spinous process of next vertebra above	Extension and contralateral rotation of spinal column	Posterior primary ramus of spinal nerves
Semispinalis cervicis	Transverse processes of T1–T5 or T6	Spinous processes of C2–C5	Extension and contralateral rotation of vertebral column	All divisions, posterior primary ramus of spinal nerves
Semispinalis thoracis	Transverse processes of T6–T10	Spinous processes of C6, C7, and T1–T4	Extension and contralateral rotation of vertebral column	Posterior primary ramus of spinal nerves

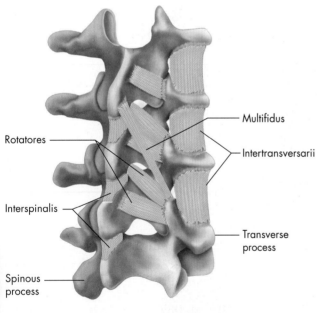

Right posterolateral view

FIG. 12.14 • Deep muscles associated with the vertebrae posterolateral view from the right.

Muscles of the thorax

The thoracic muscles are involved almost entirely in respiration (Fig. 12.15). During quiet rest, the diaphragm is responsible for breathing movements. As it contracts and flattens, the thoracic volume is increased and air is inspired to equalize the pressure. When larger amounts of air are needed, such as during exercise, the other thoracic muscles take on a more significant role in inspiration. The scalene muscles elevate the first two ribs to increase the thoracic volume. Further expansion of the chest is accomplished by the external intercostals. Additional muscles of inspiration are the levator cos tarum and the serratus posterior (Fig. 12.16). Forced expiration occurs with contraction of the internal intercostals, transversus thoracis, and subcostales. All these muscles are detailed in Table 12.4.

Chapter **12**

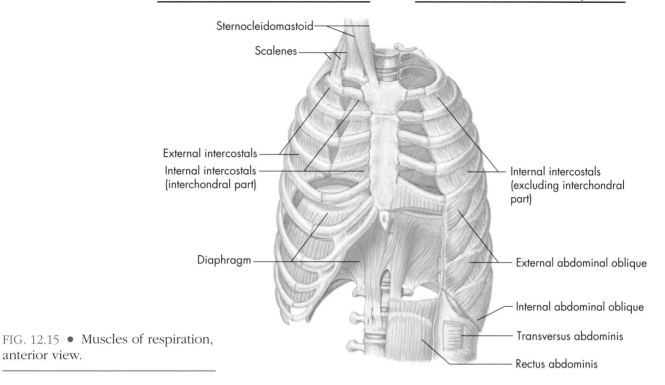

Sternocleidomastoid
Scalenes
External intercostals
Internal intercostals (interchondral part)
Diaphragm

Internal intercostals (excluding interchondral part)
External abdominal oblique
Internal abdominal oblique
Transversus abdominis
Rectus abdominis

FIG. 12.15 • Muscles of respiration, anterior view.

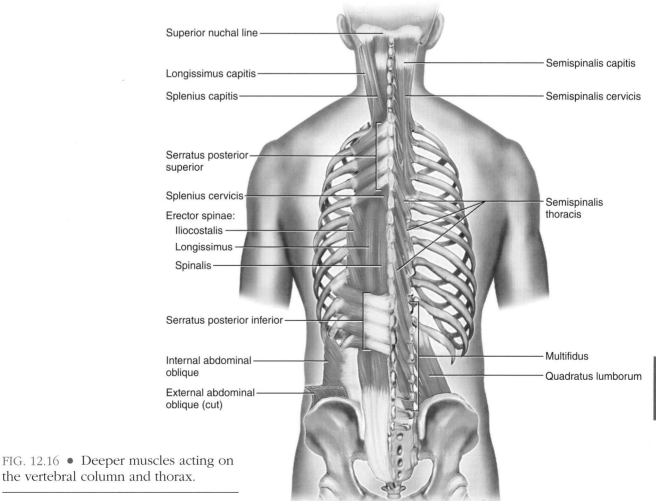

Superior nuchal line
Longissimus capitis
Splenius capitis
Serratus posterior superior
Splenius cervicis
Erector spinae:
  Iliocostalis
  Longissimus
  Spinalis
Serratus posterior inferior
Internal abdominal oblique
External abdominal oblique (cut)

Semispinalis capitis
Semispinalis cervicis
Semispinalis thoracis
Multifidus
Quadratus lumborum

FIG. 12.16 • Deeper muscles acting on the vertebral column and thorax.

Chapter
12

TABLE 12.4 • Muscles of the thorax

Muscle	Origin	Insertion	Action	Innervation
Diaphragm	Circumference of thoracic inlet from xiphoid process, costal cartilages of ribs 6–12, and lumbar vertebrae	Central tendon of diaphragm	Depresses and draws central tendon forward in inhalation, reduces pressure in thoracic cavity, and increases pressure in abdominal cavity	Phrenic nerve (C3–C5)
Internal intercostals	Longitudinal ridge on inner surface of ribs and costal cartilages	Superior border of next rib below	Elevate costal cartilages of ribs 1–4 during inhalation, depress all ribs in exhalation	Intercostal branches of T1–T11
External intercostals	Inferior border of ribs	Superior border of next rib below	Elevate ribs	Intercostal branches of T1–T11
Levator costarum	Ends of transverse processes of C7, T2–T12	Outer surface of angle of next rib below origin	Elevates ribs, lateral flexion of thoracic spine	Intercostal nerves
Subcostales	Inner surface of each rib near its angle	Medially on the inner surface of 2nd or 3rd rib below	Draws the ventral part of the ribs downward, decreasing the volume of the thoracic cavity	Intercostal nerves
Scalenus anterior	Transverse processes of C3–C6	Inner border and upper surface of 1st rib	Elevates 1st rib, flexion, lateral flexion, and contralateral rotation of cervical spine	Ventral rami of C5, C6, sometimes C4
Scalenus medius	Transverse processes of C2–C7	Superior surface of 1st rib	Elevates 1st rib, flexion, lateral flexion, and contralateral rotation of cervical spine	Ventral rami of C3–C8
Scalenus posterior	Transverse processes of C5–C7	Outer surface of 2nd rib	Elevates 2nd rib, flexion, lateral flexion, and slight contralateral rotation of cervical spine	Ventral rami of C6–C8
Serratus posterior (superior)	Ligamentum nuchae, spinous processes of C7, T1, and T2 or T3	Superior borders lateral to angles of ribs 2–5	Elevates upper ribs	Branches from anterior primary rami of T1–T4
Serratus posterior (inferior)	Spinous processes of T10–T12 and L1–L3	Inferior borders lateral to angles of ribs 9–12	Counteracts inward pull of diaphragm by drawing last 4 ribs outward and downward	Branches from anterior primary rami of T9–T12
Transversus thoracis	Inner surface of sternum and xiphoid process, sternal ends of costal cartilages of ribs 3–6	Inner surfaces and inferior borders of costal cartilages 3–6	Depresses ribs	Intercostal branches of T3–T6

Erector spinae muscles*
(sacrospinalis)   FIGS. 12.16, 12.17, 12.18

*Know as group

(e-rek´tor spi´ne) (sa´kro-spi-na´lis)

Iliocostalis

(il´i-o-kos-ta´lis): lateral layer

Longissimus

(lon-jis´i-mus): middle layer

Spinalis

(spi-na´lis): medial layer

---

*This muscle group includes the iliocostalis, the longissimus dorsi, the spinalis dorsi, and divisions of these muscles in the lumbar, thoracic, and cervical sections of the spinal column.

Origin

Iliocostalis: medial iliac crest, thoracolumbar aponeurosis from sacrum, posterior ribs 3–12
Longissimus: medial iliac crest, thoracolumbar aponeurosis from sacrum, lumbar 1–5 transverse processes and thoracic 1–5 transverse processes, cervical 5–7 articular processes
Spinalis: ligamentum nuchae, seventh cervical spinous process, thoracic 11 and 12 spinous processes, and lumbar 1 and 2 spinous processes

Insertion

Iliocostalis: posterior ribs 1–12, cervical 4–7 transverse processes Longissimus: cervical 2–6 spinous processes, thoracic

1–12 transverse processes, lower nine ribs, mastoid process Spinalis: second cervical spinous process, thoracic

5–12 spinous processes, occipital bone

Lumbar extension

Cervical extension

Lumbar lateral flexion

Cervical lateral flexion

Lumbar rotation unilaterally

Cervical rotation unilaterally

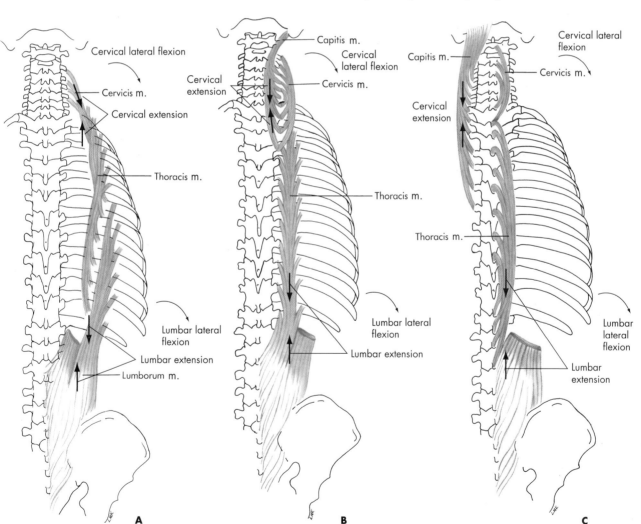

FIG. 12.17 • Erector spinae (sacrospinalis) muscle, posterior view. **A,** Iliocostalis lumborum, thoracis, and cervicis; **B,** Longissimus thoracis, cervicis, and capitis; **C,** Spinalis thoracis, cervicis, and capitis.

Chapter
12

Action

Extension, lateral flexion, and ipsilateral rotation of the spine and head

Anterior pelvic rotation

Lateral pelvic rotation to contralateral side

Palpation

Deep and difficult to distinguish from other muscles in the cervical and thoracic regions; with subject prone, palpate immediately lateral to spinous processes in lumbar region with active extension

Innervation

Posterior branches of the spinal nerves

Application, strengthening, and flexibility

The erector spinae muscles function best when the pelvis is posteriorly rotated. This lowers the origin of the erector spinae and makes it more effective in keeping the spine straight. As the spine is held straight, the ribs are raised, thus fixing the chest high and consequently making the abdominal muscles more effective in holding the pelvis up in front and flattening the abdominal wall.

An exercise known as the dead lift, employing a barbell, uses the erector spinae in extending the spine. In this exercise, the subject bends over, keeping the arms and legs straight; picks up the barbell; and returns to a standing position. In performing this type of exercise, it is very important to always use correct technique to avoid back injuries. Voluntary static contraction of the erector spinae in the standing position can provide a mild exercise and improve body posture.

The erector spinae and the various divisions may be strengthened through numerous forms of back extension exercises. These are usually done in a prone or face-down position in which the spine is already in some state of flexion. The subject then uses these muscles to move part or all of the spine toward extension against gravity. A weight may be held in the hands behind the head to increase resistance.

Maximal hyperflexion of the entire spine stretches the erector spinae muscle group. Stretch may be isolated to specific segments through specific movements. Maximal flexion of the head and cervical spine stretches the capitis and cervicis segments. Flexion combined with lateral flexion to one side accentuates the stretch on the contralateral side. Thoracic and lumbar flexion places the stretch primarily on the thoracis and lumborum segments.

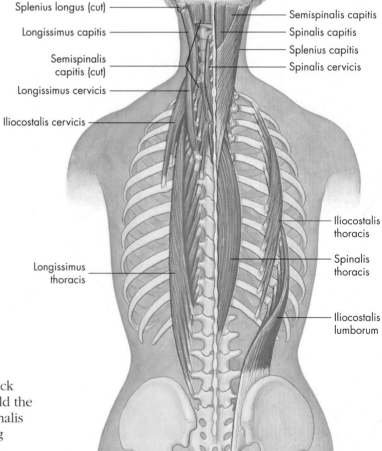

FIG. 12.18 • Muscles of the back and the neck help move the head (posterior view) and hold the torso erect. The splenius capitis and semispinalis have been cut on the left to show underlying muscles.

Chapter
12

Quadratus lumborum muscle   FIG. 12.19

(kwad-ra´tus lum-bo´rum)

Origin

Posterior inner lip of the iliac crest

Insertion

Approximately one-half the length of the lower border of the twelfth rib and the transverse process of the upper four lumbar vertebrae

Action

Lateral flexion to the ipsilateral side
Stabilizes the pelvis and lumbar spine
Extension of the lumbar spine
Anterior pelvic rotation
Lateral pelvic rotation to contralateral side

Palpation

With subject prone, just superior to iliac crest and lateral to lumbar erector spinae with isometric lateral flexion

Innervation

Branches of T12, L1 nerves

Application, strengthening, and flexibility

The quadratus lumborum is important in lumbar lateral flexion and in elevating the pelvis on the same side in the standing position. Trunk rotation and lateral flexion movements against resistance are good exercises for development of this muscle. The position of the body relative to gravity may be changed to increase resistance on this and other trunk and abdominal muscles. Left lumbar lateral flexion while in lumbar flexion stretches the right quadratus lumborum, and vice versa.

Lumbar
lateral
flexion

Lumbar
extension

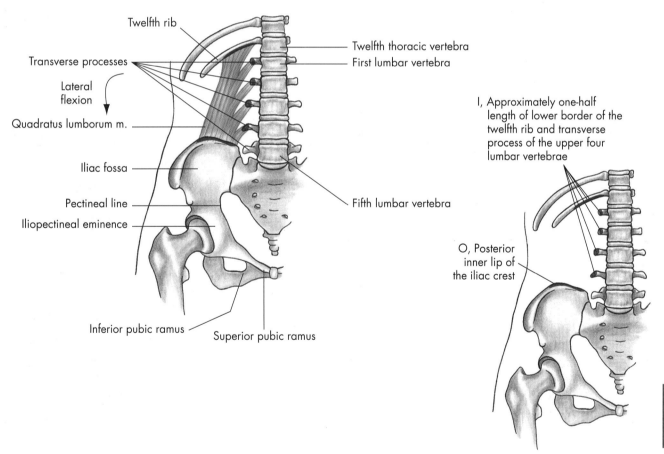

FIG. 12.19 • Quadratus lumborum muscle. O, Origin; I, Insertion.

Chapter
12

Muscles of the abdominal wall

FIGS. 12.20, 12.21, 12.22

Lumbar
flexion

Lumbar
lateral
flexion

Lumbar
rotation
unilaterally

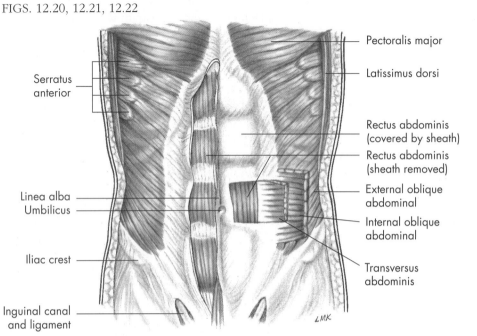

Pectoralis major

Latissimus dorsi

Rectus abdominis
(covered by sheath)

Rectus abdominis
(sheath removed)

External oblique
abdominal

Internal oblique
abdominal

Transversus
abdominis

Serratus
anterior

Linea alba
Umbilicus

Iliac crest

Inguinal canal
and ligament

FIG. 12.20 • Muscles of
the abdomen: external
oblique and rectus
abdominis. The fibrous
sheath around the rectus
has been removed on the
right side to show the
muscle within.

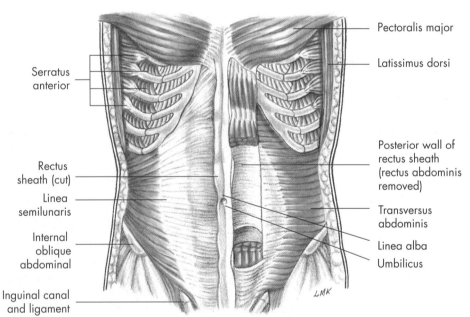

Pectoralis major

Latissimus dorsi

Posterior wall of
rectus sheath
(rectus abdominis
removed)

Transversus
abdominis

Linea alba

Umbilicus

Serratus
anterior

Rectus
sheath (cut)

Linea
semilunaris

Internal
oblique
abdominal

Inguinal canal
and ligament

FIG. 12.21 • Muscles of
the abdomen. The external
oblique has been removed
on the right side to reveal
the internal oblique. The
external and internal
obliques have been
removed on the left side
to reveal the transversus
abdominis. The rectus
abdominis has been cut to
reveal the posterior rectus
sheath.

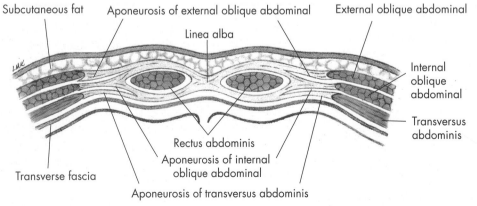

Subcutaneous fat

Aponeurosis of external oblique abdominal

Linea alba

External oblique abdominal

Internal
oblique
abdominal

Transversus
abdominis

Rectus abdominis

Aponeurosis of internal
oblique abdominal

Transverse fascia

Aponeurosis of transversus abdominis

FIG. 12.22 • Abdominal
wall above umbilicus.
The unique arrangement
of the four abdominal
muscles with their fascial
attachment in and around
the rectus abdominis
muscle is shown. With no
bones for attachments,
these muscles can be
adequately maintained
through exercise.

Rectus abdominis muscle FIG. 12.23

(rek´tus ab-dom´i-nis)

Origin

Crest of the pubis

Insertion

Cartilage of the fifth, sixth, and seventh ribs and the xiphoid process

Action

Both sides: lumbar flexion
Posterior pelvic rotation
Right side: weak lateral flexion to the right
Left side: weak lateral flexion to the left

Palpation

Anteromedial surface of the abdomen, between the rib cage and the pubic bone with isometric trunk flexion

Innervation

Intercostal nerves (T7–T12)

Application, strengthening, and flexibility

The rectus abdominis muscle controls the tilt of the pelvis and the consequent curvature of the lower spine. By rotating the pelvis posteriorly, the rectus abdominis flattens the lower back, making the erector spinae muscle more effective as an extensor of the spine and the hip flexors (the iliopsoas muscle, particularly) more effective in raising the legs.

In a relatively lean person with well-developed abdominals, three distinct sets of lines or depressions may be noted. Each represents an area of tendinous connective tissue connecting or supporting the abdominal arrangement of muscles in lieu of bony attachments. Running vertically from the xiphoid process through the umbilicus to the pubis is the **linea alba**. It divides each rectus abdominis and serves as its medial border. Lateral to each rectus abdominis is the **linea semilunaris**, a crescent, or moon-shaped, line running vertically. This line represents the aponeurosis connecting the lateral border of the rectus abdominis and the medial border of the external and internal abdominal obliques. The **tendinous inscriptions** are horizontal indentations that transect the rectus abdominis at three or more locations, giving the muscle its segmented appearance. Refer to Fig. 12.20.

There are several exercises for the abdominal muscles, such as bent-knee sit-ups, crunches, and isometric contractions. Bent-knee sit-ups with the arms folded across the chest are considered by many to be a safe and efficient exercise. Crunches are considered to be even more effective for isolating the work to the abdominals. Both of these exercises shorten the iliopsoas muscle and other hip flexors, thereby reducing their ability to generate force. Twisting to the left and right brings the oblique muscles into more active contraction. In all the above exercises, it is important to use proper technique, which involves gradually moving to the up position until the lumbar spine is actively flexed maximally and then slowly returning to the beginning position. Jerking movements using momentum should be avoided. Movement continued beyond full lumbar flexion exercises only the hip flexors, which is not usually an objective. Even though all these exercises may be helpful in strengthening the abdominals, careful analysis should occur before deciding which are indicated in the presence of various injuries and problems of the lower back.

The rectus abdominis is stretched by simultaneously hyperextending both the lumbar and the thoracic spine. Extending the hips assists in this process by accentuating the anterior rotation of the pelvis to hyperextend the lumbar spine.

Lumbar
flexion

Lumbar
lateral
flexion

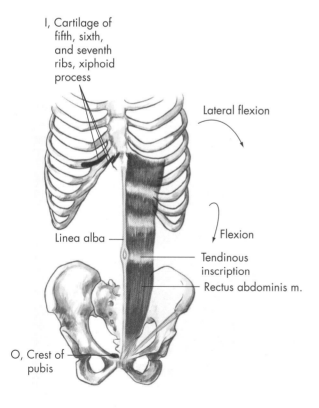

I, Cartilage of fifth, sixth, and seventh ribs, xiphoid process

Lateral flexion

Linea alba

Flexion

Tendinous inscription

Rectus abdominis m.

O, Crest of pubis

FIG. 12.23 • Rectus abdominis muscle, anterior view. O, Origin; I, Insertion.

Chapter
12

External oblique abdominal muscle   FIG. 12.24

(ek-stur´nel o-bleek´ ab-dom´i-nel)

Origin

Borders of the lower eight ribs at the side of the chest, dovetailing with the serratus anterior muscle*

Insertion

Anterior half of the crest of the ilium, the inguinal ligament, the crest of the pubis, and the fascia of the rectus abdominis muscle at the lower front

Action

Both sides: lumbar flexion

Posterior pelvic rotation

Right side: lumbar lateral flexion to the right and rotation to the left, lateral pelvic rotation to the left

Left side: lumbar lateral flexion to the left and rotation to the right, lateral pelvic rotation to the right

---

*Sometimes the origin and insertion are reversed in anatomy books. This is the result of different interpretations of which bony structure is the more movable. The insertion is considered the most movable part of a muscle.

Palpation

With subject supine, palpate lateral to the rectus abdominis between iliac crest and lower ribs with active rotation to the contralateral side

Innervation

Intercostal nerves (T8–T12), iliohypogastric nerve (T12, L1), and ilioinguinal nerve (L1)

Application, strengthening, and flexibility

Working on each side of the abdomen, the external oblique abdominal muscles aid in rotating the trunk when working independently of each other. Working together, they aid the rectus abdominis muscle in its described action. The left external oblique abdominal muscle contracts strongly during sit-ups when the trunk rotates to the right, as in touching the left elbow to the right knee. Rotating to the left brings the right external oblique into action.

Each side of the external oblique must be stretched individually. The right side is stretched by moving into extreme left lateral flexion combined with extension, or by extreme lumbar rotation to the right combined with extension. The opposite movements combined with extension stretch the left side.

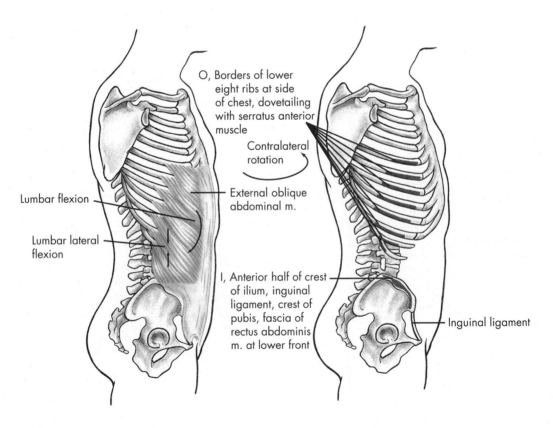

O, Borders of lower eight ribs at side of chest, dovetailing with serratus anterior muscle

Contralateral rotation

Lumbar flexion

External oblique abdominal m.

Lumbar lateral flexion

I, Anterior half of crest of ilium, inguinal ligament, crest of pubis, fascia of rectus abdominis m. at lower front

Inguinal ligament

FIG. 12.24 • External oblique abdominal muscle, lateral view. O, Origin; I, Insertion.

Internal oblique abdominal muscle

FIG. 12.25

(in-ter´nel o-bleek´ ab-dom´i-nel)

Origin

Upper half of the inguinal ligament, anterior two-thirds of the crest of the ilium, and lumbar fascia

Insertion

Costal cartilages of the eighth, ninth, and tenth ribs and the linea alba

Action

Both sides: lumbar flexion
Posterior pelvic rotation
Right side: lumbar lateral flexion to the right and rotation to the right, lateral pelvic rotation to the left
Left side: lumbar lateral flexion to the left and rotation to the left, lateral pelvic rotation to the right

Palpation

With subject supine, palpate anterolateral abdomen between iliac crest and lower ribs with active rotation to the ipsilateral side

Innervation

Intercostal nerves (T8–T12), iliohypogastric nerve (T12, L1), and ilioinguinal nerve (L1)

Application, strengthening, and flexibility

The internal oblique abdominal muscles run diagonally in the direction opposite that of the external obliques. The left internal oblique rotates to the left, and the right internal oblique rotates to the right.

In touching the left elbow to the right knee in crunches, the left external oblique and the right internal oblique abdominal muscles contract at the same time, assisting the rectus abdominis muscle in flexing the trunk to make completion of the movement possible. In rotary movements, the internal oblique and the external oblique on the opposite side always work together.

Like the external oblique, each side of the internal oblique must be stretched individually. The right side is stretched by moving into extreme left lateral flexion and extreme left lumbar rotation combined with extension. The same movements to the right combined with extension stretch the left side.

Lumbar flexion

Lumbar lateral flexion

Lumbar rotation unilaterally

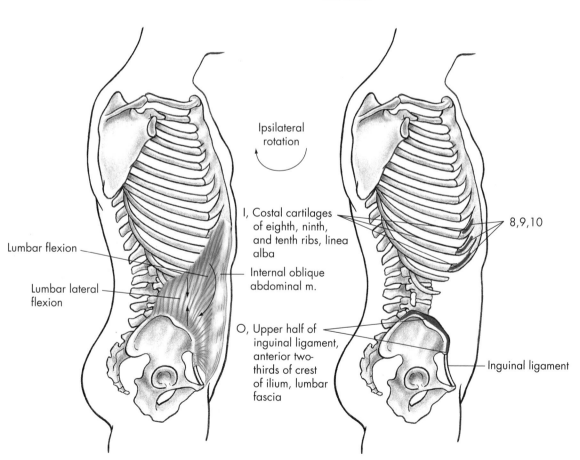

FIG. 12.25 • Internal oblique abdominal muscle, lateral view. O, Origin; I, Insertion.

Chapter
12

Transversus abdominis muscle

FIG. 12.26

(trans-vurs´us ab-dom´i-nis)

Origin

Lateral third of the inguinal ligament, inner rim of the iliac crest, inner surface of the costal cartilages of the lower six ribs, and lumbar fascia

Insertion

Crest of the pubis and the iliopectineal line
Abdominal aponeurosis to the linea alba

Action

Forced expiration by pulling the abdominal wall inward

Palpation

With subject supine, palpate anterolateral abdomen between iliac crest and lower ribs during forceful exhalation; very difficult to distinguish from the abdominal obliques

Innervation

Intercostal nerves (T7–T12), iliohypogastric nerve (T12, L1), and ilioinguinal nerve (L1)

Application, strengthening, and flexibility

The transversus abdominis is the chief muscle of forced expiration and is effective—together with the rectus abdominis, the external oblique abdominal, and the internal oblique abdominal muscles—in helping hold the abdomen flat. It, along with the other abdominal muscles, is considered by many to be key in providing and maintaining core stability. This abdominal flattening and forced expulsion of the abdominal contents are the only action of this muscle.

The transversus abdominis muscle is exercised effectively by attempting to draw the abdominal contents back toward the spine. This may be done isometrically in the supine position or while standing. A maximal inspiration held in the abdomen applies stretch.

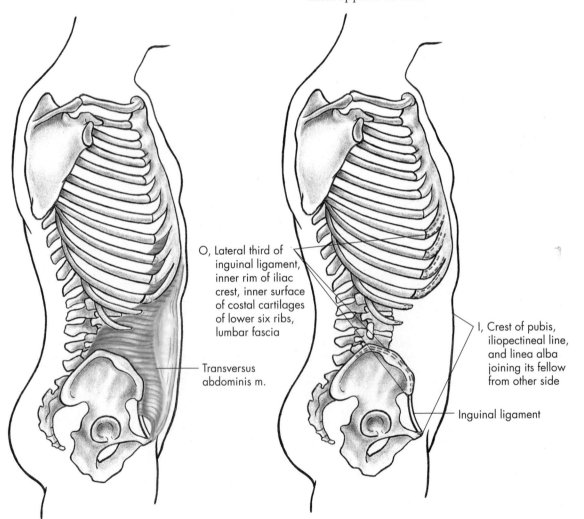

O, Lateral third of inguinal ligament, inner rim of iliac crest, inner surface of costal cartilages of lower six ribs, lumbar fascia

Transversus abdominis m.

I, Crest of pubis, iliopectineal line, and linea alba joining its fellow from other side

Inguinal ligament

FIG. 12.26 • Transversus abdominis muscle, lateral view. O, Origin; I, Insertion.

REVIEW EXERCISES

1. List the planes in which each of the following movements occurs. List the axis of rotation for each movement in each plane.
   a. Cervical flexion
   b. Cervical extension
   c. Cervical rotation
   d. Cervical lateral flexion
   e. Lumbar flexion
   f. Lumbar extension
   g. Lumbar rotation
   h. Lumbar lateral flexion
   i. Capital flexion
   j. Capital extension
2. Why is good abdominal muscular development so important? Why is this area so frequently neglected?
3. Why are weak abdominal muscles frequently blamed for lower back pain?
4. Prepare an oral or a written report on abdominal or back injuries found in the literature.
5. Research common spinal disorders such as brachial plexus neuropraxia, cervical radiculopathy, lumbosacral herniated nucleus pulposus, sciatica, spondylolysis, and spondylolisthesis. Report your findings in class.

6. **Muscle analysis chart** • Cervical and lumbar spine

Complete the chart by listing the muscles primarily involved in each movement.	
Cervical spine	
Flexion	Extension
Lateral flexion right	Rotation right
Lateral flexion left	Rotation left
Lumbar spine	
Flexion	Extension
Lateral flexion right	Rotation right
Lateral flexion left	Rotation left

Chapter
12

7. Antagonistic muscle action chart • Cervical and lumbar spine

Complete the chart by listing the muscle(s) or parts of muscles that are antagonist in their actions to the muscles in the left column.

Agonist	Antagonist
Splenius capitis	
Splenius cervicis	
Sternocleidomastoid	
Erector spinae	
Rectus abdominis	
External oblique abdominal	
Internal oblique abdominal	
Quadratus lumborum	

LABORATORY EXERCISES

1. Locate the following parts of the spine on a human skeleton and on a human subject:
   a. Cervical vertebrae
   b. Thoracic vertebrae
   c. Lumbar vertebrae
   d. Spinous processes
   e. Transverse processes
   f. Sacrum
   g. Manubrium
   h. Xiphoid process
   i. Sternum
   j. Rib cage (various ribs)

2. How and where can the following muscles be palpated on a human subject?
   a. Rectus abdominis
   b. External oblique abdominal
   c. Internal oblique abdominal
   d. Erector spinae
   e. Sternocleidomastoid
   f. Splenius cervicis
   g. Splenius capitis
   h. Quadratus lumborum

3. Contrast crunches, bent-knee sit-ups, and straight-leg sit-ups. Does having a partner to hold the feet make a difference in the ability to do the bent-knee and straight-leg sit-ups? If so, why?

4. Have a laboratory partner stand and assume a position exhibiting good posture. What motions in each region of the spine does gravity attempt to produce? Which muscles are responsible for counteracting these motions against the pull of gravity?

5. Compare and contrast the spinal curves of a laboratory partner sitting erect with those of one sitting slouched in a chair. Which muscles are responsible for maintaining good sitting posture?

6. Which exercise is better for the development of the abdominal muscles—leg lifts or sit-ups? Analyze each exercise with regard to the activity of the abdominal muscles. Defend your answer.

7. Trunk and spine exercise movement analysis chart

After analyzing each exercise in the chart, break each into two primary movement phases, such as a lifting phase and a lowering phase. For each phase, determine what trunk and spine movements occur, and then list the trunk and spine muscles primarily responsible for causing/controlling those movements. Beside each muscle in each movement, indicate the type of contraction as follows: I—isometric; C—concentric; E—eccentric.

Exercise	Initial movement phase		Secondary movement phase	
	Movement(s)	Agonist(s)—(contraction type)	Movement(s)	Agonist(s)—(contraction type)
Push-up				
Squat				
Dead lift				
Sit-up, bent knee				
Prone extension				
Rowing exercise				
Leg raises				
Stair machine				

8. Trunk and spine sport skill analysis chart

Analyze each skill in the chart, and list the movements of the trunk and spine in each phase of the skill. You may prefer to list the initial positions that the trunk and spine are in for the stance phase. After each movement, list the trunk and spine muscle(s) primarily responsible for causing/controlling the movement. Beside each muscle in each movement, indicate the type of contraction as follows: I—isometric; C—concentric; E—eccentric. It may be desirable to review the concepts for analysis in Chapter 8 for the various phases. Assume right hand/ leg dominant where applicable. Circle R or L to indicate the dominant extremity for the exercise, if appropriate.

Exercise		Stance phase	Preparatory phase	Movement phase	Follow-through phase
Baseball pitch	(R)				
	(L)				
Football punt	(R)				
	(L)				
Walking	(R)				
	(L)				

Chapter

12

Trunk and spine sport skill analysis chart (continued)

Exercise		Stance phase	Preparatory phase	Movement phase	Follow-through phase
Softball pitch	(R)				
	(L)				
Soccer pass	(R)				
	(L)				
Batting	(R)				
	(L)				
Bowling	(R)				
	(L)				
Basketball jump shot	(R)				
	(L)				

References

Clarkson HM, Gilewich GB: *Musculoskeletal assessment: joint range of motion and manual muscle strength,* ed 2, Baltimore, 1999, Lippincott Williams & Wilkins.

Day AL: Observation on the treatment of lumbar disc disease in college football players, *American Journal of Sports Medicine* 15:275, January–February 1987.

Field D: *Anatomy: palpation and surface markings,* ed 3, Oxford, 2001, Butterworth-Heinemann.

Gench BE, Hinson MM, Harvey PT: *Anatomical kinesiology,* Dubuque, IA, 1995, Eddie Bowers.

Hamilton N, Weimer W, Luttgens K: *Kinesiology: scientific basis of human motion,* ed 12, New York, 2012, McGraw-Hill.

Hislop HJ, Montgomery J: *Daniels and Worthingham's muscle testing: techniques of manual examination,* ed 8, Philadelphia, 2007, Saunders.

Lindsay DT: *Functional human anatomy,* St. Louis, 1996, Mosby.

Magee DJ: *Orthopedic physical assessment,* ed 5, Philadelphia, 2008, Saunders.

Martens MA, et al: Adductor tendonitis and muscular abdominis tendopathy, *American Journal of Sports Medicine* 15:353, July–August 1987.

Marymont JV: Exercise-related stress reaction of the sacroiliac joint, an unusual cause of low back pain in athletes, *American Journal of Sports Medicine* 14:320, July–August 1986.

Muscolino JE: *The muscular system manual: the skeletal muscles of the human body,* ed 3, St. Louis, 2010, Elsevier Mosby.

National Strength and Conditioning Association; Baechle TR, Earle RW: *Essentials of strength training and conditioning,* ed 2, Champaign, IL, 2000, Human Kinetics.

Oatis CA: *Kinesiology: the mechanics and pathomechanics of human movement,* ed 2, Philadelphia, 2008, Lippincott Williams & Wilkins.

Perry JF, Rohe DA, Garcia AO: *The kinesiology workbook,* Philadelphia, 1992, Davis.

Prentice WE: *Principles of athletic training: a competency-based approach,* ed 15, New York, 2014, McGraw-Hill.

Rasch PJ: *Kinesiology and applied anatomy,* ed 7, Philadelphia, 1989, Lea & Febiger.

Saladin KS: *Anatomy & physiology: the unity of form and function,* ed 5, New York, 2010, McGraw-Hill.

Seeley RR, Stephens TD, Tate P: *Anatomy & physiology,* ed 8, New York, 2008, McGraw-Hill.

Sieg KW, Adams SP: *Illustrated essentials of musculoskeletal anatomy,* ed 4, Gainesville, FL, 2002, Megabooks.

Stone RJ, Stone JA: *Atlas of the skeletal muscles,* ed 6, New York, 2009, McGraw-Hill.

Thibodeau GA, Patton KT: *Anatomy & physiology,* ed 9, St. Louis, 1993, Mosby.

Van De Graaff KM: *Human anatomy,* ed 6, Dubuque, IA, 2002, McGraw-Hill.

 For additional resources and a list of related websites, visit **www.mhhe.com/floyd19e**.

Worksheet Exercises

For in- or out-of-class assignments, or for testing, utilize this tear-out worksheet.

Worksheet 1

Using crayons or colored markers, draw and label on the worksheet the following muscles. Indicate the origin and insertion of each muscle with an "O" and an "I," respectively, and draw in the origin and insertion on the anterior view as applicable.

    a. Rectus abdominis

    b. External oblique abdominal

    c. Internal oblique abdominal

    d. Sternocleidomastoid

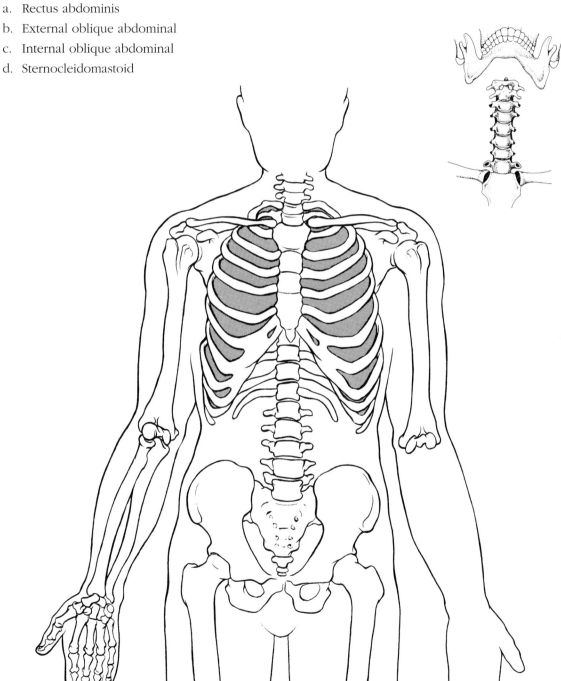

Worksheet Exercises

For in- or out-of-class assignments, or for testing, utilize this tear-out worksheet.

Worksheet 2

Using crayons or colored markers, draw and label on the worksheet the following muscles. Indicate the origin and insertion of each muscle with an "O" and an "I," respectively, and draw in the origin and insertion on the posterior view as applicable.

- a. Erector spinae
- b. Quadratus lumborum
- c. Splenius—cervicis and capitis

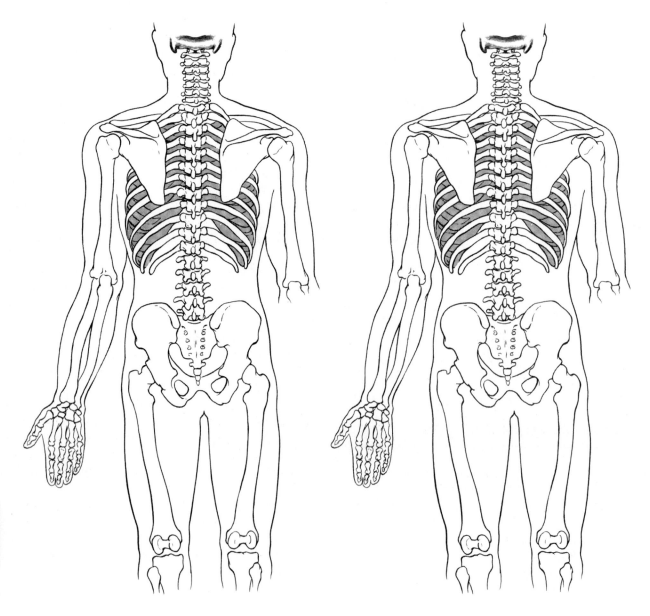

MUSCULAR ANALYSIS OF TRUNK AND LOWER EXTREMITY EXERCISES

Objectives

- To analyze an exercise to determine the joint movements and the types of contractions occurring in the specific muscles involved in those movements

- To learn to group individual muscles into units that produce certain joint movements

- To begin to think of exercises that increase the strength and endurance of individual muscle groups

- To learn to analyze and prescribe exercises to strengthen major muscle groups

- To apply the concept of the kinetic chain to the lower extremity

Online Learning Center Resources

Visit *Manual of Structural Kinesiology*'s **Online Learning Center** at **www.mhhe.com/floyd19e** for additional information and study material for this chapter, including:

- *Self-grading quizzes*
- *Anatomy flashcards*
- *Animations*
- *Related websites*

Chapter 8 presented an introduction to the analysis of exercise and activities. That chapter included only analysis of the muscles previously studied in the upper-extremity region.

Since that chapter, all the other joints and large muscle groups of the human body have been considered. The exercises and activities in this chapter concentrate more on the muscles in the trunk and lower extremity.

Strength, endurance, and flexibility of the muscles of the lower extremity, trunk, and abdominal sections are also very important in skillful physical performance and body maintenance.

The type of contraction is determined by whether the muscle is lengthening or shortening during the movement. However, muscles may shorten or lengthen in the absence of a contraction through passive movement caused by other contracting muscles, momentum, gravity, or external forces such as manual assistance and exercise machines.

A concentric contraction is a shortening contraction of the muscles against gravity or resistance, whereas an eccentric contraction is a contraction in which the muscle lengthens under tension to control the joints moving with gravity or resistance.

Contraction against gravity is also quite evident in the lower extremities. To simply stand still, isometric contractions are utilized in the hip extensors, knee extensors, and plantar flexors to prevent hip flexion, knee flexion, and dorsiflexion, respectively.

Chapter
13

The quadriceps muscle group contracts eccentrically when the body slowly lowers in a weight-bearing movement through lower-extremity action. The quadriceps functions as a decelerator to knee joint flexion in weight-bearing movements by contracting eccentrically to prevent too rapid a downward movement. One can easily demonstrate this fact by palpating this muscle group while slowly moving from a standing position to a half squat. This type of contraction involves almost as much effort as concentric contractions.

In this example involving the quadriceps, the slow descent is eccentric, and the ascent from the squatted position is concentric. If the descent were under no muscular control, it would be at the same speed as gravity, and the muscle lengthening would be passive. That is, the movement and change in length of the muscle would be both caused and controlled by gravity and not by active muscular contractions.

More and more medical and allied health professionals have been emphasizing the development of muscle groups through resistance training and circuit-training activities. Athletes and non-athletes, both male and female, need overall muscular development. Even those who do not necessarily desire significant muscle mass are advised to develop and maintain their muscle mass through resistance training. As we age, we normally tend to lose muscle mass, and as a result our metabolism decreases. This factor, combined with improper eating habits, results in unhealthful fat accumulation and excessive weight gain. Through increasing our muscle mass, we burn more calories and are less likely to gain excessive fat.

Sports participation does not ensure sufficient development of muscle groups. Also, more and more emphasis has been placed on mechanical kinesiology in physical education and athletic skill teaching. This is desirable and can help bring about more skillful performance. However, it is important to remember that mechanical principles will be of little or no value to performers without adequate strength and endurance of the muscular system, which is developed through planned exercises and activities. In the fitness and health revolution of recent decades, a much greater emphasis has been placed on exercises and activities that improve the physical fitness, strength, endurance, and flexibility of participants. This chapter continues the practice of analyzing the muscles through simple exercises, the approach begun in Chapter 8. When these techniques are practiced extensively and mastered, the individual is ready to analyze and prescribe exercises and activities for the muscular strength and endurance needed in sports activities and for healthful living.

To further assist in analyzing the muscles primarily involved in exercises, review the "Concepts for Analysis" section in Chapter 8. It would also be beneficial to utilize Appendix 5 and determine the muscles involved in the different phases using the Skill Analysis Worksheet found in the Worksheet Exercises at the end of this chapter. The worksheets together provide for analysis of up to six different phases.

Chapter

13

Abdominal curl-up   FIG. 13.1

Description

The participant lies on the back, forearms crossed and lying across the chest, with the knees flexed approximately 90 degrees and the feet about hip width apart. The hips and knees are flexed in this manner to reduce the length of the hip flexors, thereby reducing their contribution to the curl-up.

The participant performs trunk flexion up to a curl-up position, rotates the trunk to the right and points the left elbow toward the anterior right pelvis (anterior superior iliac spine), and then returns to the starting position. On the next repetition, the participant should rotate to the left instead of to the right for balanced muscular development.

Analysis

This open kinetic chain exercise is divided into four phases for analysis: (1) trunk flexion phase to curl-up position, (2) rotating to right/left phase, (3) return phase to curl-up position, and (4) return phase to starting position (Table 13.1).

FIG. 13.1 • Abdominal curl-up. **A,** Beginning relaxed position; **B,** Trunk flexion to curl-up position; **C,** Trunk flexion and right rotation curl-up position.

TABLE 13.1 • Abdominal curl-up

Joint	Trunk flexion phase to curl-up position — Action	Agonists	Rotating to right/left phase — Action	Agonists	Return phase to curl-up position — Action	Agonists	Return phase to starting position — Action	Agonists
Cervical spine	Flexion	Cervical spine flexors Sternocleido-mastoid	Mainte-nance of cervical flexion	Cervical spine flexors (isometric contraction) Sternocleido-mastoid	Mainte-nance of cervical flexion	Cervical spine flexors (isometric contraction) Sternocleido-mastoid	Extension	Cervical spine flexors (eccentric contraction) Sternocleido-mastoid
Trunk	Flexion	Trunk flexors Rectus abdominis External oblique abdominal Internal oblique abdominal	Right lumbar rotation	Right lumbar rotators (R) Rectus abdominis (L) External oblique abdominal (R) Internal oblique abdominal (R) Erector spinae	Left lumbar rotation to neutral position	Right lumbar rotators (eccentric contraction) (R) Rectus abdominis (L) External oblique abdominal (R) Internal oblique abdominal (R) Erector spinae	Extension	Trunk flexors (eccentric contraction) Rectus abdominis External oblique abdominal Internal oblique abdominal
Hip	Flexion	Hip flexors Iliopsoas Rectus femoris Pectineus	Mainte-nance of hip flexion	Hip flexors (isometric contraction) Iliopsoas Rectus femoris Pectineus	Mainte-nance of hip flexion	Hip flexors (isometric contraction) Iliopsoas Rectus femoris Pectineus	Extension	Hip flexors (eccentric contraction) Iliopsoas Rectus femoris Pectineus

Chapter
13

Alternating prone extensions FIG. 13.2

Description

The participant lies in a prone position, face-down, with the shoulders fully flexed in a relaxed position lying in front of the body. The head, upper trunk, right upper extremity, and left lower extremity are raised from the floor. The knees are kept in full extension. Then the participant returns to the starting position. On the next repetition, the head, upper trunk, left upper extremity, and right lower extremity are raised from the floor.

Analysis

This open kinetic chain exercise is separated into two phases for analysis: (1) lifting phase to raise the right upper extremity off the surface and raise the left lower extremity off the floor and (2) lowering phase to relaxed position (Table 13.2).

FIG. 13.2 • Alternating prone extensions. **A,** Beginning relaxed position; **B,** Raised position.

TABLE 13.2 • Alternating prone extensions (Superman exercise)

| Joint | Lifting phase to raise upper and lower extremities | | Lowering phase to relaxed position | |
	Action	Agonists	Action	Agonists

Shoulder	Flexion	Shoulder joint flexors
		  Pectoralis major (clavicular head or upper fibers)
		  Deltoid
		  Coracobrachialis
		  Biceps brachii
			Extension	Shoulder joint flexors (eccentric contraction)
				  Pectoralis major (clavicular head or upper fibers)
				  Deltoid
				  Coracobrachialis
				  Biceps brachii

Shoulder girdle	Adduction	Shoulder girdle adductors
		  Trapezius
		  Rhomboids
			Abduction	Shoulder girdle adductors (eccentric contraction)
				  Trapezius
				  Rhomboids

Trunk	Extension	Trunk extensors
		  Erector spinae
		  Splenius
		  Quadratus lumborum
			Flexion (return to neutral relaxed position)	Trunk and cervical spine extensors (eccentric contraction)
				  Erector spinae
				  Splenius
				  Quadratus lumborum

Hip	Extension	Hip extensors
		  Gluteus maximus
		  Semitendinosus
		  Semimembranosus
		  Biceps femoris
			Flexion (return to neutral relaxed position)	Hip extensors (eccentric contraction)
				  Gluteus maximus
				  Semitendinosus
				  Semimembranosus
				  Biceps femoris

Squat   FIG. 13.3

Description

The participant places a barbell on the shoulders behind the neck and grasps it with the palms-forward position of the hands. The participant squats down, flexing at the hips while keeping the spine in normal alignment, until the thighs are parallel to the floor. The participant then returns to the starting position. This exercise is commonly performed improperly by allowing the knees to move forward beyond the plane of the feet, which greatly increases the risk of injury. Care should be taken to ensure that the shins remain as vertical as possible during this exercise.

The feet should be parallel, with slight external rotation of the lower extremity. The knees should point over the ankles and feet without going in front of, between, or outside of the vertical plane of the feet.

Analysis

This closed kinetic chain exercise is separated into two phases for analysis: (1) lowering phase to the squatted position and (2) lifting phase to the starting position (Table 13.3). *Note:* It is assumed that no movement will take place in the shoulder joint, shoulder girdle, wrists, hands, or back, although isometric muscle activity is required in these areas to maintain proper positioning.

FIG. 13.3 • Squat. **A,** Starting position; **B,** Squatted position.

TABLE 13.3 • Squat

Joint	Lowering phase to squatted position			Lifting phase to starting position		
	Action	Agonists		Action	Agonists	
Hip	Flexion	Hip extensors (eccentric contraction) Gluteus maximus Semimembranosus Semitendinosus Biceps femoris		Extension	Hip extensors Gluteus maximus Semimembranosus Semitendinosus Biceps femoris	
Knee	Flexion	Knee extensors (eccentric contraction) Rectus femoris Vastus medialis Vastus intermedius Vastus lateralis		Extension	Knee extensors Rectus femoris Vastus medialis Vastus intermedius Vastus lateralis	
Ankle	Dorsiflexion	Plantar flexors (eccentric contraction) Gastrocnemius Soleus		Plantar flexion	Plantar flexors Gastrocnemius Soleus	

Chapter

13

Dead lift   FIG. 13.4

Description

The participant begins in hip/knee flexed position, keeping the arms, legs, and back straight, and grasps the barbell on the floor. Then a movement to the standing position is made by extending the hips. This exercise, when done improperly by allowing lumbar flexion, may contribute to low back problems. It is essential that the lumbar extensors be used more as isometric stabilizers of the low back while the hip extensors perform the majority of the lift in this exercise.

Analysis

This closed kinetic chain exercise is divided into two phases for analysis: (1) lifting phase to the hip extended/knee extended position and (2) lowering phase to the hip flexed/knee flexed starting position (Table 13.4).

FIG. 13.4 • Dead lift. **A,** Beginning hip flexed/knee flexed position; **B,** Ending hip extended/knee extended position.

A   B

TABLE 13.4 • Dead lift

Joint	Lifting phase to hip/knee extended position		Lowering phase to hip/knee flexed position	
	Action	Agonists	Action	Agonists
Wrist and hand	Flexion	Wrist and hand flexors (isometric contraction) 　Flexor carpi radialis 　Flexor carpi ulnaris 　Palmaris longus 　Flexor digitorum profundus 　Flexor digitorum superficialis 　Flexor pollicis longus	Flexion	Wrist and hand flexors (isometric contraction) 　Flexor carpi radialis 　Flexor carpi ulnaris 　Palmaris longus 　Flexor digitorum profundus 　Flexor digitorum superficialis 　Flexor pollicis longus
Trunk	Maintenance of extension	Trunk extensors (isometric contraction) 　Erector spinae (sacrospinalis) 　Quadratus lumborum	Maintenance of extension	Trunk extensors (isometric contraction) 　Erector spinae (sacrospinalis) 　Quadratus lumborum
Hip	Extension	Hip extensors 　Gluteus maximus 　Semimembranosus 　Semitendinosus 　Biceps femoris	Flexion	Hip extensors (eccentric contraction) 　Gluteus maximus 　Semimembranosus 　Semitendinosus 　Biceps femoris
Knee	Extension	Knee extensors (quadriceps) 　Rectus femoris 　Vastus medialis 　Vastus intermedius 　Vastus lateralis	Flexion	Knee extensors (quadriceps) (eccentric contraction) 　Rectus femoris 　Vastus medialis 　Vastus intermedius 　Vastus lateralis

Note: Slight movement of the shoulder joint and shoulder girdle is not being analyzed.

Isometric exercises

An exercise technique called isometrics is a type of muscular activity in which there is contraction of muscle groups with little or no muscle shortening. Though not as productive as isotonics in terms of overall strength gains, isometrics are an effective way to build and maintain muscular strength in a limited range of motion.

A few selected isometric exercises are analyzed muscularly to show how they are designed to develop specific muscle groups. Although there are varying approaches to isometrics, most authorities agree that isometric contractions should be held approximately 7 to 10 seconds for a training effect.

Abdominal contraction   FIG. 13.5

Description

The participant contracts the muscles in the anterior abdominal region as strongly as possible, with no movement of the trunk or hips. This exercise can be performed in the sitting, standing, or supine position. The longer the contraction in seconds, the more valuable the exercise will be, to a degree.

Analysis

Abdomen
   Contraction
       Rectus abdominis
       External oblique abdominal
       Internal oblique abdominal
       Transversus abdominis

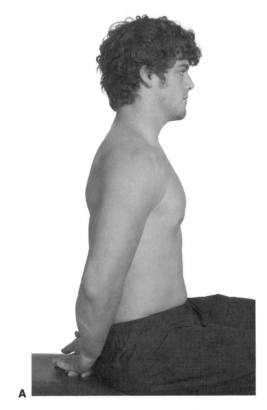
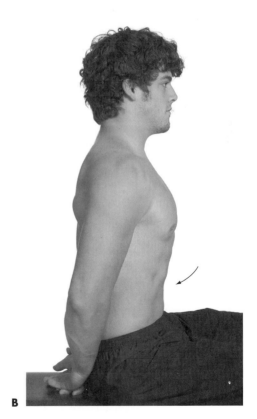

FIG. 13.5 • Abdominal contraction. **A,** Beginning position; **B,** Contracted position.

Chapter
13

Rowing exercise FIG. 13.6

Description

The participant sits on a movable seat with the knees and hips flexed close to the chest. The arms are reaching forward to grasp a horizontal bar. The legs are extended forcibly as the arms are pulled toward the chest. Then the legs and arms are returned to the starting position.

Analysis

This closed kinetic chain exercise is divided into two movements for analysis: (1) arm pull to chest/ leg push to extend knees and hip phase and (2) return phase to the starting position (Table 13.5).

A

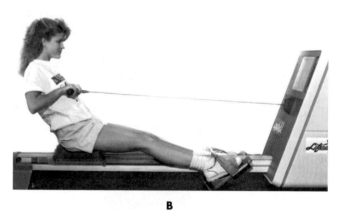

B

FIG. 13.6 • Rowing exercise machine. **A,** Starting position; **B,** Movement.

TABLE 13.5 • Rowing exercise

| Joint | Arm pull/leg push phase | | Return phase to starting position | |
	Action	Agonists	Action	Agonists
Foot and ankle	Plantar flexion	Ankle plantar flexors Gastrocnemius Soleus	Dorsiflexion	Ankle dorsiflexors Tibialis anterior Extensor hallucis longus Extensor digitorum longus Peroneus tertius
Knee	Extension	Quadriceps (knee extensors) Rectus femoris Vastus medialis Vastus intermedius Vastus lateralis	Flexion	Knee flexors (hamstrings) Biceps femoris Semitendinosus Semimembranosus
Hip	Extension	Hip extensors Gluteus maximus Biceps femoris Semimembranosus Semitendinosus	Flexion	Hip flexors Iliopsoas Rectus femoris Pectineus
Trunk	Extension	Trunk extensors Erector spinae	Flexion	Trunk flexors Rectus abdominis Internal oblique abdominal External oblique abdominal

TABLE 13.5 (continued) • **Rowing exercise**

| Joint | Arm pull/leg push phase | | Return phase to starting position | |
	Action	Agonists	Action	Agonists

Shoulder girdle	Adduction, downward rotation, and depression	Shoulder girdle adductors, downward rotators, and depressors
		  Trapezius (lower)
		  Rhomboid
		  Pectoralis minor
			Abduction, upward rotation, and elevation	Shoulder girdle adductors, downward rotators, and depressors (eccentric contraction)
				  Trapezius (lower)
				  Rhomboid
				  Pectoralis minor

Shoulder joint	Extension	Shoulder joint extensors
		  Latissimus dorsi
		  Teres major
		  Posterior deltoid
		  Teres minor
		  Infraspinatus
			Flexion	Shoulder joint extensors (eccentric contraction)
				  Latissimus dorsi
				  Teres major
				  Posterior deltoid
				  Teres minor
				  Infraspinatus

Elbow joint	Flexion	Elbow joint flexors
		  Biceps brachii
		  Brachialis
		  Brachioradialis
			Extension	Elbow joint flexors (eccentric contraction)
				  Biceps brachii
				  Brachialis
				  Brachioradialis

Wrist and hand	Flexion	Wrist and hand flexors (isometric contraction)
		  Flexor carpi radialis
		  Flexor carpi ulnaris
		  Palmaris longus
		  Flexor digitorum profundus
		  Flexor digitorum superficialis
		  Flexor pollicis longus
			Flexion	Wrist and hand flexors (isometric contraction)
				  Flexor carpi radialis
				  Flexor carpi ulnaris
				  Palmaris longus
				  Flexor digitorum profundus
				  Flexor digitorum superficialis
				  Flexor pollicis longus

REVIEW EXERCISES

1. Select, describe, and completely analyze five conditioning exercises.
2. Collect, analyze, and evaluate exercises that are found in newspapers, in magazines, and on the Internet or are observed on television.
3. Prepare a set of exercises that will ensure development of all large muscle groups in the body.
4. Select exercises from exercise books for analysis.
5. Bring to class other typical exercises for members to analyze.
6. Analyze the conditioning exercises given by your physical education teachers, coaches, and athletic trainers.
7. Consider a sport (basketball or any other sport) and develop exercises applying the overload principle that would develop all the large muscle groups used in that sport.
8. Prepare a list of exercises not found in this chapter to develop the lower-extremity and spinal muscles. Separate the list into open versus closed kinetic chain exercises.

LABORATORY EXERCISES

1. Observe children using playground equipment. Analyze muscularly the activities they are performing.
2. Visit the facility on your campus where the free weights and specific or multifunction exercise machines are located. Analyze exercises that can be done with each machine. Compare and contrast similar exercises using different exercise machines and free weights. *Note:* Manufacturers of all types of exercise apparatus have a complete list of exercises that can be performed with their machines. Secure a copy of recommended exercises and muscularly analyze each exercise.
3. Lie supine on a table with the knees flexed and hips flexed 90 degrees and the ankles in the

Chapter
13

neutral 90-degree position. Extend each joint until your knee is fully extended, your hip is flexed only 10 degrees, and your ankle is plantarflexed 10 degrees by performing each of the following movements in the order given:

- Full knee extension
- Hip extension to within 10 degrees of neutral
- Plantar flexion to 10 degrees

Analyze the movements and the muscles responsible for each movement at the hip, knee, and ankle.

4. Stand with your back and buttocks against a smooth wall and place your feet (shoulder width apart) with approximately 12 inches between your heels and the wall. Maintain your feet in position, with hips and knees each flexed approximately 90 degrees so that your thighs are parallel to the floor. Keeping your feet in place, slowly slide your back and buttocks up the wall until your buttocks are as far away from the floor as possible without moving your feet. Analyze the movements and the muscles responsible for each movement at the hip, knee, and ankle.

5. What is the difference between the two exercises in Questions 3 and 4? Can you perform the movement in Question 4 one step at a time, as you did in Question 3?

6. **Exercise analysis chart**

Analyze each exercise in the chart. Use one row for each joint involved that actively moves during the exercise. Do not include joints for which there is no active movement or joints maintained in one position isometrically.

Exercise	Phase	Joint, movement occurring	Force causing movement (muscle or gravity)	Force resisting movement (muscle or gravity)	Functional muscle group, type of contraction
Abdominal curl-up	Trunk flexion phase to curl-up position				
	Rotating to right phase				
	Return phase to curl-up position				
	Return phase to starting position				
Alternating prone extensions	Lifting phase				
	Lowering phase				

Exercise analysis chart (continued)

Exercise	Phase	Joint, movement occurring	Force causing movement (muscle or gravity)	Force resisting movement (muscle or gravity)	Functional muscle group, type of contraction
Squat	Lowering phase				
	Lifting phase				
Dead lift	Lifting phase				
	Lowering phase				
Rowing exercise	Arm pull/leg push phase				
	Return phase to starting position				

Chapter

13

References

Adrian M: Isokinetic exercise, *Training and Conditioning* 1:1, June 1991.

Altug Z, Hoffman JL, Martin JL: *Manual of clinical exercise testing, prescription and rehabilitation,* Norwalk, CT, 1993, Appleton & Lange.

Andrews JR, Wilk KE, Harrelson GL: *Physical rehabilitation of the injured athlete,* ed 3, Philadelphia, 2004, Saunders.

Ellenbecker TS, Davies GJ: *Closed kinetic chain exercise: a comprehensive guide to multiple-joint exercise,* Champaign, IL, 2001, Human Kinetics.

Fahey TD: *Athletic training: principles and practices,* Mountain View, CA, 1986, Mayfield.

Logan GA, McKinney WC: *Anatomic kinesiology,* ed 3, New York, 1982, McGraw-Hill.

Matheson O, et al: Stress fractures in athletes, *American Journal of Sports Medicine* 15:46, January–February 1987.

National Strength and Conditioning Association; Baechle TR, Earle RW: *Essentials of strength training and conditioning,* ed 2, Champaign, IL, 2000, Human Kinetics.

Northrip JW, Logan GA, McKinney WC: *Analysis of sport motion: anatomic and biomechanic perspectives,* ed 3, New York, 1983, McGraw-Hill.

Powers SK, Howley ET: *Exercise physiology: theory and application of fitness and performance,* ed 7, New York, 2009, McGraw-Hill.

Prentice WE: *Rehabilitation techniques in sports medicine,* ed 5, New York, 2011, McGraw-Hill.

Steindler A: *Kinesiology of the human body,* Springfield, IL, 1970, Charles C Thomas.

Torg JS, Vegso JJ, Torg E: *Rehabilitation of athletic injuries: an atlas of therapeutic exercise,* Chicago, 1987, Year Book.

Wirhed R: *Athletic ability and the anatomy of motion,* ed 3, St. Louis, 2006, Mosby Elsevier.

 For additional resources and a list of related websites, visit **www.mhhe.com/floyd19e**.

Chapter

13

Worksheet Exercises

For in- or out-of-class assignments, or for testing, utilize this tear-out worksheet.

Skill analysis worksheet

Using the techniques taught in this chapter and in Chapter 8, analyze the joint movements and muscles used in each phase of movement for selected skills in a given sport or physical activity. For each phase, list the initial and subsequent joint position and the joint motion that occurs with the approximate degrees of movement. List the muscles utilized in the appropriate cell as to whether contracting concentrically to *cause* movement or eccentrically to *control* movement or isometrically to *prevent* movement or maintain the position. For passive movements that may occur, place a dash across all three contraction types.

Kinesiology skill analysis table

Phase		Toes	TT/ST	Ankle	Knee	Hip	Lumbar	Cervical	Shou Gir	Elbow	Rad/ulna	Wrist	Fingers
	Position												
	Movement/deg												
	Concentric												
	Eccentric												
	Isometric												
	Position												
	Deg/move												
	Agon/concen												
	Eccentric												
	Isometric												
	Position												
	Deg/move												
	Agon/concen												
	Eccentric												
	Isometric												

Chapter

13

Worksheet Exercises

For in- or out-of-class assignments, or for testing, utilize this tear-out worksheet.

Skill analysis worksheet

Using the techniques taught in this chapter and in Chapter 8, analyze the joint movements and muscles used in each phase of movement for selected skills in a given sport or physical activity. For each phase, list the initial and subsequent joint position and the joint motion that occurs with the approximate degrees of movement. List the muscles utilized in the appropriate cell as to whether contracting concentrically to *cause* movement or eccentrically to *control* movement or isometrically to *prevent* movement or maintain the position. For passive movements that may occur, place a dash across all three contraction types.

Kinesiology skill analysis table

Phase		Toes	TT/ST	Ankle	Knee	Hip	Lumbar	Cervical	Shou Gir	Elbow	Rad/ulna	Wrist	Fingers
	Position												
	Movement/deg												
	Concentric												
	Eccentric												
	Isometric												
	Position												
	Deg/move												
	Agon/concen												
	Eccentric												
	Isometric												
	Position												
	Deg/move												
	Agon/concen												
	Eccentric												
	Isometric												

Appendix

Appendix 1

Range of motion for diarthrodial joints of the upper extremity

Joint	Type	Motion	Range
Sternoclavicular	Arthrodial	Protraction	Moves anteriorly 15°
		Retraction	Moves posteriorly 15°
		Elevation	Moves superiorly 45°
		Depression	Moves inferiorly 5°
		Upward rotation	45°
		Downward rotation	5°
Acromioclavicular	Arthrodial	Protraction-retraction	20°–30° rotational and gliding motion
		Elevation-depression	20°–30° rotational and gliding motion
		Upward rotation–downward rotation	20°–30° rotational and gliding motion
Scapulothoracic	Not a true synovial joint; all movement totally dependent on AC and SC joints	Abduction-adduction	25° total range
		Upward rotation–downward rotation	60° total range
		Elevation-depression	55° total range
Glenohumeral	Enarthrodial	Flexion	90°–100°
		Extension	40°–60°
		Abduction	90°–95°
		Adduction	0° prevented by trunk, 75° anterior to trunk
		Internal rotation	70°–90°
		External rotation	70°–90°
		Horizontal abduction	45°
		Horizontal adduction	135°
Elbow	Ginglymus	Extension	0°
		Flexion	145°–150°
Radioulnar	Trochoid	Supination	80°–90°
		Pronation	70°–90°

Range of motion for diarthrodial joints of the upper extremity *(continued)*

Joint	Type	Motion	Range
Wrist	Condyloid	Flexion	70°–90°
		Extension	65°–85°
		Abduction	15°–25°
		Adduction	25°–40°
Thumb carpometacarpal	Sellar	Flexion	15°–45°
		Extension	0°–20°
		Adduction	0°
		Abduction	50°–70°
Thumb metacarpophalangeal	Ginglymus	Extension	0°
		Flexion	40°–90°
Thumb interphalangeal	Ginglymus	Flexion	80°–90°
		Extension	0°
2nd, 3rd, 4th, and 5th metacarpophalangeal joints	Condyloid	Extension	0°–40°
		Flexion	85°–100°
		Abduction	Variable 10°–40°
		Adduction	Variable 10°–40°
2nd, 3rd, 4th, and 5th proximal interphalangeal joints	Ginglymus	Flexion	90°–120°
		Extension	0°
2nd, 3rd, 4th, and 5th distal interphalangeal joints	Ginglymus	Flexion	80°–90°
		Extension	0°

Appendix 2
Range of motion for diarthrodial joints of the spine and lower extremity

Joint	Type	Motion	Range
Cervical	Arthrodial except atlantoaxial joint, which is trochoid	Flexion	80°
		Extension	20°–30°
		Lateral flexion	35°
		Rotation unilaterally	45°
Lumbar	Arthrodial	Flexion	45°
		Extension	45°
		Lateral flexion	45°
		Rotation unilaterally	60°
Hip	Enarthrodial	Flexion	130°
		Extension	30°
		Abduction	35°
		Adduction	0°–30°
		External rotation	50°
		Internal rotation	45°
Knee *For internal and external rotation to occur, the knee must be flexed approximately 30° or more.*	Ginglymus (trochoginglymus)	Extension	0°
		Flexion	140°
		Internal rotation	30°
		External rotation	45°
Ankle (talocrural)	Ginglymus	Plantar flexion	50°
		Dorsal flexion	15°–20°
Transverse tarsal and subtalar	Arthrodial	Inversion	20°–30°
		Eversion	5°–15°
Great toe metatarsophalangeal	Condyloid	Flexion	45°
		Extension	70°
		Abduction	Variable 5°–25°
		Adduction	Variable 5°–25°
Great toe interphalangeal	Ginglymus	Flexion	90°
		Extension	0°
2nd, 3rd, 4th, and 5th metatarsophalangeal joints	Condyloid	Flexion	40°
		Extension	40°
		Abduction	Variable 5°–25°
		Adduction	Variable 5°–25°
2nd, 3rd, 4th, and 5th proximal interphalangeal joints	Ginglymus	Flexion	35°
		Extension	0°
2nd, 3rd, 4th, and 5th distal interphalangeal joints	Ginglymus	Flexion	60°
		Extension	30°

Appendix 3

Commonly used exercises for strengthening selected muscles

Some exercises may be more or less specific for certain muscles. In some cases, certain exercises are designed to emphasize specific portions of a particular muscle more than other portions. Some exercises may be modified slightly to further emphasize or deemphasize certain muscles or portions of muscles. In addition to the muscles listed, numerous other muscles in surrounding joints or other parts of the body may be involved by contracting isometrically to maintain appropriate position of the body for the muscles listed to carry out the exercise movement. Appropriate strength and endurance of these stabilizing muscles is essential for correct position and execution of the listed exercises. Finally, these exercises may be documented by different names by different authorities. Illustration and documentation of the proper techniques and indications for these exercises is beyond the scope of this text. Please consult several strength and conditioning texts for further details.

Upper extremity		
Muscle groups	Muscles	Exercise
Scapula abductors	Serratus anterior Pectoralis minor	Front press Dumbbell flys Dumbbell press Bench press Close-grip bench press Incline press Push-ups Incline dumbbell press Pec deck flys Cable crossover flys
Scapula adductors	Rhomboid Trapezius, lower fibers Trapezius, middle fibers	Bent-over lateral raises Pec deck rear delt laterals Seated rows Bent rows T-bar rows Dead lifts Sumo dead lifts
Scapula upward rotators	Serratus anterior Trapezius, lower fibers Trapezius, middle fibers	Dumbbell press Dumbbell flys One-arm dumbbell press Lateral raises Upright rows
Scapula downward rotators	Pectoralis minor Rhomboid	Parallel bar dips Dumbbell pullovers Barbell pullovers Chin-ups Reverse chin-ups Lat pulldowns Back lat pulldowns Close-grip lat pulldowns Straight-arm lat pulldowns One-arm dumbbell rows
Scapula elevators	Rhomboid Levator scapulae Trapezius, upper fibers Trapezius, middle fibers	Upright rows Barbell shrugs Dumbbell shrugs Machine shrugs Dead lifts
Scapula depressors	Pectoralis minor Trapezius, lower fibers	Parallel bar dips Body dips

Upper extremity		
Muscle groups	Muscles	Exercise
Shoulder flexors	Deltoid, anterior fibers Deltoid, middle fibers Pectoralis major, clavicular fibers Coracobrachialis	Arm curls Triceps dips One-arm dumbbell press Front raises Low pulley front raises One-dumbbell front raises Barbell front raises
Shoulder extensors	Latissimus dorsi Teres major Triceps brachii, long head Pectoralis major, sternal fibers Deltoid, posterior fibers Infraspinatus Teres minor	Dumbbell pullovers Barbell pullovers Reverse chin-ups Close-grip lat pulldowns Straight-arm lat pulldowns One-arm dumbbell rows
Shoulder abductors	Deltoid, anterior fibers Deltoid, posterior fibers Deltoid, middle fibers Pectoralis major, clavicular fibers Supraspinatus	Back press Front press Dumbbell press One-arm dumbbell press Lateral raises Side-lying lateral raises Low pulley lateral raises Upright rows Nautilus lateral raises Upright rows
Shoulder adductors	Pectoralis major Latissimus dorsi Teres major Subscapularis Coracobrachialis	Triceps dips Parallel bar dips Chin-ups Lat pulldowns Back lat pulldowns
Shoulder internal rotators	Pectoralis major Latissimus dorsi Teres major Subscapularis	Triceps dips Side-lying internal rotations Standing internal rotations at 90 degrees abduction
Shoulder external rotators	Infraspinatus Teres minor	Side-lying external rotations Standing external rotations at 90 degrees abduction

Upper extremity		
Muscle groups	Muscles	Exercise
Shoulder horizontal abductors	Latissimus dorsi Infraspinatus Teres minor Deltoid, middle fibers Deltoid, posterior fibers	Bent-over lateral raises Low pulley bent-over lateral raises Pec deck rear delt laterals Bent rows T-bar rows
Shoulder horizontal adductors	Pectoralis major Coracobrachialis	Dumbbell flys Triceps dips Dumbbell press Bench press Close-grip bench press Incline press Decline press Push-ups Incline dumbbell press Incline dumbbell flys Pec deck flys Cable crossover flys Seated rows
Elbow flexors	Biceps brachii Brachialis Brachioradialis	Arm curls concentration curls Hammer curls Low pulley curls High pulley curls Barbell curls Machine curls Preacher curls Reverse barbell curls Chin-ups Reverse chin-ups Lat pulldowns Back lat pulldowns Close-grip lat pulldowns Seated rows One-arm dumbbell rows Bent rows T-bar rows Upright rows
Elbow extensors	Triceps brachii Triceps brachii, lateral head Triceps brachii, long head Triceps brachii, medial head Anconeus	Barbell pullovers Bench press Close-grip bench press Decline press Dumbbell press Dumbbell pullovers Dumbbell triceps extensions Front press Incline dumbbell press Incline press One-arm dumbbell triceps extensions One-arm reverse pushdowns Parallel bar dips Pushdowns Push-ups Reverse pushdowns Seated dumbbell triceps extensions Seated ez-bar triceps extensions Triceps dips Triceps extensions Triceps kickbacks

Upper extremity		
Muscle groups	Muscles	Exercise
Wrist flexors	Flexor carpi radialis Palmaris longus Flexor carpi ulnaris Flexor digitorum superficialis Flexor digitorum profundus	Wrist curls
Wrist extensors	Extensor carpi radialis longus Extensor carpi radialis brevis Extensor carpi ulnaris	Reverse barbell curls Reverse wrist curls Reverse pushdowns
Finger flexors	Hand intrinsics Flexor digitorum profundus Flexor digitorum superficialis Flexor pollicis longus	Ball squeezes Putty squeezes Rice bucket grips Wrist curls Dead lifts
Finger extensors	Extensor digitorum Extensor digiti minimi Extensor indicis	Reverse barbell curls Reverse wrist curls Rubber band stretches

Lower extremity		
Muscle groups	Muscles	Exercise
Hip flexors	Rectus femoris Iliopsoas Pectineus Tensor fasciae latae	Crunches Sit-ups Gym ladder sit-ups Calves over bench sit-ups Incline bench sit-ups Specific bench sit-ups Machine crunches Incline leg raises Leg raises Hanging leg raises
Hip extensors	Gluteus maximus Biceps femoris, long head Semitendinosus Semimembranosus	Stiff-legged dead lifts Dead lifts Back extensions Dumbbell squats Squats Front squats Power squats Angled leg press Good mornings Lunges Cable back kicks Machine hip extensions Floor hip extensions Bridging Prone arches
Hip abductors	Gluteus medius Gluteus maximus Tensor fascia latae	Cable hip abductions Standing machine hip abductions Floor hip abductions Seated machine hip abductions

Lower extremity		
Muscle groups	Muscles	Exercise
Hip adductors	Adductor magnus Adductor longus Adductor brevis Gracilis	Sumo dead lifts Power squats Cable adductions Machine adductions
Hip external rotators	Gluteus maximus Piriformis Gemellus superior Gemellus inferior Obturator externus Obturator internus Quadratus femoris	Hip turn-outs Body turn-aways
Knee extensors	Vastus medialis Vastus intermedius Rectus femoris Vastus lateralis	Leg extensions Dead lifts Sumo dead lifts Dumbbell squats Squats Front squats Angled leg press Power squats Hack squats Lunges
Knee flexors	Semitendinosus Biceps femoris, long head Biceps femoris, short head Semimembranosus Gastrocnemius, lateral head Gastrocnemius, medial head	Standing leg curls Seated leg curls Lying leg curls
Ankle dorsiflexors	Tibialis anterior Extensor hallucis longus Extensor digitorum longus Peroneous tertius	Towel pulls Elastic band pulls
Ankle plantar flexors	Gastrocnemius, lateral head Soleus Gastrocnemius, medial head	Standing calf raises One-leg toe raises Donkey calf raises Seated calf raises Seated barbell calf raises
Transverse tarsal/ subtalar inversion	Tibialis anterior Tibialis posterior Flexor digitorum longus Flexor hallucis longus	Towel drags Elastic band turn-ins
Transverse tarsal/ subtalar eversion	Extensor digitorum longus Peroneus longus Peroneus brevis Peroneus tertius	Towel drags Elastic band turn-outs
Toe extensors	Extensor hallucis longus Extensor digitorum longus	Towel pulls Elastic band pulls
Toe flexors	Flexor digitorum longus Flexor hallucis longus Foot intrinsics	Towel curls Marble pickups Pencil pickups

Cervical spine and trunk		
Muscle groups	Muscles	Exercise
Cervical extensors	Splenius cervicis Splenius capitus Trapezius, upper fibers	Dead lifts Neck extensions
Cervical flexors	Sternocleidomastoid	Chin tucks Sit-ups
Cervical rotators	Sternocleidomastoid Splenius cervicis Splenius capitus	Machine neck rotations
Trunk extensors	Erector spinae	Back extensions Alternating prone extensions Prone arches
Trunk flexors	Rectus abdominis External oblique abdominal Internal oblique abdominal	Dead lifts Crunches Crunch twists Sit-ups Gym ladder sit-ups Calves over bench sit-ups Incline bench sit-ups Specific bench sit-ups High pulley crunches Machine crunches Incline leg raises Leg raises Hanging leg raises
Trunk rotators	External oblique abdominal Internal oblique abdominal	Crunches Crunch twists Sit-up twists Gym ladder sit-ups Calves over bench sit-ups Incline bench sit-ups Specific bench sit-ups High pulley crunches Machine crunches Incline leg raises Leg raises Hanging leg raises Broomstick twists Machine trunk rotations
Trunk lateral flexors	External oblique abdominal Internal oblique abdominal Quadratus lumborum Rectus abdominis	Crunches Crunch twists Sit-ups Gym ladder sit-ups Calves over bench sit-ups Incline bench sit-ups Specific bench sit-ups High pulley crunches Machine crunches Incline leg raises Leg raises Hanging leg raises Broomstick twists Dumbbell side bends Roman chair side bends

Appendix 4

Etymology of commonly used terms in kinesiology

Below are some of the most commonly used terms in naming the muscles, bones, and joints as well as some additional terms utilized in explaining their function. This etymology is provided in order to better understand the origin and historical development of these terms and to provide a more meaningful background as to how these terms came to be used in the study of the body and its movement today.

abdominis   Latin: belly

abductor   Latin: abducere

acromion   Greek: akron, extremity + omus, shoulder

adductor   Latin: adducere, adduct-, to bring to, contract

amphiarthrodial   Greek: ampho, both + arthron, joint + eidos, form, shape

anconeus   Greek: agkon, elbow

antebrachial   Latin: ante, before + brachium, arm

antecubital   Latin: ante, before + cubitum, elbow

anterior   Latin: comparative of ante, before

appendicular   Latin: appendere, to hang to

arthrodial   Greek: arthron, joint + eidos, form, shape

axillary   Latin: axilla

axon   Greek: axon, axis

biceps   Latin: two-headed, bi-, two; caput, head

brachialis   Latin: brachialis, brachial, arm

brachii   Latin: bracchium, arm

brachioradialis   Latin: bracchium, arm + radialis, radius

brevis   Latin: (adj.) short, low, little, shallow

buccal   Latin: cheek

bursa   Greek: a leather sack

calcaneus   Latin: calcaneus, heel, from Latin calcaneum, from calx, calc

cancellous   Latin: cancellus, lattice

capitate, capitis   Latin: caput, head

carpal   Latin: carpalis, from carpus, wrist

carpus, carpi   Latin: from Greek karpos, wrist

caudal   Latin: caudalis, tail

celiac   Greek: koilia, belly

cephalic   Greek: kephale, head

cerebellum   Latin: little brain

cerebrum   Latin: cerebrum, brain

cervicis   Latin: cervix, neck

clavicle   French: clavicule, collarbone; Latin: clavicula, little key

coccyx   Greek: kokkyz, cuckoo

colli   Latin: collare, necklace, band or chain for the neck, from collum, the neck

concentric   Latin: con, together with + centrum, center

condyle   Greek: kondylos, knuckle

condyloidal   Greek: kondylos, knuckle + eidos, form, shape

coracobrachialis   Greek: from coracoid korax, crow; eidos, form; Greek: brachion + Latin, radial

coracoid   Greek: korax, raven + eidos, form, shape

coronal   Greek: korone, crown

coronoid   Greek: korone, something curved, kind of crown + eidos, form, shape

cortical   Latin: rind

costal   Latin: costa, rib

coxal   Latin: coxa, hip

cranial   Latin: cranialis, cranium; Greek: kranion

crest   Latin: crista, crest

crural   Latin: cruralis, pertaining to leg or thigh

cubital   Latin: cubitum, elbow

cuboid   Greek: kybos, cube

cuneiforms   Latin: cuneus, wedge + forma, form

deltoid, deltoidius   Latin: deltoides: Greek deltoeides, triangular: delta, delta + -oeides, -oid

dendrite   Greek: dendrites, pertaining to a tree

derma   Greek: derma, skin

dermatome   Greek: derma, skin + tome, incision

diaphysis   Greek: diaphysis, a growing through

diarthrodial   Greek: dis, two + arthron, joint + eidos, form, shape

digital, digitorum   Latin: digitus, finger or toe

distal   Latin: distare, to be distant

dorsal   Latin: dorsalis, dorsualis, of the back, from dorsum, back

dorsi   Latin: dorsi, genitive of dorsum, back

eccentric   Greek: ek, out + kentron, center

enarthrodial   Greek: en, in + arthron, joint + eidos, form, shape

endosteum   Greek: endon, within + osteon, bone

epiphyseal   Greek: epi, above + phyein, to grow

epiphysis   Greek: a growing upon

erector   Latin: erigere, to erect

extensor   Latin: extendere, to stretch out

external, externus   Latin: externus, outside, outward

fabella   Latin: faba, little bean

facet   French: facette, small face

fasciae   Latin: fascia, band

femoris   Latin: femur, thigh, genitive of femur, thigh

femur   Latin: thigh

fibers   Latin: fibra, a fiber, filament, of uncertain origin, perhaps related to Latin: filum, thread

fibrous   Latin: fibra, fiber

fibula   Latin: clasp, brooch

flexor   Latin: bender

foramen   Latin: hole

fossa   Latin: ditch

fovea   Latin: pit

frontal   Latin: frontem (nom, frons), forehead, literally that which projects

fusiform   Latin: fusus, spindle + forma, shape

gaster   Greek: gaster, belly

gastrocnemius   Greek: gaster, belly + kneme, leg

gemellus   Latin: twin

ginglymus   Greek: ginglymos, hinge

gluteus   Greek: gloutos, buttock

gomphosis   Greek: bolting together

goniometer   Greek: gonia, angle + metron, measure

gracilis   Latin: graceful

greater   Middle English: grete; Old English: great, thick, coarse; French: grand, which is from Latin: magnus

hallucis   Latin: hallex, large toe

hamate   Latin: hamatus, hooked

head   Latin: from caput

humerus   Latin: a misspelling borrowing umerus, shoulder

hyaline   Greek: hyalos, glass

iliacus   Latin: ilium, flank

iliocostalis   Latin: ilium, flank + costa, rib

ilium   Latin: ilium, groin, flank, variant of Latin: ilia

indicis   Latin: forefinger, pointer, sign, list

inferior   Latin: inferior, lower

infraspinatus, infraspinous   Latin: infra, below + spina, spine

inguinal   Latin: inguinalis, groin

insertion   Latin: in, into + serere, to join

intermediate   Latin: intermediates, lying between; Latin: intermedius, that which is between; from inter, between + medius, in the middle

intermedius   Latin: inter, between, mediare, to divide that which is between

internal, internus   Latin: internus, within

interossei   Latin: inter, between + os, bone

interspinalis   Latin: inter, between + spina, spine

intertransversarii   Latin: inter, between + transverses, cross-direction

ischium   Greek: ischion, hip joint

isokinetic   Greek: isos, equal + kinesis, motion

isometric   Greek: isos, equal + metron, measure

isotonic   Greek: isos, equal + tonus, tone

kinematic   Greek: kinematos, movement

kinesiology   Greek: kinematos, movement + logos, word, reason

kinesthesia   Greek: kinematos, movement + aesthesis, sensation

kyphosis   Greek: humpback

latae   Latin: latus, side

lateral, lateralis   Latin: lateralis, belonging to the side

latissimus   Latin: latissimus, superlative of latus, wide

lesser   Middle English: lesse; Latin: minor

levator   Latin: levator, lifter

lever   Latin: levare, to raise

linea   Latin: linea, line

longissimus   Latin: longest, very long

longus   Latin: long

lordosis   Greek: lordosis, bending

lumbar, lumborum   Latin: lumbus, loin

lumbricals   Latin: lumbus, loin; referred to vermiform; Latin: vermis, worm + forma, form

lunate   Latin: lunatus, past participle of lunare, to bend like a crescent, from luna, moon

magnus   Latin: great

major   Middle English: majour; Latin: major

malleolus   Latin: malleolus, little hammer

mammary   Latin: mamma, breast

mandible   Latin: mandibula, jaw; Latin: mandere, to chew

manubrium   Latin: handle, from manus, hand

margin   Latin: marginalis, border

maxilla   Latin: upper jaw, of mala, jaw, cheekbone

maximus   Latin: greatest

meatus   Latin: meatus, passage

medial, medialis   Latin: medialis, of the middle; Latin: medius, middle

medius   Latin: middle

mental   Latin: mentum, chin

metacarpal   Greek: meta, after, beyond, over + Latin: carpalis, from carpus, wrist

metatarsals   Greek: meta, after, beyond, over + tarsos, flat surface

middle   Old English: middle; Latin: medium

minimus, minimi   Latin: minimum, smallest

minor   Latin: lesser, smaller, junior

multifidus   Latin: multus, many + clefts or segments

muscle   Latin: musculus

myo   Greek: mys, muscle

myotome   Greek: mys, muscle + tome, incision

nasal   Latin: nasus, nose

navicular   Latin: navicula, boat, diminutive of navis, ship

neural   Latin: neuralis, nerve

neuron   Greek: neuron, nerve, sinew

notch   French: noche, indention, depression

nuchal   Latin: nape (back) of the neck

oblique, obliquus   Latin: obliquus, slanted

obturator   Latin: obturare, to close

occiput   Latin: occiput (gen. occipitis), back of the skull, from ob, against, behind + caput, head

olecranon   Greek: elbow

omos   Greek: omos, shoulder

opponens   Latin: opponentem (nom. opponens), prp. of opponere, oppose, object to, set against

oral   Latin: oralis, mouth

orbital   Latin: orbita, track

origin   Latin: origo, beginning

osseous   Latin: osseus, bony

otic   Greek: otikos, ear

palmar   Latin: palma, palm of the hand

palmaris   Latin: palma, palm

patella   Latin: pan, kneecap

pectineus   Latin: pectin, comb

pectoralis   Latin: pectoralis, from pectus, pector-, breast; Middle English, French, Latin: pectorale, breastplate, from neuter of pectoralis

pedal   Latin: pedalis, foot

pennate   Latin: penna, feather

perineal   Greek: perinaion, perineum

periosteum   Greek: peri, around + osteon, bone

peroneus   Greek: perone, brooch

phalanges   A plural of phalanx Greek: phalangos, finger or toe bone

phalanx   Greek: phalangos, finger or toe bone

piriformis   Latin: piriformis, pear shaped

pisiform   Latin: pisa, pea + forma, form

plantae   Latin: planta, sole of the foot

pollicis   Latin: from pollex, thumb, big toe

popliteus   Latin: poples, ham of the knee

posterior   Latin: comparative of posterus, coming after, from post, afterward

process   Latin: processus, going before

profundus   Latin: deep, bottomless, vast

pronator   Latin: pronare, to bend forward

proximal   Latin: proximitatem (nom. proximitas), nearness, vicinity, from proximus, nearest

psoas   Greek: psoa, muscle of the loin

pubis   Latin: (os) pubis, bone of the groin

quadratus   Latin: quadratus, square

quinti   Latin: quintus, fifth

radialis   Latin: radialis, radius, beam of light

radiate   Latin: radiare, to emit rays

radius   Latin: beam of light

ramus   Latin: branch

rectus   Latin: rectus, straight

rhomboids, rhomboidus   From the word rhombus; Latin: flatfish, magician's circle; from Greek: rhombos, rhombus

rotatores   Latin: rotare, to rotate

sacrum   Latin: os sacrum, sacred bone

sartorius   Latin: sartor, tailor

scaphoid   Latin: skiff, boat shaped + eidos, form

scapulae, scapula   Latin: shoulder; Latin: scapulae, the shoulder blades

scoliosis   Greek: scoliosis, crookedness

sellar   Latin: Turkish saddle

semimembranosus   Latin: semi, half + membrane, membrane

semispinalis   Latin: semi, half + spina, spine

semitendinosus   Latin: semi, half + tendere, to stretch

serratus   Latin: serratus, saw-shaped, from serra, saw

sesamoid   Latin: sesamoides, resembling a grain of sesame in size or shape

sinus   Latin: curve, hollow

soleus   Latin: solea, sandal

somatic   Greek: soma, body

sphincter   Greek: sphincter, hand

spinae   Latin: thorn

spinal   Latin: spinalis, spine

splenius   Greek: splenion, bandage

sternocleidomastoid   Greek: sternon, chest + kleis, key + mastos, breast + eidos, form

sternum   Greek: sternon, chest, breast, breastbone

styloid   Anglo-Saxon: stigan, to rise + eidos, form

subscapularis   Latin: sub, beneath + scapulae, shoulder blades

sulcus Latin: groove

superficialis   Latin: superficies, of or pertaining to the surface

superior   Latin: superiorem, higher

supinator   Latin: reflectere, to bend back

supraspinatus, supraspinous   Latin: supra, above + spina, spine

sural   Latin: sura, calf

suture   Latin: sutura, a seam

symphysis   Greek: symphysis, growing together

synarthrodial   Greek: syn, together + arthron, joint + eidos, form, shape

synchondrosis   Greek: syn, together + chondros, cartilage + osis, condition

syndesmosis   Greek: syndesmos, ligament + osis, condition

synovial   Latin: synovia, joint fluid

talus   Latin: ankle, anklebone, knucklebone

tarsal   Greek: tarsalis, ankle

temporal   Latin: temporalis, of time, temporary, from tempus (temporis), time, season, proper time or season

tendon   Latin: tendo, tendon

tensor   Latin: tendere, to stretch

teres   Latin: rounded

tertius   Latin: third

thoracic   Greek: thorax, chest

tibia   Latin: shinbone

tibialis   Latin: tibia, pipe, shinbone

transverse   Latin: transverses, oblique

trapezium   Greek: trapezion, a little table

trapezius   Latin: trapezium, trapezium, from the shape of the muslces paired

trapezoid   Greek: trapezoeides, table shaped

triceps   Latin: three-headed; tri-, tri- + caput, head

triquetrum   Latin: neuter of triquetrus, three-cornered

trochanter   Greek: trokhanter, to run

trochlear   Greek: trokhileia, system of pulleys

tubercle   Latin: turberculum, a little swelling

tuberosity   Latin: tuberositas, tuberosity

ulna   Latin: elbow

ulnaris   Latin: ulna, elbow

umbilical   Latin: umbilicus, naval

vastus   Latin: immense, extensive, huge

vertical   Latin: verticalis, overhead, vertex, highest point

visceral   Latin: viscera, body organs

volar   Latin: vola, sole, palm

xiphoid   Greek: xiphos, sword + eidos, form, shape

zygoma   Latin: zygoma, zygomat-, from Greek zugoma, bolt, from zugoun, to join

Appendix 5

Determining if a muscle (or muscle group) is contracting and, if so, the contraction type (isometric, concentric, or eccentric)

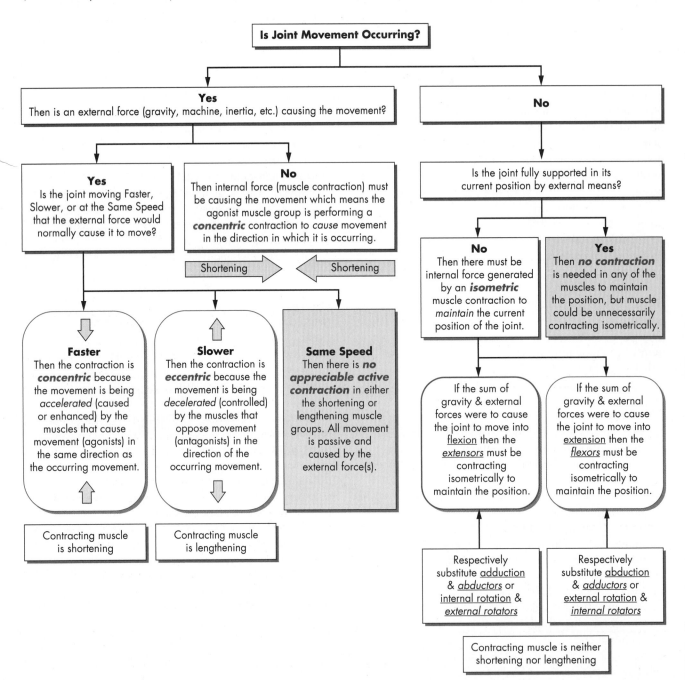

Glossary

abduction Lateral movement away from the midline of the trunk, as in raising the arms or legs to the side horizontally.

acceleration The rate of change in velocity.

accessory motion The actual change in relationship between the articular surface of one bone relative to another, characterized as roll, spin, and glide.

action potential Electrical signal transmitted from the brain and spinal cord through axons to the muscle fibers in a particular motor unit providing the stimulus to contract.

active insufficiency Point reached when a muscle becomes shortened to the point that it cannot generate or maintain active tension.

active tension Tension in muscles that is generated via an active contraction of the respective muscle fibers in that muscle.

adduction Movement medially toward the midline of the trunk, as in lowering the arms to the side or legs back to the anatomical position.

afferent nerves Nerves that bring impulses from receptors in the skin, joints, muscles, and other peripheral aspects of the body to the central nervous system.

aggregate muscle action Muscles working together in groups rather than independently to achieve given joint motions.

agonist A muscle or muscle group that is described as being primarily responsible for a specific joint movement when contracting.

all or none principle States that regardless of the number involved, the individual muscle fibers within a given motor unit will fire and contract either maximally or not at all.

amphiarthrodial (amphiarthrosis) joints Joints that functionally allow only a very slight amount of movement such as synchondrosis (e.g., costochondral joint of the ribs with sternum), syndesmosis (e.g., distal tibiofibular), and symphysis (e.g., symphysis pubis) joints.

amplitude Range of muscle fiber length between maximal and minimal lengthening.

anatomical position The position of reference in which the subject is in the standing position, with feet together and palms of hands facing forward.

angle Bend or protruding angular projection of a bone such as superior and inferior angle of scapula.

angle of pull The angle between the muscle insertion and the bone on which it inserts.

angular displacement The change in location of a rotating body.

angular motion Motion involving rotation around an axis.

antagonist A muscle or muscle group that counteracts or opposes the contraction of another muscle or muscle group.

anterior axillary line A line parallel to the mid-axillary line which passes through the anterior axillary skinfold.

anteroposterior axis The axis that has the same directional orientation as the sagittal plane of motion and runs from front to back at a right angle to the frontal plane of motion. Also known as the sagittal or AP axis.

anteversion Abnormal or excessive rotation forward of a structure, such as femoral anteversion.

aponeurosis A tendinous expansion of dense fibrous connective tissue, sheet- or ribbonlike in appearance and resembling a flattened tendon, which serves as a fascia to bind muscles together or as a means of connecting muscle to bone.

appendicular skeleton The appendages, or the upper and lower extremities, and the shoulder and pelvic girdles.

arthrodial joints Joints in which bones glide on each other in limited movement, as in the bones of the wrist (carpal) or the bones of the foot (tarsal).

arthrokinematics Motion between the actual articular surfaces of the bones at a joint.

arthrosis Joint or articulation between two or more bones.

axial skeleton The skull, vertebral column, ribs, and sternum.

axis of rotation The point in a joint about which a bone moves or turns to accomplish joint motion.

axon An elongated projection that transmits impulses away from the neuron cell body.

balance The ability to control equilibrium, either static or dynamic.

biarticular muscles Those muscles that, from origin to insertion, cross two different joints, allowing them to perform actions at each joint.

bilateral Relating to the right and left sides of the body or of a body structure such as the right and left extremities.

biomechanics The study of mechanics as it relates to the functional and anatomical analysis of biological systems, especially humans.

bipennate A type of pennate muscle with fibers running obliquely on both sides from a central tendon, such as the rectus femoris and flexor hallucis longus.

border or margin   Edge or boundary line of a bone such as lateral and medial border of scapula.

brachial plexus   Group of spinal nerves composed of cervical nerves 5 through 8, along with thoracic nerve 1; supplies motor and sensory function to the upper extremity and most of the scapula.

cancellous bone   Spongy, porous bone that lies under cortical bone.

cardinal plane   Specific planes that divide the body exactly into two halves.

carpal tunnel   A three-sided arch, concave on the palmar side and formed by the trapezium, trapezoid, capitate, and hamate. It is spanned by the transverse carpal and volar carpal ligaments creating a tunnel.

carpal tunnel syndrome   A condition characterized by swelling and inflammation with resultant increased pressure in the carpal tunnel, which interferes with normal function of the median nerve, leading to reduced motor and sensory function of its distribution; particularly common with repetitive use of the hand and wrist in manual labor and clerical work such as typing and keyboarding.

carrying angle   In the anatomical position, the angle formed by the forearm deviating laterally from the arm, typically 5 to 15 degrees.

cartilaginous joints   Joints joined together by hyaline cartilage or fibrocartilage, allowing very slight movement, such as synchondrosis and symphysis.

center of gravity   The point at which all of the body's mass and weight are equally balanced or equally distributed in all directions.

center of rotation   The point or line around which all other points in the body move.

central nervous system (CNS)   The cerebral cortex, basal ganglia, cerebellum, brain stem, and spinal cord.

cervical plexus   Group of spinal nerves composed of cervical nerves 1 through 4; generally responsible for sensory and motor function from the upper part of the shoulders to the back of the head and front of the neck.

circumduction   Circular movement of a bone at the joint, as in movement of the hip, shoulder, or trunk around a fixed point. Combination of flexion, extension, abduction, and adduction.

closed kinetic chain   When the distal end of an extremity is fixed, preventing movement of any one joint unless predictable movements of the other joints in the extremity occur.

coefficient of friction   The ratio between the force needed to overcome friction over the force holding the surfaces together.

collagen   A protein in the body that forms fibrous connective tissues such as ligaments, tendons, cartilage, bone, and skin. Its elongated fibrils provide strength and flexibility to these tissues.

concentric contraction   A contraction in which there is a shortening of the muscle that causes motion to occur at the joints it crosses.

concurrent   Movement pattern allowing the involved biarticular muscle to maintain a relatively consistent length because of the same action at both of its joints.

condyle   Large, rounded projection that usually articulates with another bone, such as the medial or lateral condyle of the femur.

condyloid joint   Type of joint in which the bones permit movement in two planes without rotation, as in the wrist between the radius and the proximal row of the carpal bones or the second, third, fourth, and fifth metacarpophalangeal joints.

contractility   The ability of muscle to contract and develop tension or internal force against resistance when stimulated.

contraction phase   In a single muscle fiber contraction, it is the phase following the latent perion in which the muscle fiber actually begins shortening; lasts about 40 milliseconds.

core training   Strengthening and conditioning that focuses on the diaphragm, transversus abdominis, lumbar multifidus, and the muscles of the pelvic floor as well as the rectus abdominis, external obliques, internal obliques, and erector spinae.

coronal axis   Runs from side to side through the body and is at a right angle to the sagittal plane of motion. Also known as the frontal or lateral axis.

cortex   Diaphyseal wall of long bones, formed from hard, dense compact bone.

cortical bone   Harder, more compact bone that forms the outer bony surface of the diaphysis.

countercurrent   Movement pattern resulting from opposite actions occurring simultaneously at both joints of a biarticular muscle resulting in substantial shortening of the biarticular muscle.

cranial nerves   The group of 12 pairs of nerves originating from the undersurface of the brain and exiting from the cranial cavity through skull openings; they supply specific motor and sensory function to the head and face.

crest   Prominent, narrow, ridgelike projection of bone, such as the iliac crest of the pelvis.

curvilinear motion   Motion along a curved line.

Davis's law   States that ligaments, muscle, and other soft tissue when placed under appropriate tension will adapt over time by lengthening and conversely when maintained in a loose or shortened state over a period of time will gradually shorten.

dendrite   One or more branching projections from the neuron cell body that transmit impulses to the neuron and cell body.

depression   Inferior movement of the shoulder girdle, as in returning to the normal position from a shoulder shrug.

dermatome   A defined area of skin supplied by a specific spinal nerve.

dexter   Relating to, or situated to the right or on the right side of something.

diagonal abduction   Movement by a limb through a diagonal plane away from the midline of the body such as in the hip or glenohumeral joint.

diagonal adduction   Movement by a limb through a diagonal plane toward and across the midline of the body such as in the hip or glenohumeral joint.

diagonal or oblique axis   Axis that runs at a right angle to the diagonal plane. As the glenohumeral joint moves from diagonal abduction to diagonal adduction in overhand throwing, its axis runs perpendicular to the plane through the humeral head.

diagonal plane   A combination of more than one plane. Less than parallel or perpendicular to the sagittal, frontal, or transverse plane. Also known as oblique plane.

diaphysis   The long cylindrical portion or shaft of long bones.

diarthrodial (diarthrosis) joints   Freely movable synovial joints containing a joint capsule and hyaline cartilage and lubricated by synovial fluid.

displacement   A change in position or location of an object from its original point of reference.

distal   Farthest from the midline or point or reference; the fingertips are the most distal part of the upper extremity.

distance   The path of movement; refers to the actual sum length of units of measurement traveled.

dislocating component   When the angle of pull is greater than 90 degrees, the force pulls the bone away from its joint axis, thereby increasing joint distraction forces.

dorsal (dorsum)   Relating to the back, being or located near, on, or toward the back, posterior part, or upper surface of; also relating to the top of the foot.

dorsiflexion (dorsal flexion)   Flexion movement of the ankle resulting in the top of foot moving toward the anterior tibia.

duration   An exercise variable usually referring to the number of minutes per exercise bout.

dynamic equilibrium   Occurs when all of the applied and inertial forces acting on the moving body are in balance, resulting in movement with unchanging speed or direction.

dynamics   The study of mechanics involving systems in motion with acceleration.

eccentric contraction   A contraction in which the muscle lengthens in an attempt to control the motion occurring at the joints that it crosses, characterized by the force of gravity or applied resistance being greater than the contractile force.

eccentric force   Force that is applied in a direction not in line with the center of rotation of an object with a fixed axis. In objects without a fixed axis, it is an applied force that is not in line with the object's center of gravity.

efferent nerves   Nerves that carry impulses to the outlying regions of the body from the central nervous system.

elasticity   The ability of muscle to return to its original length following stretching.

elastin   A protein in the body that forms connective tissue. It has a highly elastic quality and will return to its original state after stress, whether compressed or stretched.

electromyography (EMG)   A method utilizing either surface electrodes or fine wire/needle electrodes to detect the action potentials of muscles and provide an electronic readout of the contraction intensity and duration.

elevation   Superior movement of the shoulder girdle, as in shrugging the shoulders.

enarthrodial joint   Type of joint that permits movement in all planes, as in the shoulder (glenohumeral) and hip joints.

endochondral bones   Long bones that develop from hyaline cartilage masses after the embryonic stage.

endosteum   Dense, fibrous membrane covering the inside of the cortex of long bones.

epicondyle   Projection located above a condyle, such as the medial or lateral epicondyle of the humerus.

epiphyseal plate   Thin cartilage plate separating the diaphysis and epiphysis during bony growth; commonly referred to as growth plate.

epiphysis   The end of a long bone, usually enlarged and shaped to join the epiphysis of an adjacent bone, formed from cancellous or trabecular bone.

equilibrium   State of zero acceleration in which there is no change in the speed or direction of the body.

eversion   Turning of the sole of the foot outward or laterally, as in standing with the weight on the inner edge of the foot.

extensibility   The ability of muscle to be stretched back to its original length following contraction.

extension   Straightening movement resulting in an increase of the angle in a joint by moving bones apart, as when the hand moves away from shoulder during extension of the elbow joint.

external rotation   Rotary movement around the longitudinal axis of a bone away from the midline of the body. Also known as rotation laterally, outward rotation, and lateral rotation.

extrinsic muscles   Muscles that arise or originate outside of (proximal to) the body part on which they act.

facet   Small flat or shallow bony articular surface such as the articular facet of a vertebra.

fascia   Sheet or band of fibrous connective tissue that envelops, separates, or binds together parts of the body such as muscles, organs, and other soft tissue structures of the body.

fibrous joints   Joints joined together by connective tissue fibers and generally immovable, such as gomphosis, sutures, and syndesmosis.

fibular   Relating to the fibular (lateral) side of the lower extremity.

first-class lever   A lever in which the axis (fulcrum) is between the force and the resistance, as in the extension of the elbow joint.

flat muscles   A type of parallel muscle that is usually thin and broad, with fibers originating from broad, fibrous, sheetlike aponeuroses such as the rectus abdominus and external oblique.

flexion   Movement of the bones toward each other at a joint by decreasing the angle, as in moving the hand toward the shoulder during elbow flexion.

follow-through phase   Phase that begins immediately after the climax of the movement phase, in order to bring about negative acceleration of the involved limb or body segment; often referred to as the deceleration phase. The velocity of the body segment progressively decreases, usually over a wide range of motion.

foramen   Rounded hole or opening in bone, such as the foramen magnum in the base of the skull.

force   The product of mass times acceleration.

force arm   The perpendicular distance between the location of force application and the axis. The shortest distance from the axis of rotation to the line of action of the force. Also known as the moment arm or torque arm.

force couple   Occurs when two or more forces are pulling in different directions on an object, causing the object to rotate about its axis.

force magnitude   Amount of force usually expressed in newtons.

fossa   Hollow, depressed, or flattened surface of bone, such as the supraspinous fossa or iliac fossa.

fovea   Very small pit or depression in bone, such as the fovea capitis of the femur.

frequency   An exercise variable usually referring to the number of times exercise is conducted per week.

friction   Force that results from the resistance between the surfaces of two objects moving upon one another.

frontal plane   Plane that bisects the body laterally from side to side, dividing it into front and back halves. Also known as the lateral or coronal plane.

fundamental position   Reference position essentially the same as the anatomical position, except that the arms are at the sides and the palms are facing the body.

fusiform muscles   A type of parallel muscle with fibers shaped together like a spindle with a central belly that tapers to tendons on each end, such as the brachialis and the biceps brachii.

gaster   The central, fleshy, contractile portion of the muscle that generally increases in diameter as the muscle contracts.

ginglymus joint   Type of joint that permits a wide range of movement in only one plane, as in the elbow, ankle, and knee joints.

glide (slide) (translation)   A type of accessory motion characterized by a specific point on one articulating surface coming in contact with a series of points on another surface.

Golgi tendon organ (GTO)   A proprioceptor, sensitive to both muscle tension and active contraction, found in the tendon close to the muscle tendon junction.

gomphosis   A type of immovable articulation, as of a tooth inserted into its bony socket.

goniometer   Instrument used to measure joint angles or compare the changes in joint angles.

goniometry   Measuring the available range of motion in a joint or the angles created by the bones of a joint.

ground reaction force   The force of the surface reacting to the force placed on it, as in the reaction force between the body and the ground when running across a surface.

hamstrings   A common name given to the group of posterior thigh muscles: biceps femoris, semitendinosus, and semimembranosus.

head   Prominent, rounded projection of the proximal end of a bone, usually articulating, such as the humeral or femoral head.

heel-strike   First portion of the walking or running stance phase characterized by landing on the heel with the foot in supination and the leg in external rotation.

horizontal abduction   Movement of the humerus in the horizontal plane away from the midline of the body.

horizontal adduction   Movement of the humerus in the horizontal plane toward the midline of the body.

hyaline cartilage   Articular cartilage; covers the end of bones at diarthrodial joints to provide a cushioning effect and reduce friction during movement.

hyperextension   Extension beyond the normal range of extension.

impingement syndrome   Occurs when the tendons of the rotator cuff muscles, particularly the supraspinatus and infraspinatus, become irritated and inflamed as they pass through the subacromial space between the acromion process of the scapula and the head of the humerus, typically resulting in pain, weakness, and loss of movement.

impulse   The product of force and time.

inertia   Resistance to action or change; resistance to acceleration or deceleration. Inertia is the tendency for the current state of motion to be maintained, regardless of whether the body segment is moving at a particular velocity or is motionless.

innervation   The supplying of a muscle, organ, or body part with nerves.

insertion   The distal attachment or point of attachment of a muscle farthest from the midline or center of the body, generally considered the most movable part.

instantaneous center of rotation   The center of rotation at a specific instant in time during movement.

intensity   An exercise variable usually referring to a certain percentage of the absolute maximum that a person can sustain.

internal rotation   Rotary movement around the longitudinal axis of a bone toward the midline of the body. Also known as rotation medially, inward rotation, and medial rotation.

interneurons   Central or connecting neurons that conduct impulses from sensory neurons to motor neurons.

intrinsic muscles   Muscles that are entirely contained within a specified body part; usually refers to the small, deep muscles found in the foot and hand.

inversion   Turning of the sole of the foot inward or medially, as in standing with the weight on the outer edge of the foot.

irritability   The property of muscle being sensitive or responsive to chemical, electrical, or mechanical stimuli.

isokinetic   Type of dynamic exercise usually using concentric and/or eccentric muscle contractions in which the speed (or velocity) of movement is constant and muscular contraction (usually maximal contraction) occurs throughout the movement.

isometric contraction   A type of contraction with little or no shortening of the muscle resulting in no appreciable change in the joint angle.

isotonic Contraction occurring in which there is either shortening or lengthening in the muscle under tension; also known as a dynamic contraction, and classified as being either concentric or eccentric.

joint capsule Sleevelike covering of ligamentous tissue surrounding diarthrodial joints.

joint cavity The area inside the joint capsule of diarthrodial or synovial joints.

kinematics The description of motion, including consideration of time, displacement, velocity, acceleration, and space factors of a system's motion.

kinesiology The science of movement, which includes anatomical (structural) and biomechanical (mechanical) aspects of movement.

kinesthesis The awareness of the position and movement of the body in space; sense that provides awareness of bodily position, weight, or movement of the muscles, tendons, and joints.

kinetic friction The amount of friction occurring between two objects that are sliding upon one another.

kinetics The study of forces associated with the motion of a body.

Krause's end bulbs A proprioceptor sensitive to touch and thermal changes found in the skin, subcutaneous tissue, lip and eyelid mucosa, and external genitals.

kyphosis Increased curving of the spine outward or backward in the sagittal plane.

latent period In a single muscle fiber contraction, it is the brief period of a few milliseconds following the stimulus before the contraction phase begins.

lateral axis Axis that has the same directional orientation as the frontal plane of motion and runs from side to side at a right angle to the sagittal plane of motion. Also known as the frontal or coronal axis.

lateral epicondylitis A common problem quite frequently associated with gripping and lifting activities that usually involves the extensor digitorum muscle near its origin on the lateral epicondyle; commonly known as tennis elbow.

lateral flexion Movement of the head and/or trunk laterally away from the midline; abduction of spine.

law of acceleration A change in the acceleration of a body occurs in the same direction as the force that caused it and is directly proportional to the force causing it and inversely proportional to the mass of the body.

law of reaction For every action there is an opposite and equal reaction.

lever A rigid bar (bone) that moves about an axis.

ligament A type of tough connective tissue that attaches bone to bone to provide static stability to joints.

line Ridge of bone less prominent than a crest, such as the linea aspera of the femur.

linea alba Tendinous division and medial border of the rectus abdominis running vertically from the xiphoid process through the umbilicus to the pubis.

linear displacement The distance that a system moves in a straight line.

linear motion Motion along a line; also referred to as translatory motion.

linea semilunaris Lateral to the rectus abdominis, a crescent, or moon-shaped, line running vertically that represents the aponeurosis connecting the lateral border of the rectus abdominis and medial border of the external and internal abdominal obliques.

lordosis Increased curving of the spine inward or forward in the sagittal plane.

lumbar kyphosis A reduction of its normal lordotic curve, resulting in a flat-back appearance.

lumbar plexus Group of spinal nerves composed of L1 through L4 and some fibers from T12, generally responsible for motor and sensory function of the lower abdomen and the anterior and medial portions of the lower extremity.

mass The amount of matter in a body.

maximal stimulus A stimulus strong enough to produce action potentials in all of the motor units of a particular muscle.

meatus Tubelike passage within a bone, such as the external auditory meatus of the temporal bone.

mechanical advantage The advantage gained through the use of machines to increase or multiply the applied force in performing a task; enables a relatively small force to be applied to move a much greater resistance; determined by dividing the load by the effort.

mechanics The study of physical actions of forces; can be subdivided into statics and dynamics.

medial epicondylitis An elbow problem associated with the medial wrist flexor and pronator group near their origin on the medial epicondyle; frequently referred to as golfer's elbow.

median Relating to, located in, or extending toward the middle, situated in the middle, mesial.

medullary cavity Marrow cavity between the walls of the diaphysis, containing yellow or fatty marrow.

Meissner's corpuscles A proprioceptor sensitive to fine touch and vibration found in the skin.

mid-axillary line A line running vertically down the surface of the body passing through the apex of the axilla (armpit).

mid-clavicular line A line running vertically down the surface of the body passing through the midpoint of the clavicle.

mid-inguinal point A point midway between the anterior superior iliac spine and the public symphysis.

midsagittal (median) plane Cardinal plane that bisects the body from front to back, dividing it into right and left symmetrical halves.

midstance Middle portion of the walking or running stance phase characterized by pronation and internal rotation of the foot and leg; may be divided into loading response, midstance, and terminal stance.

mid-sternal line A line running vertically down the surface of the body passing through the middle of the sternum.

momentum The quality of motion, which is equal to mass times velocity.

motor neurons Neurons that transmit impulses away from the brain and spinal cord to muscle and glandular tissue.

motor unit   Consists of a single motor neuron and all of the muscle fibers it innervates.

movement phase   The action part of a skill, sometimes known as the acceleration, action, motion, or contact phase. Phase in which the summation of force is generated directly to the ball, sport object, or opponent, and is usually characterized by near-maximal concentric activity in the involved muscles.

multiarticular muscles   Those muscles that, from origin to insertion, cross three or more different joints, allowing them to perform actions at each joint.

multipennate muscle   A type of pennate muscle that has several tendons with fibers running diagonally between them, such as the deltoid.

muscle spindle   A proprioceptor sensitive to stretch and the rate of stretch that is concentrated primarily in the muscle belly between the fibers.

myotatic or stretch reflex   The reflexive contraction that occurs as a result of the motor neurons of a muscle being activated from the CNS secondarily to a rapid stretch occurring in the same muscle; the knee jerk or patella tendon reflex is an example.

myotome   A muscle or group of muscles supplied by a specific spinal nerve.

neuromuscular junction   Connection between the nervous system and the muscular system via synapses between efferent nerve fibers and muscle fibers.

neuron   Nerve cell that is the basic functional unit of the nervous system responsible for generating and transmitting impulses.

neuron cell body   Portion of a neuron containing the nucleus but not including the axon and dendrites.

neutralizers   Muscles that counteract or neutralize the action of other muscles to prevent undesirable movements; referred to as neutralizing, they contract to resist specific actions of other muscles.

nonrotary component (horizontal component)   Component (either stabilizing or dislocating) of muscular force acting parallel to the long axis of the bone (lever).

notch   Depression in the margin of a bone such as the trochlear and radial notch of the ulna.

open kinetic chain   When the distal end of an extremity is not fixed to any surface, allowing any one joint in the extremity to move or function separately without necessitating movement of other joints in the extremity.

opposition   Diagonal movement of the thumb across the palmar surface of the hand to make contact with the hand and/or fingers.

origin   The proximal attachment or point of attachment of a muscle closest to the midline or center of the body, generally considered the least movable part.

osteoblasts   Specialized cells that form new bone.

osteoclasts   Specialized cells that resorb new bone.

osteokinematic motion   Motion of the bones relative to the three cardinal planes, resulting from physiological movements.

Pacinian corpuscles   A proprioceptor sensitive to pressure and vibration found in the subcutaneous, submucosa,

subserous tissues around joints, external genitals, and mammary glands.

palmar flexion   Flexion movement of the wrist in the sagittal plane with the volar or anterior side of the hand moving toward the anterior side of the forearm.

palpation   Using the sense of touch to feel or examine a muscle or other tissue.

parallel muscles   Muscles that have their fibers arranged parallel to the length of the muscle, such as flat, fusiform, strap, radiate, or sphincter muscles.

parasagittal plane   Planes parallel to the midsagittal plane.

passive insufficiency   State reached when an opposing muscle becomes stretched to the point where it can no longer lengthen and allow movement.

passive tension   Tension in muscles that is due to externally applied forces and is developed as a muscle is stretched beyond its normal resting length.

pennate muscles   Muscles that have their fibers arranged obliquely to their tendons in a manner similar to a feather, such as unipennate, bipennate, and multipennate muscles.

periodization   The intentional variance of overload through a prescriptive reduction or increase in a training program to bring about optimal gains in physical performance.

periosteum   The dense, fibrous membrane covering the outer surface of the diaphysis.

peripheral nervous system (PNS)   Portion of the nervous system containing the sensory and motor divisions of all the nerves throughout the body except those found in the central nervous system.

pes anserinus   Distal tendinous expansion formed by the sartorius, gracilis, and semitendinosus and attaching to the anteromedial aspect of the proximal tibia below the level of the tibial tuberosity.

physiological movement   Normal movements of joints such as flexion, extension, abduction, adduction, and rotation, accomplished by bones moving through planes of motion about an axis of rotation at the joint.

plane of motion   An imaginary two-dimensional surface through which a limb or body segment is moved.

plantar   Relating to the sole or undersurface of the foot.

plantar flexion   Extension movement of the ankle, resulting in the foot and/or toes moving away from the body.

plica   An anatomical variant of synovial tissue folds that may be irritated or inflamed with injuries or overuse of the knee.

posterior axillary line   A line parallel to the mid-axillary line which passes through the posterior axillary skinfold.

preparatory phase   Skill analysis phase, often referred to as the cocking or wind-up phase, used to lengthen the appropriate muscles so that they will be in position to generate more force and momentum as they concentrically contract in the next phase.

primary mover   Muscles that contribute significantly to causing a specific joint movement when contracting concentrically.

process   Prominent projection of a bone, such as the acromion process of the scapula or the olecranon process of the humerus.

pronation   Internally rotating the radius so that it lies diagonally across the ulna, resulting in the palm-down position of the forearm; term also refers to a combination of ankle dorsiflexion, subtalar eversion, and forefoot abduction (toe-out).

proprioception   Feedback relative to the tension, length, and contraction state of muscle, the position of the body and limbs, and movements of the joints provided by internal receptors located in the skin, joints, muscles, and tendons.

protraction   Forward movement of the shoulder girdle away from the spine; abduction of the scapula.

proximal   Nearest to the midline or point of reference; the forearm is proximal to the hand.

Q angle (quadriceps angle)   The angle at the patella formed by the intersection of the line of pull of quadriceps with the line of pull of the patella tendon.

quadriceps   A common name given to the four muscles of the anterior aspect of the thigh: rectus femoris, vastus medialis, vastus intermedius, and vastus lateralis.

radial   Relating to the radial (lateral) side of the forearm or hand.

radial deviation (radial flexion)   Abduction movement at the wrist of the thumb side of the hand toward the forearm.

radiate muscles   A type of parallel muscle with a combined arrangement of flat and fusiform muscle in that they originate on broad aponeuroses and converge onto a tendon such as the pectoralis major or trapezius. Also described sometimes as being triangular, fan-shaped, or convergent.

ramus   Part of an irregularly shaped bone that is thicker than a process and forms an angle with the main body such as the superior and inferior ramus of pubis.

range of motion (ROM)   The specific amount of movement possible in a joint.

reciprocal inhibition   Activation of the motor units of the agonists, causing a reciprocal neural inhibition of the motor units of the antagonists, which allows them to subsequently lengthen under less tension. Also referred to as reciprocal innervation.

recovery phase   Skill analysis phase used after follow-through to regain balance and positioning to be ready for the next sport demand.

rectilinear motion   Motion along a straight line.

recurvatum   Bending backward, as in knee hyperextension.

reduction   Return of the spinal column to the anatomic position from lateral flexion; spine adduction.

relaxation phase   In a single muscle fiber contraction, it is the phase following the contraction phase in which the muscle fiber begins relaxing; lasts about 50 milliseconds.

reposition   Diagonal movement of the thumb as it returns to the anatomical position from opposition with the hand and/or fingers.

resistance   Component of the lever that is typically being attempted to be moved, usually referred to as load, weight, or mass.

resistance arm   The distance between the axis and the point of resistance application.

retinaculum   Fascial tissue that retains tendons close to the body in certain places such as around joints like the wrist and ankle.

retraction   Backward movement of the shoulder girdle toward the spine; adduction of the scapula.

retroversion   Abnormal or excessive rotation backward of a structure, such as femoral retroversion.

roll (rock)   A type of accessory motion characterized by a series of points on one articular surface contacting with a series of points on another articular surface.

rolling friction   The resistance to an object rolling across a surface, such as a ball rolling across a court or a tire rolling across the ground.

rotation   Movement around the axis of a bone, such as the turning inward, outward, downward, or upward of a bone.

rotary component (vertical component)   Component of muscular force acting perpendicular to the long axis of the bone (lever).

rotator cuff   Group of muscles intrinsic to the glenohumeral joint, consisting of the subscapularis, supraspinatus, infraspinatus, and teres minor, that is critical in maintaining dynamic stability of the joint.

Ruffini's corpuscles   A proprioceptor sensitive to touch and pressure found in the skin, subcutaneous tissue of fingers, and collagenous fibers of the joint capsule.

sacral plexus   Group of spinal nerves composed of L4, L5, and S1 through S4, generally responsible for motor and sensory function of the lower back, pelvis, perineum, posterior surface of the thigh and leg, and dorsal and plantar surfaces of the foot.

sagittal plane   Plane that bisects the body from front to back, dividing it into right and left symmetrical halves. Also known as the anteroposterior, or AP plane.

scalar   Mathematical quantities are described by a magnitude (or numerical value) alone such as speed, length, area, volume, mass, time, density, temperature, pressure, energy, work, and power.

scaption   Movement of the humerus away from the body in the scapula plane. Glenohumeral abduction in a plane halfway between the sagittal and frontal plane.

scapula line   A line running vertically down the posterior surface of the body passing through inferior angle of the scapula.

scapular plane   In line with the normal resting position of the scapula as it lies on the posterior rib cage, movements in the scapular plane are in line with the scapular which is at an angle of 30 to 45 degrees from the frontal plane.

scoliosis   Lateral curving of the spine.

second-class lever   A lever in which the resistance is between the axis (fulcrum) and the force (effort), as in plantarflexing the foot to raise up on the toes.

sellar joints   Type of reciprocal reception that is found only in the thumb at the carpometacarpal joint and permits ball-and-socket movement, with the exception of rotation.

sensory neurons   Neurons that transmit impulses to the spinal cord and brain from all parts of the body.

sesamoid bones   Small bones embedded within the tendon of a musculotendinous unit that provide protection as well as improve the mechanical advantage of musculotendinous units as in the patella.

shin splints   Slang term frequently used to describe an often chronic condition in which the tibialis posterior, tibialis anterior, and extensor digitorum longus muscles are inflamed, typically a tendinitis of one or more of these structures.

sinister   Relating to, or situated to the left or on the left side of something.

sinus   Cavity or hollow space within a bone, such as the frontal or maxillary sinus.

somatic nerves (voluntary)   Afferent nerves, which are under conscious control and carry impulses to skeletal muscles.

speed   How fast an object is moving, or the distance an object travels in a specific amount of time.

sphincter muscle   A type of parallel muscle that is a technically endless strap muscle with fibers arranged to surround and close openings upon contraction, such as the orbicularis oris. Also referred to as circular muscles.

spin   A type of accessory motion characterized by a single point on one articular surface rotating clockwise or counterclockwise about a single point on another articular surface.

spinal cord   The common pathway between the central nervous system and the peripheral nervous system.

spinal nerves   The group of 31 pairs of nerves that originate from the spinal cord and exit the spinal column on each side through openings between the vertebrae. They run directly to specific anatomical locations, form different plexuses, and eventually become peripheral nerve branches.

spine (spinous process)   Sharp, slender projection of a bone, such as the spinous process of a vertebra or spine of the scapula.

stability   The resistance to a change in the body's acceleration; the resistance to a disturbance of the body's equilibrium.

stabilizers   Muscles that surround the joint or body part and contract to fixate or stabilize the area to enable another limb or body segment to exert force and move; known as fixators, they are essential in establishing a relatively firm base for the more distal joints to work from when carrying out movements.

stabilizing component   When the angle of pull is less than 90 degrees, the force pulls the bone toward its joint axis, thereby increasing joint compression forces.

stance phase   Skill analysis phase that allows the athlete to assume a comfortable and balanced body position from which to initiate the sport skill; emphasis is on setting the various joint angles in the correct positions with respect to one another and to the sport surface.

static equilibrium   The body at complete rest or motionless.

static friction   The amount of friction between two objects that have not yet begun to move.

statics   The study of mechanics involving the study of systems that are in a constant state of motion, whether at rest with no motion or moving at a constant velocity without acceleration. Involves all forces acting on the body being in balance, resulting in the body being in equilibrium.

strap muscles   A type of parallel muscle with fibers uniform in diameter and arranged with essentially all fibers in a long parallel manner, such as the sartorius.

stretch-shortening cycle   An active stretch via an eccentric contraction of a muscle followed by an immediate concentric contraction of the same muscle.

submaximal stimuli   Stimuli that are strong enough to produce action potentials in multiple motor units, but not all motor units of a particular muscle.

subthreshold stimulus   Stimulus not strong enough to cause an action potential and therefore does not result in a contraction.

sulcus (groove)   Furrow or groovelike depression on a bone, such as the intertubercular (bicipital) groove of the humerus.

summation   When successive stimuli are provided before the relaxation phase of the first twitch is complete allowing the subsequent twitches to combine with the first to produce a sustained contraction generating greater tension than a single contraction would produce on its own.

supination   Externally rotating the radius to where it lies parallel to the ulna, resulting in the palm-up position of the forearm; term is also used in referring to the combined movements of inversion, adduction, and internal rotation of the foot and ankle.

suture   Line of union between bones, such as the sagittal suture between the parietal bones of the skull.

swing   Phase of gait that occurs when the foot leaves the ground and the leg moves forward to another point of contact.

syndesmosis joint   Type of joint held together by strong ligamentous structures that allow minimal movement between the bones, such as the coracoclavicular joint and the inferior tibiofibular joint.

synergist   Muscles that assist in the action of the agonists but are not primarily responsible for the action; known as guiding muscles, they assist in refined movement and rule out undesired motions.

synergists (helping)   Muscles that have an action common to each other, but also have actions antagonistic to each other; they help another muscle move the joint in the desired manner and simultaneously prevent undesired actions.

synergists (true)   Muscles that contract to prevent an undesired joint action of the agonist and have no direct effect on the agonist action.

synovial joints   Freely movable diarthrodial joints containing a joint capsule and hyaline cartilage and lubricated by synovial fluid.

tendinous inscriptions   Horizontal indentations that transect the rectus abdominus at three or more locations, giving the muscle its segmented appearance.

tendon   Fibrous connective tissue, often cordlike in appearance, that connects muscles to bones and other structures.

tetanus   When stimuli are provided at a frequency high enough that no relaxation can occur between muscle contractions.

third-class lever   A lever in which the force (effort) is between the axis (fulcrum) and the resistance, as in flexion of the elbow joint.

threshold stimulus   When the stimulus is strong enough to produce an action potential in a single motor unit axon and all of the muscle fibers in the motor unit contract.

tibial   Relating to the tibial (medial) side of the lower extremity.

toe-off   Last portion of the walking or running stance phase characterized by the foot returning to supination and the leg returning to external rotation.

torque   Moment of force; the turning effect of an eccentric force.

transverse plane   Plane that divides the body horizontally into superior and inferior halves; also known as the axial or horizontal plane.

treppe   A staircase effect phenomenon of muscle contraction that occurs when rested muscle is stimulated repeatedly with a maximal stimulus at a frequency that allows complete relaxation between stimuli, the second contraction produces a slightly greater tension than the first, and the third contraction produces greater tension than the second.

triceps surae   The gastrocnemius and soleus together; triceps referring to the heads of the medial and lateral gastrocnemius and the soleus; surae referring to the calf.

trochanter   A very large bony projection, such as the greater or lesser trochanter of the femur.

trochoidal joint   Type of joint with a rotational movement around a long axis, as in rotation of the radius at the radioulnar joint.

tubercle   A small, rounded, bony projection, such as the greater and lesser tubercles of the humerus.

tuberosity   A large, rounded, or roughened, bony projection, such as the radial tuberosity or tibial tuberosity.

ulnar   Relating to the ulnar (medial) side of the forearm or hand.

ulnar deviation (ulnar flexion)   Adduction movement at the wrist of the little finger side of the hand toward the forearm.

uniarticular muscles   Those muscles that, from origin to insertion, cross only one joint, allowing them to perform actions only on the single joint that they cross.

unipennate muscles   A type of pennate muscle with fibers that run obliquely from a tendon on one side only, such as the biceps femoris, extensor digitorum longus, and tibialis posterior.

valgus   Outward angulation of the distal segment of a bone or joint, as in knock-knees.

varus   Inward angulation of the distal segment of a bone or joint, as in bowlegs.

vector   Mathematical quantity described by both a magnitude and a direction such as velocity, acceleration, direction, displacement, force, drag, momentum, lift, weight, and thrust.

velocity   Includes the direction and describes the rate of displacement.

ventral   Relating to the belly or abdomen, on or toward the front, anterior part of.

vertebral line   A line running vertically down through the spinous processes of the spine.

vertical axis   Axis that runs straight down through the top of the head and spinal column and is at a right angle to the transverse plane of motion. Also known as the longitudinal or long axis.

visceral nerves (involuntary)   Nerves that carry impulses to the heart, smooth muscles, and glands; referred to as the autonomic nervous system.

volar   Relating to the palm of the hand or the sole of the foot.

Wolff's law   States that bone in a healthy individual will adapt to the loads it is placed under. When a particular bone is subjected to increased loading, the bone will remodel itself over time to become stronger to resist that particular type of loading.

Credits

Photo Credits

CHAPTER 1: Fig. 1.1 (both photos) © The McGraw-Hill Companies, Inc./Eric Wise, photographer; **1.2 (both photos)** The McGraw-Hill Companies, Inc./Joe DeGrandis, photographer; **1.11** Jim Wehtje/ Getty Images; **1.19, 1.20, 1.21** Courtesy of R.T. Floyd; **CHAPTER 2: 2.19, 2.24** Courtesy of R.T. Floyd; **2.23** Courtesy of Lisa Floyd; **CHAPTER 3: 3.16** Courtesy of Nancy Hamilton; **CHAPTER 4: 4.4 (both photos), 4.6** Courtesy of Lisa Floyd; **4.5** Courtesy of Britt Jones; **CHAPTER 5: 5.7** Courtesy of Britt Jones; **5.8, 5.9** Courtesy of Lisa Floyd; **CHAPTER 6: 6.4** Courtesy of William E. Prentice; **6.7** Courtesy of Britt Jones; **6.10, 6.12** Courtesy of Lisa Floyd; **CHAPTER 7: 7.7** Courtesy of Britt Jones; **CHAPTER 8: 8.2A, 8.3, 8.4, 8.5, 8.10, 8.11** Courtesy of Britt Jones; **8.2B–D, 8.6, 8.8, 8.9** Courtesy of R.T. Floyd; **8.7** Courtesy of Lisa Floyd; **CHAPTER 9: 9.9, 9.10** Courtesy of Britt Jones; **CHAPTER 10: 10.5, p. 292** Courtesy of Britt Jones; **CHAPTER 11: 11.7, p. 328** Courtesy of Britt Jones; **CHAPTER 12: 12.8** Courtesy of Britt Jones; **CHAPTER 13: 13.1, 13.2, 13.5** Courtesy of Britt Jones; **13.3, 13.4** Courtesy of R.T. Floyd; **13.6** Courtesy of Ron Carlberg.

Illustration Credits

CHAPTER 1: Fig. 1.3 Anthony CP, Kolthoff NJ: *Textbook of anatomy and physiology,* ed 9, St. Louis, 1975, Mosby; **1.4** Linda Kimbrough; **1.5, 1.17** Booher JM, Thibodeau GA: *Athletic injury assessment,* ed 4, Dubuque, IA, 2000, McGraw-Hill; **1.6, 1.18** R.T. Floyd; **1.7, 1.8, p. 33, p. 34** Van de Graaff KM: *Human anatomy,* ed 6, Dubuque, IA, 2002, McGraw-Hill; **1.9** Booher JM, Thibodeau GA: *Athletic injury assessment,* ed 4, New York, 2000, McGraw-Hill; Shier D, Butler J, Lewis R: *Hole's human anatomy & physiology,* ed 9, New York, 2002, McGraw-Hill; Seeley RR, Stephens TD, Tate P: *Anatomy & physiology,* ed 7, New York, 2006, McGraw-Hill; **1.10, 1.12, p. 29** Shier D, Butler J, Lewis R: *Hole's human anatomy and physiology,* ed 9, Dubuque, IA, 2006, McGraw-Hill; **1.13** Seeley RR, Stephens TD, Tate P: *Anatomy & physiology,* ed 7, New York, 2006, McGraw-Hill; **1.14, 1.15** Saladin KS: *Human Anatomy,* ed 4, New York, 2014, McGraw-Hill; **1.16** Seeley R, Stephens TD, Tate P: *Anatomy and physiology,* ed 6, Dubuque, IA, 2000, McGraw-Hill; **1.22, 1.23** Prentice WE: *Rehabilitation techniques in sports medicine,* ed 4, New York, 2004, McGraw-Hill; **Table 1.7** Modified by R.T. Floyd from Exercise Pro by BioEx Systems Inc, Smithville, TX; **CHAPTER 2: 2.1, 2.2, p. 67, p. 68** Saladin KS: *Anatomy and physiology: the unity of form and function,* ed 4, New York, 2007, McGraw-Hill; **2.3** Shier D, Butler J, Lewis R: *Hole's human anatomy and physiology,* ed 11, New York, 2007, McGraw-Hill; **2.4** Luttgens K, Hamilton N: *Kinesiology: scientific basis of human motion,* ed 10, New York, 2002, McGraw-Hill, **2.5** Ernest W. Beck; **2.6** Booher JM, Thibodeau GA: *Athletic injury assessment,* ed 4, Dubuque, IA, 2000, McGraw-Hill; **2.7, 2.12** Seeley RR, Stephens TD, Tate P: *Anatomy & physiology,* ed 8, New York, 2008, McGraw-Hill; **2.8** Mader SS: *Biology,* ed 9, New York, 2007, McGraw-Hill; **2.9, 2.13** Raven, PH, Johnson GB, Losos JB, Mason KA, Singer SR: *Biology,* ed 8, New York, 2008, McGraw-Hill; **2.10** Shier D, Butler J, Lewis R: *Hole's human anatomy & physiology,* ed 9, New York, 2002, McGraw-Hill; **2.11** Powers SK, Howley ET: *Exercise physiology: theory and applications to fitness and performance,* ed 7, New York, 2009, McGraw-Hill; **2.14, 2.17** Seeley RR, Stephens TD, Tate P: *Anatomy & physiology,* ed 7, New York, 2006, McGraw-Hill; **2.15, 2.16** Powers SK, Howley ET: *Exercise physiology: theory and application to fitness and performance,* ed 4, New York, 2001, McGraw-Hill; **2.18, 2.20** R.T. Floyd; **2.21** Prentice WE: *Principles of athletic training: a competency based approach,* ed 15, New York, 2014, McGraw-Hill; **2.22** Hall SJ: *Basic biomechanics,* ed 3, New York, 2003, McGraw-Hill; **Table 2.1** Modified from Saladin, KS: *Anatomy & physiology: the unity of form and function,* ed 4, New York, 2007, McGraw-Hill; and Seeley RR, Stephens TD, Tate P: *Anatomy & physiology,* ed 7, New York, 2008, McGraw-Hill. **CHAPTER 3: 3.1, 3.2, 3.3** Booher JM, Thibodeau GA; *Athletic injury assessment,* ed 4, New York, 2000, McGraw-Hill; Hall SJ: *Basic biomechanics,* ed 4, New York, 2003, McGraw-Hill; **3.4–3.11, 3.14, 3.17, 3.19, Table 3.1, p. 85** R.T. Floyd; **3.12, 3.15, 3.18** Hamilton N, Luttgens K: *Kinesiology: scientific basis of human motion,* ed 10, New York, 2002, McGraw-Hill; **3.13** Hall SJ: *Basic Biomechanics,* ed 6, New York, 2012, McGraw-Hill; **CHAPTER 4: 4.1, 4.3A, 4.13, 4.15** Linda Kimbrough; **4.2, 4.7** Seeley RR, Stephens TD, Tate P: *Anatomy & physiology,* ed 8, New York, 2008, McGraw-Hill; **4.3B** Shier D, Butler J, Lewis R: *Hole's human anatomy and physiology,* ed 9, New York, 2002, McGraw-Hill; **4.6** Hall SJ: *Basic biomechanics,* ed 3, Dubuque, IA, 1999, WCB/McGraw-Hill; **4.8, 4.9** Seeley RR, Stephens TD, Tate P: *Anatomy & physiology,* ed 6, Dubuque, IA, 2003, McGraw-Hill; **4.10–4.12, 4.14** Ernest W. Beck; **p. 94, p. 100, pp. 102–106** Modified by R.T. Floyd from Exercise Pro by BioEx Systems Inc, Smithville, TX; **CHAPTER 5: 5.1, 5.3, 5.4, 5.18, 5.19, 5.20, 5.24, 5.25** Linda Kimbrough; **5.2, 5.5** Saladin KS: *Anatomy & physiology: the unity of form and function,* ed 4, New York, 2007, McGraw-Hill; **5.6** Booher JM, Thibodeau GA; *Athletic injury assessment,* ed 4, Dubuque, IA, 2000, McGraw-Hill; **5.10, 5.11** Shier D, Butler J, Lewis R: *Hole's human anatomy and physiology,* ed 11, New York, 2007, McGraw-Hill; **5.14, 5.15, 5.17** Ernest W. Beck; **5.12, 5.13** Van de Graaff KM: *Human anatomy,* ed 6, Dubuque, IA. 2002, McGraw-Hill; **5.16** Shier D, Butler J, Lewis R: *Hole's human anatomy and physiology,* ed 12, New York, 2010, McGraw-Hill; **5.21** Seeley RR, Stephens TD, Tate P: *Anatomy and physiology,* ed 6, Dubuque, IA, 2003, McGraw-Hill; **5.22, 5.23** Ernest W. Beck with inserts by Linda Kimbrough; **pp. 118–119, pp. 126–127, p. 129, pp. 131–137** Modified by R.T. Floyd from Exercise Pro by BioEx Systems Inc, Smithville, TX; **CHAPTER 6: 6.1, 6.3B, 6.17–6.24** Linda Kimbrough; **6.2A-B** Saladin KS: *Anatomy & Physiology,* ed 4, New York, 2007, McGraw-Hill; **6.2C** Seeley RR, Stephens TD, Tate P: *Anatomy & physiology,* ed 7, New York, 2006, McGraw-Hill; **6.2D** Shier D, Butler J, Lewis R: *Hole's human anatomy and physiology,* ed 9, New York, 2002, McGraw-Hill; **6.3A–C, 6.15, 6.16** Van De Graaff KM: *Human anatomy,* ed 6, New York, 2002, McGraw-Hill; **6.3D** Jason Alexander; **6.5, 6.6** Booher JM, Thibodeau GA: *Athletic injury assessment,* ed 4, Dubuque, IA, 2000, McGraw-Hill; **6.8** Dail NW, Agnew TA, Floyd RT: *Kinesiology for manual therapies,* ed 1, New York, 2011, McGraw-Hill; **6.9** Saladin

Index

Page numbers in *italics* refer to tables and illustrations.

inferomedial, 4
infraspinatus muscle, 118, *119*, 120, 123, 134, *134*
innervation, 40
insertion of muscles, 40–41
instantaneous center of rotation, 79, *79*
intensity of exercise, 212
intercostal muscles, 346, *348*
intercostal nerves, 340
intermuscular septa, 241
internal oblique abdominal muscles, 337, *339, 355, 355*
internal rotation
  defined, 22
  of hip, *235*, 236
  of knee joint, 278, *278*
  of shoulder joint, *115*, 116, *117*
interneurons, 49, *50*
interphalangeal (IP) joint, 173, 297
interspinalis muscles, 345, *346*
interspinous ligaments, 333
intertransversarii muscles, 345, *346*
intertransverse ligaments, 333
intervertebral disks, *333–334, 334*
intrinsic muscles
  defined, 40
  of foot, 319, *320–323*
  of hand, 181, 199–200, *199–201*
inversion of ankle and foot joints, 23, 300, *300*
inward tilt of shoulder girdle, 93
IP (interphalangeal) joint, 173, 297
ipsilateral, 4
irregular bones, 11, *11*
irritability of muscles, 38
ischial tuberosity, 230
ischiofemoral ligament, 232
ischium, 229
isokinetics, 43
isometric contractions, 41, *42–43*, 363, 386
isometric exercises, 214, 369
isotonic contractions, 41, *42–43*

J

joint capsules, 17
joint cavities, 17
joint-isolation exercises, 210
joints, 15–27. *See also* ankle and foot joints; elbow and radioulnar joints; hip joint and pelvic girdle; knee joint; shoulder joint; *specific joints and joint types*
  accessory motions, 26–27, *26–27*
  classifications of, 15–20, *16*
  of fingers, 172
  icons for, 24, *24–26*
  movement terminology, 21–23, *22*
  physiological movements, 26
  range of motion, methods for measuring, 20–21
  of shoulder girdle, 89–90, *90–91*, 92
  stability and mobility, factors affecting, 17, 19–20
  of thumb, 173
  of toes, 297
  of trunk and spinal column, 333–335, *333–335*

K

kinematics, 70
kinesiology. *See also* structural kinesiology

defined, 1, 69
  etymology of commonly used terms in, 383–385
kinesthesis, 51–55, *53–55*
kinetic chain activities, *210–211*, 210–212
kinetic friction, 81, *81*
kinetics, 70
knee jerk reflex, 53, *54*
knee joint, 273–287
  bones of, 273–274, *274*
  icons for, *25*
  joint characteristics, 275–277, *276–277*
  ligaments of, 275, *276*
  movements of, 278, *278*
  muscles of, 278–279, *279–287, 282–287*
  nerves of, 281
  range of motion of, 277, *277*, 379
Krause's end-bulbs, 52, 55, *55*
kyphosis, 5, 94, 330

L

latent periods, 56
lateral, defined, 4
lateral axis, 6, *6–7*
lateral collateral ligament (LCL), 275
lateral epicondylitis, 150
lateral flexion of spine, 23, 336, *336–337*
lateral plane, 5, *6–7*
lateral rotation of pelvis, 236, *236–237*
lateral rotator muscles, 240, 241, 266, *266*
lateral tilt of shoulder girdle, 93
latissimus dorsi muscle, 118, *119*, 120, 123, *127*, 127–128
latissimus pulls, 128–130, 132, 220, *220*
laws of motion, 80–81, *80–81*, 84
LCL (lateral collateral ligament), 275
leg curls, 259–261
length–tension relationship, 57–58, *59*
levator costarum muscle, 346, *348*
levator scapulae muscle, 94, *95–96*, 97, 100, *100*
levers, 70–77
  classification of, 70–73, *71–72*
  defined, 70
  factors affecting use of, 73–77, *74–77*
  first-class, 70, 71, *71–72*, 73, *74*
  mechanical advantage of, 71
  second-class, 70, *71–72*, 73, *75*
  third-class, 71, *72*, 73, *75*
ligaments
  of ankle and foot joints, 296, 297
  of elbow and radioulnar joints, 145–147
  of fingers, 173, *174*
  of hip joint and pelvic girdle, 231–232, *233*
  joint stability and mobility affected by, 17–18, 20
  of knee joint, 275, *276*
  of trunk and spinal column, 333–334
  of wrist and hand joints, 173, *174*
ligamentum flavum, 333
ligamentum nuchae, 333–334
linea alba, 353
linear displacement, 79
linear motion, 79
linea semilunaris, 353
lines of pull, *47*, 47–48
long bones, 11, *11–13*
longissimus muscle, 337, 349–350, *349–350*
longitudinal arches, 297, *299*
longitudinal axis, *6–7, 7*

long thoracic nerve, 97
longus capitis muscle, 342, *342*
longus colli muscles, 345, *346*
lordosis, 5, 94, 330, 334
lower extremities. *See* ankle and foot joints; hip joint and pelvic girdle; knee joint
lower-extremity and trunk exercises, 363–371, 381–382
  abdominal contractions, 369, *369*
  abdominal curl-ups, 365, *365*
  alternating prone extensions, 366, *366*
  dead lifts, 350, 368, *368*
  importance of, 363, 364
  rowing exercise, 370, *370–371*
  squats, 367, *367*
lumbar kyphosis, 330
lumbar nerves, 48–49, *49*
lumbar spine, *26*, 335, *335*
lumbosacral plexus, 48, *49, 52, 247*, 247–248
lumbrical muscles, 199, 200, *200*, 319, *322*
lunate bone, 170

M

malleoli, 295
manubrium, 330
mass, 80
mastoid process, 330
maximal stimulus, 56
MCL (medial collateral ligament), 275
MCP (metacarpophalangeal) joints, 172, 173, *175*
mechanical advantage, 70, 71, 77, *77–78*
mechanical loading, 83, *83*
mechanics, defined, 70. *See also* biomechanics
medial, defined, 4
medial collateral ligament (MCL), 275
medial epicondylitis, 150
medial tilt of shoulder girdle, 93
median, defined, 4
median nerve, 154, *154*
median plane, 5
mediolateral axis, 6, *6–7*
medullary, 11
Meissner's corpuscles, 52, 55, *55*
menisci, 275, *276*
metacarpophalangeal (MCP) joints, 172, 173, *175*
metatarsals, 295–296
metatarsophalangeal joints, 295, 297
mid-axillary line, 3
mid-clavicular line, 3
mid-inguinal point, 3
midsagittal plane, 5
midstance, 293
mid-sternal line, 3
military press, 217, *217*
mobile wad of three, 157
momentum, 83
motion
  laws of, 80–81, *80–81*, 84
  measurement of, 79–80
  planes of, 5–6, *6–7*
motion phase, 209
motor control, 20
motor neurons, 49, *50*
motor units, 55–56, *55*
movement phase, 209

movements
accessory motions, 26–27, *26–27*
of ankle and foot joints, 22–23, 300, *300–301*
of elbow and radioulnar joints, 23, 148, *148–149*
of fingers, 173, 174, *175*
of hip joint and pelvic girdle, *234–237,* 235–236
icons for, 24, *24–26*
of knee joint, 278, *278*
passive, 41
phases of, 208–209
physiological, 26
of shoulder girdle, 23, *92–93,* 92–94
of shoulder joint, 23, *115–117,* 116
terminology for, 21–23, *22*
of thumb, 23, 173, 174, *175*
of toes, 300, *300*
of trunk and spinal column, 23, 336, *336–337*
upper-extremity exercises, analysis of, 208–209
voluntary movement, neural control of, 48–49, *49–53*
of wrist and hand joints, 23, 173–174, *174–175*
multiarticular muscles, 62
multiaxial ball-and-socket joints. *See* enarthrodial joints
multifidus muscle, 345, *346*
multipennate muscles, 38
muscle fibers, 56
muscles, 35–62. *See also* contraction/action of muscles; neuromuscular concepts; *specific muscles and muscle types*
of abdominal wall, 337, 340, *352–356,* 353–356
of ankle and foot joints, 302–304, *302–306,* *308–318,* 308–319, *320–323*
development of, 213. *See also* exercises and activities
of elbow and radioulnar joints, 150, *150–153,* 155–162, *155–162*
features of, 35
of fingers, 171, 176, 181, 193–195, 199–200
functions of, 35, 44–46
for head and neck movement, 337, 340, 341–344, *341–344*
of hip joint and pelvic girdle, 238–241, *238–246,* 250, 250–266, *252–266*
joint stability and mobility affected by, 20
of knee joint, 278–279, *279–287,* 282–287
neural control of voluntary movement, 48–49, *49–53*
nomenclature, 35–37
proprioception and kinesthesis, 51–55, *53–55*
shape and fiber arrangement classifications, 38, *39*
of shoulder girdle, 94, *95–96, 98,* 98–104, *100–104*
of shoulder joint, 118–122, *118–122,* *124–127,* 124–135, *129–135*
terminology, 38, 40–41
of thorax, 337, 340, 346–351, *347–351*
of thumb, 171, 176, 181, 196–199
tissue properties of, 38
of toes, 319
of trunk and spinal column, 337–340, *338–356,* 341–356
of vertebral column, 340, 345, *345–346*

views of, *36–37*
of wrist and hand joints, 176, *176–182,* 180–181, 184–200, *184–200*
muscle spindles, 52–53, *53–55,* 60
musculocutaneous nerve, 123, *123*
myotatic reflex, 53
myotomes, 49

N

navicular, 295
neck muscles, 337, 340, 341–344, *341–344*
nerves and nervous system. *See also specific nerves*
of ankle and foot joints, 307, *307*
central nervous system (CNS), 48, 49, 53
components of, 48–49, *49–53*
of elbow and radioulnar joints, 154, *154*
of fingers, 183
of hip joint and pelvic girdle, 247–249, *247–249*
of knee joint, 281
peripheral nervous system (PNS), 48
of shoulder girdle, 97, *97*
of shoulder joint, 123, *123*
of trunk and spinal column, 48–49, *49–52,* 340
voluntary movement, role in, 48–49, *49–53*
of wrist and hand joints, 183, *183*
neuromuscular concepts, 55–62
active and passive insufficiency, 62, *63*
all or none principle, 56
angle of pull, 60–61, *61*
concurrent and countercurrent movement patterns, 62, *62*
fiber types, 56
force–velocity relationship, 58–59, *59*
length–tension relationship, 57–58, *59*
motor units, 55–56, *55*
reciprocal inhibition or innervation, 60, *60*
stretch-shortening cycle, 59
tension development, factors affecting, 56–57, *57–58*
neuromuscular junction, 55
neuron cell bodies, 49
neurons, 49, *50*
neutralizer muscles, 45
Newton's laws of motion, 80–81, *80–81,* 84
nonrotary components, 60
nucleus pulposus, 334

O

oblique abdominal muscles, 337, 354–355, *354–355*
oblique axis, 7, *7*
oblique plane, 5–6, *7*
obliquus capitis inferior muscle, 342, *342*
obliquus capitis superior muscle, 342, *342*
obturator externus muscle, 241, *246, 266, 266*
obturator internus muscle, 241, *246, 266, 266*
obturator nerve, 248, *249*
open kinetic chain activities, *210–211,* 210–212
opponens digiti minimi muscle, 199, 200, *201*
opponens pollicis muscle, 199, *200*
opposition of thumb, 23, 174, *175*
origin of muscles, 40
os coxae, 229
osteoarthritis, 232

osteoblasts, 13
osteoclasts, 13
osteokinematic motions, 26
osteology, 9–10. *See also* bones; skeletal system
outward tilt of shoulder girdle, 93
overhead press, 217, *217*
overload principle, 212
ovoid (condyloidal) joints, 18, 172

P

Pacinian corpuscles, 52, 53, *55*
palmar, defined, 4
palmar flexion of wrist and hand, 23, 174
palmar interossei muscles, 199, *200*
palmaris brevis muscle, 199, *201*
palmaris longus muscle, 176, *176,* 181, 185, *185*
palpation, 46–47
parallel muscles, 38
parasagittal planes, 5
passive insufficiency, 62, *62*
passive movement, 41
passive tension, 58
patella, 231, 273–274, *274*
patellar tendon reflex, 53, *54*
patellofemoral joint. *See* knee joint
PCL (posterior cruciate ligament), 275
pectineus muscle, 241, *243, 254, 254*
pectoralis major muscle, 118, *118,* 120, 123, 125–126, *125–126*
pectoralis minor muscle, 94, *95–96,* 97, 103, *103*
pectoral nerves, 97, 123
pelvic bone, 230, *231*
pelvic girdle. *See* hip joint and pelvic girdle
pennate muscles, 38
periodization, 212
periosteum, 11
peripheral nervous system (PNS), 48
peroneal nerve, 249, 281, 307, *307*
peroneus brevis muscle, 303, 304, *306,* 311, *311*
peroneus longus muscle, 303, 304, *306,* 310, *310*
peroneus tertius muscle, 302, 304, *306,* 312, *312*
pes anserinus, 279
phalanges, 171, 295, 297. *See also* fingers; toes
physical fitness. *See* exercises and activities
physiological movements, 26
PIP (proximal interphalangeal) joints, 172, 297
piriformis muscle, 241, *246, 266, 266*
pisiform bone, 170, 171
pitching, 208, 209, *209*
pivot (trochoidal) joints, 19, 147
plane (arthrodial) joints, 18, 90
planes of motion, 5–6, *6–7*
plantar, defined, 4
plantar fascia, 297
plantar fasciitis, 297
plantar flexion of ankle and foot joints, 23, 300, *300*
plantar interossei muscles, 319, *322*
plantaris muscle, 303, 304
plantar nerve, 307
plica, 277
plyometric training, 60